INTRODUCTION TO SOCIAL NEUROSCIENCE

INTRODUCTION TO SOCIAL NEUROSCIENCE

Stephanie Cacioppo and John T. Cacioppo

PRINCETON UNIVERSITY PRESS
PRINCETON AND OXFORD

Published by Princeton University Press
41 William Street, Princeton, New Jersey 08540
6 Oxford Street, Woodstock, Oxfordshire OX20 1TR

press.princeton.edu

Library of Congress Cataloging-in-Publication Data

Names: Cacioppo, Stephanie, 1974– author. | Cacioppo, John T., 1951– author.
Title: Introduction to social neuroscience / Stephanie Cacioppo and John T. Cacioppo.
Description: Princeton : Princeton University Press, [2020] |
Includes bibliographical references and index.
Identifiers: LCCN 2019028877 (print) | LCCN 2019028878 (ebook) |
ISBN 9780691167275 (hardback) | ISBN 9780691189178 (ebook)
Subjects: LCSH: Neurosciences—Social aspects. | Social psychology.
Classification: LCC RC343 .C23 2020 (print) |
LCC RC343 (ebook) | DDC 616.89—dc23
LC record available at https://lccn.loc.gov/2019028877
LC ebook record available at https://lccn.loc.gov/2019028878

British Library Cataloging-in-Publication Data is available

Editorial: Alison Kalett and Lauren Bucca
Production Editorial: Natalie Baan
Text and Cover Design: Carmina Alvarez
Production: Jacquie Poirier
Publicity: Julia Hall and Matthew Taylor
Copyeditor: Erin Hartshorn

Cover art: Brain illustrations courtesy of authors

This book has been composed in Palatino LT Std and Helvetica Neue LT Std

Printed on acid-free paper. ∞

Printed in Malaysia

1 3 5 7 9 10 8 6 4 2

CONTENTS

List of Illustrations ix

1. AN INTRODUCTION TO SOCIAL NEUROSCIENCE 1
   1.1 The Evolution of Social Behaviors 1
   1.2 The Social Brain of the Desert Locust 2
   1.3 Neuroscience and Social Neuroscience 6
   1.4 What Makes Us Human? 8
   1.5 Doctrine of Multilevel Analysis and the Golden Triangle 14
   1.6. Concluding Remarks 18

2. SOCIAL CONNECTIONS MATTER 21
   2.1 Salutary Social Connections 22
   2.2 Measuring Objective and Perceived Social Isolation 24
   2.3 The Evolutionary Theory of Loneliness 28
   2.4 Pathways and Consequences of Social Isolation 31
      2.4.1 Theoretical pathways linking loneliness to mortality in the modern world 33
   2.5 Perceived Social Isolation and the Brain 48
   2.6 Concluding Remarks 52

3. THE SOCIAL BRAIN 54
   3.1 Methods for the Study of the Social Brain 55
      3.1.1 Neuroimaging Methods 56
      3.1.2 Epigenetic Processes and Gene Expressions of the Social Brain 63
   3.2 Evolution of the Social Brain 65
      3.2.1 The Emergence of *Homo sapiens* 65
      3.2.2 The Social Brain Hypothesis 71
   3.3 Development of the Social Brain in Infancy 74
   3.4 Development of the Social Brain with Salutary Relationships 77
   3.5 Concluding Remarks 84

4. CONNECTING FORCES 86
4.1 Emotional Contagion, Affective Perspective Taking, and Empathy 87
4.2 Imitation and Identification 95
4.3 Mentalizing 97
4.3.1 Egocentrism Perspective 97
4.3.2 Simulation Perspective 99
4.3.3 The Theory of Mind Perspective 105
4.3.4 Multiple Pathways 107
4.4 Social Learning 108
4.5 Concluding Remarks 109

5. SOCIAL PERCEPTION: READING THE FACE 112
5.1 Face Perception 112
5.2 Static Signals 118
5.3 Slow Signals 121
5.4 Artificial Signals 123
5.5 Rapid Signals 124
5.5.1 Expressions of Emotion 125
5.5.2 Patient and Animal Research 128
5.5.3 Negative Moral Emotions 131
5.6 Concluding Remarks 132

6. SOCIAL DECEPTION: READING THE EYES 134
6.1 Deceptive Expressions 135
6.2 Reading the Eyes 138
6.3 Gaze Direction 139
6.4 Production and Detection of Deception 145
6.4.1 Functional Organization of the Central Nervous System 146
6.4.2 Deceptive Signals and Expressions Revisited 148
6.5 Integration of Physical, Contextual, and Behavioral Influences 149
6.6 Concluding Remarks 150

7. GROUP PROCESSES 152
7.1 Interspecific and Intraspecific Competition 153
7.2 Reciprocity 154
7.3 Ingroups and Outgroups 157
7.4 The Neuroscience of Prejudice and Stereotyping 159
7.4.1 Stereotyping 160
7.4.2 Prejudice and Discrimination 162
7.4.3 Mitigating Bias 164

7.5 Punitive Altruism and Cooperation 166
7.6 Concluding Remarks 169

8. SOCIAL INFLUENCE 171
8.1 Conformity 172
8.2 Obedience to Authority 176
8.3 Persuasion 179
8.3.1 Evolutionary and Cultural Contributions to Persuasion 179
8.3.2 Communication and Persuasion 183
8.4 Concluding Remarks 187

9. SALUTARY SOCIAL CONNECTIONS 189
9.1 From “Me” to “We” 189
9.2 Passionate Romantic Love 192
9.3 Other Types of Love and Biological Drives 199
9.4 The Speed of Love in the Human Brain 204
9.5 Romantic Rejection 206
9.6 Cognitive Benefits of Love 207
9.6.1 Benefits of Love on Embodied Cognition 208
9.6.2 Benefits of Love on Social Cognition 209
9.7 Concluding Remarks 211

APPENDIX A
The Cacioppo Evolutionary Theory of Loneliness (ETL) 213

APPENDIX B
The Passionate Love Scale 217

References 225

Index 275

ILLUSTRATIONS

Figures

Figure 1.1. A. The desert locust (*Schistocera gregaria*). B. Half-brains of a solitarious locust and gregarious locust in frontal view to the same scale. 4
Figure 1.2. The gyrification (or wrinkling) of the cerebral cortex varies across species. 11
Figure 1.3. The four lobes of the cerebral hemisphere. 12
Figure 1.4. The golden triangle of social neuroscience research. 16
Figure 1.5. Comparison of spatiotemporal resolution and penetration depth of functional neuroimaging and neurophysiological techniques. 18
Figure 1.6. Mean number of scientific articles published per year on social neuroscience. 20
Figure 2.1. Partner preference tests for partner vs. stranger. 27
Figure 2.2. Illustrative antecedents of loneliness and eight interrelated pathways linking loneliness and premature mortality. 29
Figure 2.3. Functional connectivity in the resting brain reveals robust associations between loneliness and increased brain-wide functional connectivity. 31
Figure 2.4. Schematic of illustrative neural components of the network underlying the neural adjustments to loneliness. 49
Figure 3.1. Methods in social neuroscience. 55
Figure 3.2. Four core networks of the social brain. 57
Figure 3.3. Visualization of evolution of brains in primates based on the surface of the cerebral cortex. 66
Figure 3.4. Comparison of arcuate fasciculus projections in humans and nonhuman primates. 68
Figure 3.5. Alternative hypotheses for the evolution of large brains in primates. 69
Figure 3.6. Schematic representation of brain areas involved in emotion. 70
Figure 3.7. In anthropoid primates, size of the mean social group increases with relative neocortex volume. 71
Figure 3.8. fMRI experiments on loneliness. 82

Figure 4.1. Schematic representation of the development of empathy. 89
Figure 4.2. Schematic representation of proposed cognitive and neural mechanisms of emotional empathy and cognitive empathy/perspective taking. 92
Figure 4.3. Schematic representation of temporal brain dynamics involved in empathy and understanding of intentional vs. accidental harmful actions. 93
Figure 4.4. Cortical and subcortical areas involved in moral cognition. 94
Figure 4.5. Illustration of human-animal imitation. 100
Figure 4.6. Schematic representation of two conditions presented during tennis study. 101
Figure 4.7. Schematic representation of the human mirror neuron system. 104
Figure 5.1. The cranial nerves of the human brain. 113
Figure 5.2. Illustration of inverted faces. 115
Figure 5.3. Medial and lateral view of the human brain showing the mentalizing network, the mirror neuron system, and the biological motion detection network. 116
Figure 5.4. Fusiform gyrus. 116
Figure 5.5. Fusiform face area (FFA). 117
Figure 5.6. Facial expression of trust. 119
Figure 5.7. Expressions of hospitability vs. hostility. 121
Figure 5.8. Example of the omega-shaped wrinkle Charles Darwin described. 122
Figure 5.9. Examples of basic facial expressions. 127
Figure 5.10. Examples of basic emotions and their visual facial characteristics. 127
Figure 5.11. Examples of emotions perceived in the eyes. 128
Figure 5.12. Amygdalae. 129
Figure 5.13. Illustration of some of the monkey brain areas with face-selective cells. 130
Figure 6.1. A. Fake smile. B. Genuine smile. 136
Figure 6.2. Eye movements observed during a fake or genuine smile. 136
Figure 6.3. Facial expressions of fear. 139
Figure 6.4. Schematic representation of the classical thalamo-cortical visual pathway. 140
Figure 6.5. Variation of eye size and width of exposed eyeball across different species. 141
Figure 6.6. Examples of the eye-white stimuli and amygdala recruitment in response to fearful vs. happy eye-whites. 142
Figure 6.7. Example of photograph as desirable stimulus of lust or romantic love. 143

Figure 6.8. Hierarchical organization of the CNS with components of systems. 147
Figure 7.1 Women in tug-of-war in India. 154
Figure 7.2. Schematic representation of direct reciprocity. 155
Figure 7.3 Schematic representation of indirect reciprocity. 156
Figure 7.4 Schematic representation of network reciprocity. 157
Figure 7.5. Brain structures underlying components of intergroup stereotyping. 161
Figure 7.6. Several interactive brain structures underlie components of a prejudiced response. 162
Figure 7.7. Brain structures supporting the regulation of intergroup responses. 165
Figure 8.1. Example of foraging in the wild. 174
Figure 8.2. Illustration of conformity task to identify which of three lines was seen previously. 175
Figure 8.3. Time line from handwriting to social media. 182
Figure 8.4 Elaboration Likelihood Model (ELM) with both peripheral route and central route to persuasion. 184
Figure 9.1. Divorce rates for men and women by state. 190
Figure 9.2. The 12 main brain areas that are systematically activated in response to a beloved. 193
Figure 9.3. Three systems of the love brain network: affect system, reward/motivation system, and cognitive system. 194
Figure 9.4. A. Seghier's framework to account for the multiple functions of the angular gyrus. B. Schematic illustration of the interplay between the angular gyrus and other distributed subsystems. 196
Figure 9.5. Inclusion of the other in the self scale. 197
Figure 9.6. Brain regions differentially activated by love and desire. 202
Figure 9.7. Brain reconstruction of brain ischemic lesion located in the anterior insula of patient. 203

Tables

Table 2.1. Illustrative human and animal studies of the interrelated pathways associated with increases in loneliness 32
Table 4.1. Characteristics of core emotional processes 88
Table 4.2. Five types of learning 108
Table 8.1. Summary of theoretical and empirical differences between information and normative conformity 173
Table 8.2. Distribution of breakoff points 177
Table 9.1. Neisser's five constructs of the self 191
Table 9.2. Features of romantic love similar to some features of obsessive compulsive behaviors 210

Boxes

Box 1.1. Use It or Lose It 5
Box 1.2. Integration of Human and Animal Research 10
Box 2.1. Brief Loneliness Scale 27
Box 2.2. Gene Regulation by the HPA Axis 40
Box 3.1. Major Milestones in Human Evolution 55
Box. 3.2. Evolution of Laughter from Great Apes to Humans 70
Box 3.3. Human Brain Development: A Gray Matter 76
Box 3.4. Oxytocin and Pair-Bonding 78
Box 4.1. Discriminating Self from Other 90
Box 4.2. Evidence of a Dedicated Brain Network in the Macaque Monkey for the Processing of Social Interactions 98
Box 5.1. Familiar Faces Area in Primates 118
Box 6.1. Visual Attention and Courtship in Animals 144
Box 6.2. Visual Attention and Joint Attention 145
Box 7.1. Hyperscanning Connecting Brains in (and out of) the Laboratory 167
Box 9.1. Neuroimaging of Familiar Beloved Smells in Dogs 195
Box 9.2. The Angular Gyrus and Its Involvement in Self-Expansion 198
Box 9.3. Neuroimaging of Attachment Styles 201

INTRODUCTION TO SOCIAL NEUROSCIENCE

1

AN INTRODUCTION TO SOCIAL NEUROSCIENCE

The brain is the most complex organ in the known universe, and specifying the neural mechanisms underlying social structures and interactions has become one of the grand challenges for the neurosciences to address in the twenty-first century.[1] Such an endeavor is both exciting and daunting because it necessitates the integration of theories, methods, and data across levels of organization from multiple disciplines and social species. To meet this challenge, the field of social neuroscience has grown dramatically as an interdisciplinary science. Our goal in this book is to introduce you to this field.

1.1 The Evolution of Social Behaviors

Our journey begins with the question of how did social behaviors evolve? Social behaviors can be classified according to the fitness consequences for the actor and its social partners. Four types of social behaviors that are found in species ranging from bacteria to humans are:[2] (1) mutual benefit—a social behavior that benefits all involved in the interaction; (2) selfishness—a behavior that benefits the actor at the expense of the other(s) involved in the interaction; (3) altruism—a behavior that is costly for the actor but that benefits other(s); and (4) spite—a behavior that is costly for the actor and the other(s).[2]

Social behaviors that fall under the category of mutual benefit or selfishness have direct effects on the fitness of the actor and, therefore, are favored through natural selection. Social behaviors that fall under the category of altruism and spite reduce the fitness of the actor, but the same evolutionary processes can select for these behaviors when certain conditions set forth in Hamilton's rule are met.

According to Hamilton's rule, altruistic behaviors are favored when the cost to the actor is smaller than the product of the benefit to the other(s) and the relatedness of the other(s) to the actor, where genetic relatedness describes the genetic similarity between two individuals, relative to a reference population. For instance, positive relatedness means that two individuals share more genes than average, and negative relatedness means two individuals share fewer genes than average. By the same logic, spiteful behaviors are favored when the cost to the actor is smaller than the product of the cost to the other(s) and the negative relatedness of the other(s) to the actor, or when the mutually costly behaviors represent a cost to the actor that is smaller than the product of the benefit for a third party and the relatedness of the third party to the actor.[2,3]

It also follows from Hamilton's rule that genetic relatedness, which can be signaled by factors such as kin recognition, may play a larger role in the evolution of social behaviors that fall under the categories of altruism or spite than in the evolution of social behaviors that fall under the categories of mutual benefit and selfishness.

The evolutionary principles favoring social behaviors are the same across species, and social behaviors evolved long before the appearance of humans. The human brain shares many design features with those of other organisms, and both comparative studies and animal models play an important role in revealing the secrets of brain function. The human brain also differs from that of other species. Apparently uniquely, the human brain contemplates the history of the earth, the reach of the universe, the origin of our species, the genetic blueprint of life, and the physical basis of our own unique mental existence.[4] Two observations that arise repeatedly in our journey through social neuroscience are that (1) there are conserved neural, hormonal, cellular, and molecular mechanisms underlying social behavior; and (2) social connections (e.g., kinship), social complexity (e.g., possible interaction partners), and social and cultural learning are driving forces behind the evolution of the remarkable capacities of the human brain.

1.2 The Social Brain of the Desert Locust

Social species are so characterized because they interact frequently with members of their own species (or what is termed "conspecifics") to form structures (i.e., patterns of interaction such as pair bonds, mother-infant attachments, and teams) that extend beyond the individual. Frequent interactions with conspecifics introduces complexities, demands, challenges, dangers, opportunities, and benefits not faced by nonsocial species. As a result, not only does the brain underlie social processes and structures, but social structures and processes can influence brain function and structure. These influences are generally thought to occur over generations through evolutionary processes such as natural selection. However, the influence of the social environment on brain structures and function can also be seen within a lifetime.

At any single moment in time, an individual member of a social species may vary in terms of its position along a continuum of social integration (salubrious social connections and bonds) to social isolation (e.g., exclusion, neglect). Where an organism falls along this continuum can be studied longitudinally in natural settings or manipulated experimentally in the laboratory to investigate the causal effects of the social context. Research has shown that where an organism falls along this continuum can influence brain structures and functions.[5] We survey a number of such influences in chapter 2 but consider the desert locust (*Schistocerca gregaria*) as a case in point (figure 1.1).

The desert locust is found in Africa, Asia, and the Middle East. It is a voracious insect, eating its weight in fruit, leaves, seeds, flowers, stems, bark, and shoots. At 2 grams, the daily consumption of an individual insect is insignificant. However, in their gregarious state, desert locusts form swarms of fast flying insects numbering as many as 50 billion and consuming up to 200,000 tons of food per day. Swarms of desert locust have ravaged crops and spawned famine for centuries.

What makes the desert locust of special interest here is that it can switch back and forth between an asocial state and a social state. The asocial state is the more typical condition, during which period the locust generally avoids conspecifics. Under specifiable conditions (e.g., stimulation indicative of swarming), the locusts transform from a solitary state to a gregarious (social) state, at which point the brains of these locusts grow approximately 30% larger presumably to accommodate the additional information-processing demands of their now more complicated social environment (figure 1.1).[6]

The deprivation of these social connections leads to a return to the asocial state, along with a consequent reduction in brain volume. Importantly, the brains of these locusts do not grow generally, but rather the growth is in brain regions that are particularly important in the swarming phase (box 1.1).

Social processes in humans were once thought to have been incidental to human learning and cognition. However, there is growing evidence that this is not the case, and instead that the social complexities

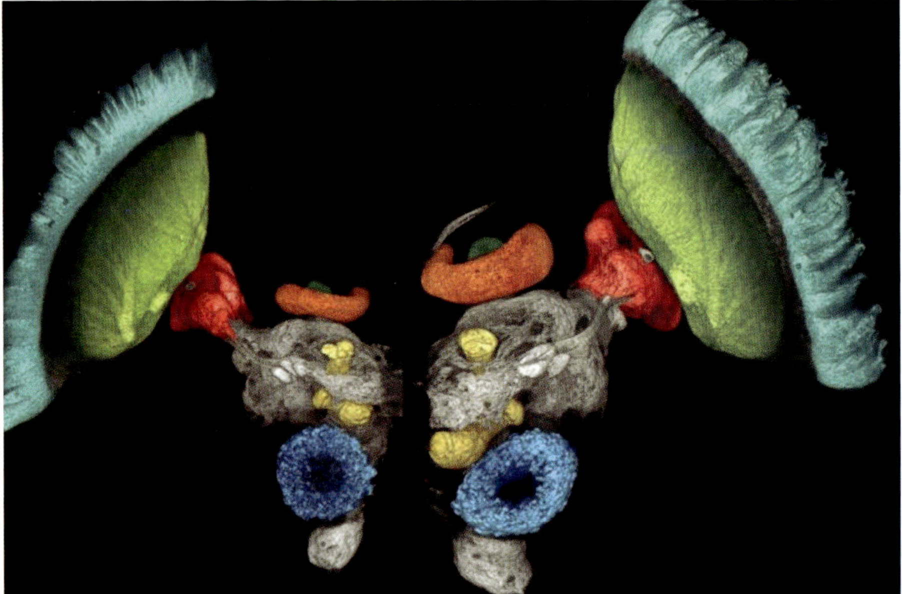

FIGURE 1.1. A. Top panel. The desert locust (*Schistocera gregaria*). iStock.com /MaYcal. B. Bottom panel. Half-brains of a solitarious locust (left) and gregarious locust (right) in frontal view to the same scale (scale bar, 1 mm). The locusts were of near-identical body size. This image shows the visual neuropiles of the optic lobe that include the medulla (lime), the lamina (cyan), and the lobula (red). Adapted from Ott and Rogers.[6] From Fig. 1 of Swidbert R. Ott and Stephen M. Rogers. Gregarious desert locusts have substantially larger brains with altered proportions compared with the solitarious phase. *Proceedings of the Royal Society B: Biological Sciences* http://doi.org/10.1098/rspb.2010.0694.

BOX 1.1. Use It or Lose It

Not the entire brain increases in size when submitted to complex social interactions. Because the brain is energetically expensive, it has been posited that specific brain regions should enlarge only when needed to meet functional demands.[46] In other words, the size of a neural region is related to its functional significance. If social connections/isolation follows the rule of "use it or lose it," regional neuroanatomical adjustments should occur contingent on the demands of social versus isolated living conditions. Consistent with this reasoning, experimental studies of social isolation or solitary states on brain size indicate that the effects are not uniform across the brain but instead are most evident in brain regions that reflect differences in the functional demands of solitary versus social living for that particular species. For instance, the gregarious locust has a larger midbrain to optic lobe ratio, and within both the visual and olfactory systems higher multimodal integration **centers** are disproportionately larger than the primary sensory neuropils.[6,48] The central complex, an important multimodal sensory and sensorimotor integration center, is also considerably larger in gregarious locusts. Despite the solitary desert locust having a smaller brain overall, the solitary locust has disproportionally large primary visual and olfactory neuropils, putatively due to the increased individual predation risk and the need for the solitary locust to detect visual stimuli at a greater distance.[48] Similar reductions in regional brain size in socially isolated animals have been found in other animals, including *Drosophila melanogaster*,[49] several species of honeybees,[(see 5 for review)] and mammals.[27] For instance, Technau[49] showed that socially isolated adult female wild-type Kapelle *Drosophila melanogaster* have fewer mushroom body fibers than do members of a control group—the mushroom bodies in *D. melanogaster* are involved in olfactory learning, multisensory integration, and memory.[(see 50 for review)] Other animals, like mice, that rely heavily on tactile inputs from whiskers have an enlarged sensory cortex. On the other hand, bats, which rely a lot on echolocation, have a large auditory cortex, and the highly visual short-tailed opossums have a large visual cortex.[27,51]

and demands of primate species have contributed to the evolution of the neocortex and to various aspects of human cognition.[7,8] For instance, cross-species comparisons have revealed that the evolution of large and metabolically expensive brains is more closely associated with social than ecological complexity.[9] Moreover, although human toddlers and chimpanzees have similar cognitive skills for engaging and interacting in the physical world, toddlers show more sophisticated cognitive skills than chimpanzees for engaging the social world.[10] We further address this topic in chapter 3.

1.3 Neuroscience and Social Neuroscience

The human brain is a surprisingly recent evolutionary development. If we compressed the 4.5 billion year–long history of the Earth into a 24-hour period, the first single-cell organisms would have emerged around 18 hours ago, the first simple nervous systems separating animals from plants would have emerged around 3.75 hours ago, the first brain would have emerged about 2.67 hours ago, the first hominid brain would have emerged less than 2.5 minutes ago, and the current model of the human brain would have emerged less than 3 seconds ago.

Despite the long evolutionary heritage, the human brain is not the most impressive looking structure. The average human brain measures about 140 millimeters (5.5″) wide, 170 millimeters (6.6″) long, and 90 millimeters (3.6″) high, and weighs about 1,300–1,400 grams (3 pounds). Yet the human brain is the most complex organ in the known universe. It consists of around 86 billion neurons, and each neuron is estimated to form around 5,000 synapses with other neurons, forming approximately 430 trillion synaptic connections for information transfer. If we were to develop a machine that could count all these connections at a rate of 1 per second, it would take more than 13.5 million years to complete the count for a single human brain. Moreover, these structures and transfers remain modifiable across the life span based in part on the environmental demands placed on the brain, including the demands placed on it through interactions with others.

The brain is the central organ of perceiving, identifying, and adapting to social and physical stressors via multiple interacting mediators from the cell surface to the cytoskeleton to epigenetic regulation and non-genomic mechanisms.[11,12] The brain has evolved to determine what is threatening to it, and to respond or adapt to the potential threat with a remarkable plasticity. By elucidating the underlying mechanisms of plasticity and vulnerability of the brain, social neuroscience provides a basis for understanding the efficacy of interventions for a broad variety of social disorders.[11–16] Social stressors cause an imbalance of neural circuitry that may alter one's cognitive or emotional state. This imbalance, in turn, affects systemic physiology via neuroendocrine, autonomic, immune, and metabolic mediators[14,17]. While acute vigilance or hyperattention to potential social threats may be adaptive, the chronic surveillance of the environment may be maladaptive and require intervention with a combination of pharmacological and behavioral therapies, as is the case for chronic loneliness.[17–19] While prevention is key, the plasticity of the brain gives hope for therapies that take into consideration individual differences, gender differences, and brain-body interactions.[11,12,17]

The scientific study of the structure and function of human brain plasticity is so complex that it requires a variety of basic, clinical, and applied disciplines.[20] It also requires comparative research across species as well as studies of healthy people, patients, and animal models to cover the terrain. Although scientific investigations of structure and function go hand in hand, differences in emphasis exist in this scientific frontier. The emphasis in some fields is on identifying constituent structures at different levels of organization, such as neuroanatomy. The emphasis on others is weighted more toward understanding the function of the brain and nervous system, such as the complementary fields of behavioral, cognitive, and social neuroscience.

Behavioral neuroscience, the oldest of these perspectives, replaced the black box between a stimulus and a response in behaviorism with the brain. Accordingly, the brain was viewed as an instrument of sensation and response, with representative topics of study including perception, learning, motivation, homeostasis, biological rhythms, and reproduction.

Cognitive neuroscience, which emerged in the early 1990s, grew out of the cognitive sciences to view the brain as the classic computer, with an operating system; input devices that were designed for selective input; output devices of various types; methods of representing, transforming, manipulating, and storing information; software programs that permitted incoming information to be combined with stored information to produce adaptable responses; and so forth. Accordingly, representative topics of study included attention, representations, memory systems, reasoning, decision making, executive functioning, and response inhibition and response selection.

Social neuroscience, which also emerged in the early 1990s, represents yet another broad perspective on brain function.[21] In social neuroscience, the human brain is regarded not as an isolated computer but metaphorically akin to a smart phone—computationally powerful, mobile, and broadband connected. The connection with other such devices—and sites that have been shaped or visited by other devices—is what makes our phones so powerful and so special, and the same is the case for the human brain.

The functions that are highlighted by this perspective go beyond the solitary computer to include the connections and coordination among interconnected computing devices as well as the structures and processes that were developed in the service of these devices (e.g., the existence and culture of social media). The brain functions that immediately come into focus from this perspective include communication, social perception and recognition, impression formation, imitation, empathy, competition, cooperation, pair-bonding, mother-infant attachment, bi-parental caregiving, social learning, status hierarchies, norms and cultures, social learning, conformity, contagion, social networks, societies, and culture.

The existence of connections between computing devices leaves them vulnerable to various forms of malware, including malicious software such as computer viruses, ransomware, spyware, Trojan horses, worms, adware, and scareware. The human brain is no different. Scientific investigations of the social brain have shown that humans are capable of altruism and salutary relationships and they are capable of deceptive, exploitive, and malicious interactions and relationships. Investigations discussed in this book are beginning to illuminate the biological mechanisms underlying salutary social interactions and relationships as well as protective mechanisms to reduce vulnerability to hostile interactions and exploitive interactions and relationships.

Behavioral, cognitive, and social neuroscience may look at the same construct or behavior but do so from different perspectives and interests. For instance, from the perspective of cognitive neuroscience, language is a system for the representation and processing of information within the brain; from the perspective of social neuroscience, language is a system for information exchange between brains, a system that promotes communication and coordination across discrete and sometimes distant organisms. This illustrates how each of these perspectives can provide important, *complementary* perspectives for understanding brain function.

In sum, social species are so characterized because through social recognition and interaction they form structures that extend beyond any individual member of the species. Social structures and processes differ across species but have evolved hand in hand with neural, hormonal, cellular, and genomic mechanisms because the consequent capacities and behaviors—such as communication, mutual aid, and mutual protection—helped these organisms survive, reproduce, and leave a genetic legacy. Social neuroscience is defined as the study of the neural, hormonal, cellular, and genomic mechanisms underlying social structures and processes. An important goal of social neuroscience is to identify these biological mechanisms and to specify the transduction pathways between neural and social structures and processes.

1.4 What Makes Us Human?

The question "what makes us human?" typically means, how are we different from other species? The debate over what differentiates humans from other species has a venerable history. Charles Darwin reasoned that the difference in mind between humans and the higher animals, great as it is, is one of degree and not of kind. For most of the twentieth century, research emphasized the similarities between the mind, brain, and biology of human and nonhuman animals, demonstrating that we are not unique in our use of language, tools, cultures, syntax, or even teachers.

"What makes us human?" here means what in our evolutionary past has contributed to the human brain and nervous system. In this section, we introduce the human brain from this perspective, and we elaborate on the evolution of the human brain in chapter 3.

Although humans are a unique species, the human body and brain share many design features with those of other organisms. Many of the structures and associated functions of the human brain and body are related to antecedents in other animals (box 1.2). These similarities are not always evident because selective evolutionary pressures may produce a discontinuity in the form or function of a structure across species. As neuroscientist Michael Gazzaniga[22] noted: "Just as gases can become liquids, which can become solids, phase shifts occur in evolution, shifts so large in their implications that it becomes almost impossible to think of them as having the same components" (p. 3).

The human brain has evolved yet differs from nonhuman brains in more fundamental ways than simply the size of the brain. Differences have been found in gene expression across neocortical layers of the human, in contrast to the nonhuman primate, brain, suggesting substantial neocortical reorganization.[23,24] The predominant neural circuit underlying sensory-motor hierarchies in nonhuman primates, for instance, may have yielded to a form that spans the cortex, develops late, and promotes intermodal integration, abstract representation, manipulation, and storage of information.[25] Moreover, astrocytes, glia cells, neuronal synapses, and morphology of cortical minicolumns are not the same in all animals but instead show an evolutionary expansion to support increased computational capacities across regions of the brain.[23,26–28] As the behavior of species becomes more complex, more room is needed for the increase in the number of cells and intracellular connections in the brain. Real estate within the cranial vault is precious, so an evolutionary adaptation is an increased convolution (wrinkling) of the cerebral cortex[29] (figure 1.2). Each of these solutions to the need for computational power emerged in the mammalian brain long before the appearance of humans, and each contributes to what makes us human today.

The vertebrate brain is composed of three major components: (1) the hindbrain, the evolutionarily oldest part of the brain, which includes

BOX 1.2. Integration of Human and Animal Research

The basic structure of various brain systems has been conserved in vertebrate species throughout evolutionary time. There are not only similarities across vertebrate brains in neural structures, but also in the systems that control gene activity and in the neurochemicals that influence neuronal functions (e.g., glutamate, gamma-aminobutyric acid [GABA], norepinephrine, dopamine, serotonin, corticotropin releasing factor, oxytocin, vasopressin, endorphins).[23,52,53]

These similarities make animal models an important source of information about brain structures and brain function in the neurosciences, and these models provide an opportunity to study ancient aspects of social motivations (e.g., pair-bonding, response to isolation, parental nurturing) through experiments that include techniques such as optogenetics, electrophysiology, and gene manipulations that are not possible in humans. Neuroimaging techniques in humans are more focused on the role of cortical structures, whereas animal models involving rodents are more focused on the role of evolutionarily older subcortical structures in social behavior. Neurobiologists Damian Stanley and Ralph Adolphs from the California Institute of Technology summarized the relative strengths and weaknesses of four animal groups commonly used in social neuroscience.[54]

Differences in neural structures have also evolved across vertebrate species based on their unique needs and adaptations, which makes comparative studies across species possible. For instance, rodents, compared to humans, rely on olfactory cues for information about their environment, and their olfactory bulbs are relatively large and evolved. Humans, in contrast, rely more on visual cues, and their visual cortices and associated regions are relatively large and evolved compared to those in rodents. The oxytocin receptors in prairie voles are located within the dopamine-rich area of the striatum, whereas the oxytocin receptors in montane voles are not. These differences in receptor distributions are associated with differences in pair-bonding (prairie voles form pair bonds, montane voles do not), and experimental studies demonstrated pair-bonding depends on oxytocin acting on receptors within the striatum.[55]

The integration of knowledge from human and animal studies is especially important in social neuroscience, with findings in human studies providing insights for animal experimentation, and animal studies providing insights into the molecular, cellular, and circuit-level attributes of the social brain.

areas such as the cerebellum, pons, and medulla; (2) the midbrain, which includes areas such as the tectum (superior and inferior colliculi) and tegmentum (red nucleus, periaqueductal gray, and substantia nigra); and (3) the forebrain, generally the evolutionarily newest part of the brain, which includes areas such as the cerebral cortex, amygdala, thalamus, and hypothalamus.

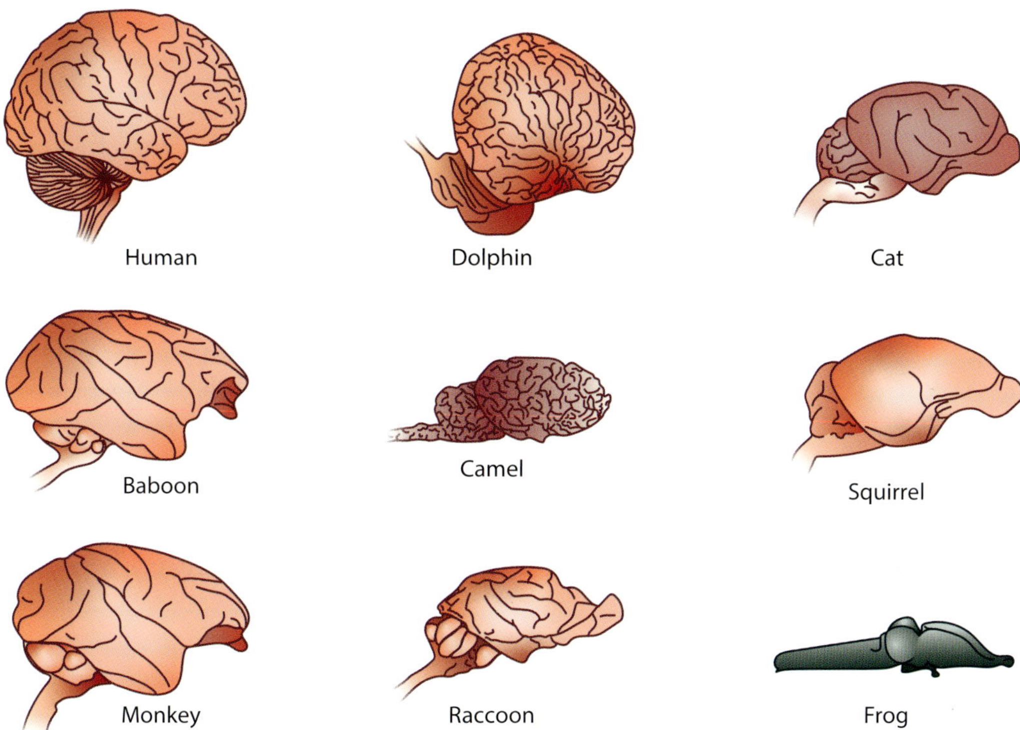

FIGURE 1.2. The gyrification (or wrinkling) of the cerebral cortex varies across species. Gyrification permits more brain mass to fit within a given volume. From https://serendipstudio.org/exchange/brains.

The cerebral cortex has two hemispheres (the right hemisphere and the left hemisphere), each in appearance the mirror image of the other. The surface of each of the cerebral hemispheres is characterized by gyri (ridges), sulci (shallow grooves between gyri), and fissures (deep grooves between gyri), which give the cerebral cortex its wrinkled appearance.

Major gyri and sulci divide each hemisphere into four lobes (figure 1.3): (1) the frontal lobe, the anterior portion of the cerebral cortex, which includes motor areas, supplementary, and premotor areas, areas involved in aspects of mentalizing and self-representation, language (e.g., Broca's area), planning, and executive functioning; (2) the parietal lobe, the middle portion of the cerebral cortex, which includes the somatosensory associative areas that interpret sensations as well as brain areas involved in visuospatial attention, mathematics, language comprehension, and abstract constructs (e.g., aspects of self-representation); (3) the temporal lobe, a lateral portion of the cerebral cortex, which includes the primary and associative auditory areas as well as areas

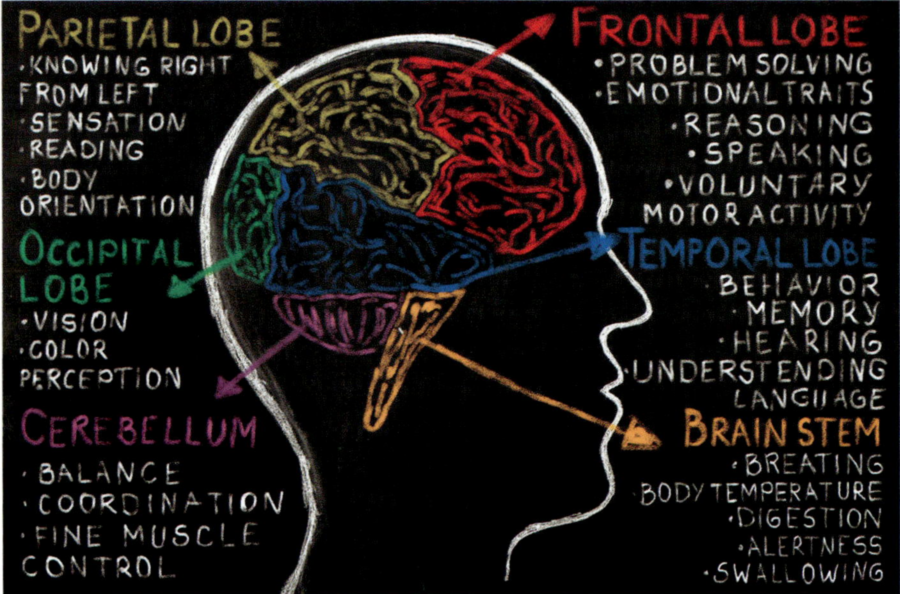

FIGURE 1.3. Each cerebral hemisphere consists of four lobes: the frontal lobe (red), parietal lobe (yellow), temporal lobe (blue), and occipital lobe (green). The cerebellum (purple) and brainstem (orange) are also shown here. Each lobe can have various functions (white). iStock.com/eli_asenova.

involved in memory, aspects of mentalizing, and social perception; and (4) the occipital lobe, the posterior portion of the cerebral cortex, which includes primary and associative visual areas. The left and right hemispheres are connected by a wide bundle of nerves called the corpus callosum.

The two hemispheres are central to a feat of the brain that appears to have emerged uniquely in humans. One function performed by the left hemisphere is to interpret events in a way that forms a coherent narrative. The interpretive function of the left hemisphere integrates information from the two hemispheres and associated operations of the brain to create this narrative.[30,31] These narratives may be confabulations—invalid narratives fabricated by the left hemisphere—but the resulting conscious experience is that the narrative is an accurate explanation for the observed events.[22] It is this left brain function that also provides for our unique ability to process vast amounts of contingently true information, which contributes to our unique social, cognitive, and environmental adaptability.

The notion that the left hemisphere functions like an Interpreter that fabricates a story—in many cases an erroneous story—to create a coherent personal narrative of events—past, present, and future—is contrary to the ancient notion that one can trust the accuracy of one's own conscious reasoning and interpretations. Indeed, this lay notion seems so beyond reproach that long ago common sense was given the mantle of being axiomatic—self-evident and unquestionable. And yet research on human brain function has shown that what seems self-evident or commonsense about one's own thoughts and behaviors cannot be taken as true any more than the commonsense notion that the sun and the stars circle around the earth can be taken as true.

In a classic series of experimental investigations, psychologists Richard Nisbett and Timothy Wilson[32] found that "when people attempt to report on their cognitive processes, that is, on the processes mediating the effects of a stimulus on a response, they do not do so on the basis of any true introspection" (p. 231). The results of their studies—and many that have since been conducted—indicate that self-reports of processes of this form tend to be invalid because (1) people can be unaware of the existence of a stimulus that influences a response, (2) they can be unaware of the existence of the response, (3) they can be unaware that the stimulus has affected the response, (4) they may have developed a false belief about the cause for a response, and (5) they may have overgeneralized a belief about a cause for a response. In short, these self-reports reflect the operation of the Interpreter within us trying to make sense of the world.

Self-reports of current conscious states or behaviors depend on our willingness to report what we are experiencing at a given moment in time. These are not equivalent to self-reports of the narratives or explanations we generate for our conscious states or behaviors—a conscious state that typically reflects the work of the Interpreter. For instance, when people are asked to rate what they feel after having been exposed to an advertisement, they are able to report on their conscious state. (Whether or not they are *willing* to report this state accurately is a different question.) However, when people are asked to rate how their exposure to an advertisement influenced what they feel, they typically feel able to do so but their reports depend on the operation of the Interpreter and are generally invalid. Nisbett and Wilson showed that even when such reports are accurate, the accuracy is the result of a lay theory of the effects of some stimulus being correct, not the result of people having access to the process on which they are reporting. People are willing to say more than they can know in large part because the brain functions to form coherent narratives from past and present events, resulting in people often having no clue about what they think they know versus what they can actually know. This issue will emerge throughout the book.

All the ways in which our brains differ from other brains and with what consequences is a story that has only begun to be written. What is clear is that the final story will not be self-evident and that, given the importance of our interactions with others, it is a story that social neuroscience will likely have a significant hand in writing.

1.5 Doctrine of Multilevel Analysis and the Golden Triangle

Social organisms are constituted at various levels of organization. Although not an exhaustive list, these levels include: (1) individual cells (e.g., nerve cells), which are an important unit of structure in living organisms and may serve a specific function(s); (2) tissues (e.g., nerves), which are made up of cells that are similar in structure and work together to perform a specific function(s); (3) organs (e.g., brain), which consist of tissues that work together to perform a specific function(s); (4) organ systems (e.g., nervous system), which consist of two or more organs that work together to perform a specific function(s); (5) individual organisms (e.g., humans), which refers to the entirety of a living thing that carries out basic processes such as taking in materials, releasing energy from food, releasing wastes, growing, sensing and responding to the environment, and reproducing; (6) dyads (e.g., mother-infant), referring to a pair of organisms who together perform social processes such as communication, coordination, mutual aid and protection, bonding, imitation, nurturance, cooperation, social influence, and social learning; (7) groups, which refers to sets of three of more organisms who operate together to perform social processes; and (8) society, which refers to a large group involved in persistent social interaction who share institutions and a distinctive set of beliefs and knowledge (e.g., norms, practices, and behaviors) transmitted through social learning (i.e., culture).

Mapping across systems and levels (from genome to societies and cultures) requires interdisciplinary expertise, comparative studies, innovative methods, and integrative conceptual analyses.[33,34] The *doctrine of multilevel analysis* in social neuroscience provides one such framework for the scientific investigation of social structures or processes across multiple levels of organization.[21] The doctrine includes three principles for formulating and interpreting investigations along the continuum of organizational levels.

The first is the *principle of multiple determinism*, which specifies that social behaviors may have multiple antecedents (causes) within or across levels of organization. For instance, one might consume a considerable quantity of pizza in an effort to remedy a low blood-sugar condition (biological determinant) or to win a food-eating contest (social determinant). Biological responses can also be multiply determined. Although immune response was once thought to reflect only physiological responses to pathogens or tissue damage, a more complete understanding of immunity has led to demonstrations of how a person's perceptions of his or her close personal relationships may impact inflammation and immunity.[35] Psychosocial stress, operating through the brain's perception of the meaning of events, can also increase proinflammatory cytokine production in the absence of infection or injury. Animal research has revealed related findings in mice: exposure of mice to two weeks of social isolation enhances tumor liver metastasis in part via its suppressive effect on the immune system of the host.[36] An important implication of the principle of multiple determinism is that *comprehensive* theories require a consideration of multiple factors, often from various levels of organization—for example, from the biological through the organismal to the social level.

The *principle of nonadditive determinism* specifies that properties of the whole are not always readily predictable by the simple sum of the (initially recognized) properties of the parts.[21] For instance, the behavior of nonhuman primates was examined following the administration of amphetamine or placebo.[37] No clear pattern emerged until each primate's position in the social hierarchy was considered. When this social factor was taken into account amphetamine was found to increase dominant behavior in primates high in the social hierarchy and to increase submissive behavior in primates low in the social hierarchy. A strictly biological (or social) analysis, regardless of the sophistication of the measurement technology, might not have unraveled the orderly relationship that existed.

Finally, the tendency is to think that biological factors determine social structures and processes, but the *principle of reciprocal determinism* specifies that there can be mutual influences among biological and social factors.[21] For example, it is unsurprising that the level of testosterone in nonhuman male primates has been shown to promote sexual behavior. However, it has also been shown that the availability of receptive females in a colony influences the level of testosterone in the male primates.[38,39] That is, the causal pathway between the biological and social level is bidirectional or reciprocal. There are numerous other such examples, such as maternal behavior altering the expression of genes in female rodent infants through a process of deoxyribonucleic

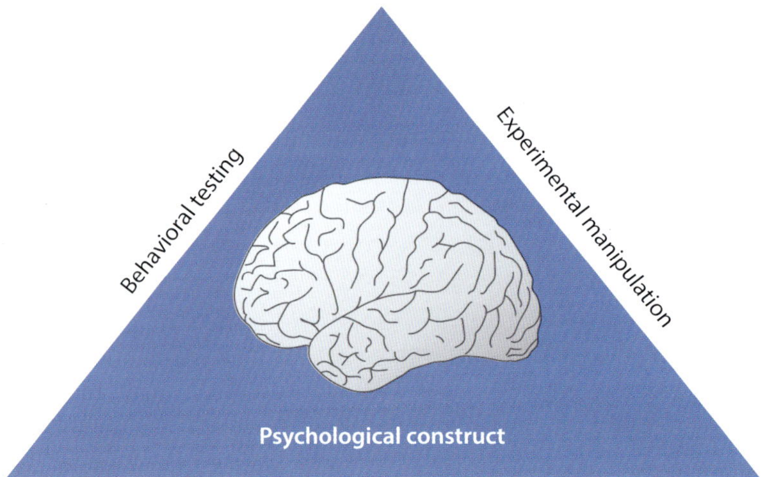

FIGURE 1.4. The golden triangle of social neuroscience research. This equilateral triangle represents the equal importance of three angles (behavioral testing, experimental manipulation, and physiological measures) to perform sound research on brain–behavior relationships. Based on Cacioppo & Cacioppo.[42]

acid (DNA) methylation, and the genes altered in this way later affecting the maternal behavior of these female rodents.[40] That is, the effects of social and biological processes can be reciprocal. One important implication is that comprehensive accounts of social behavior may not be achieved if the biological, cognitive, or social level of organization is considered unnecessary or irrelevant.

These principles point to the importance of multilevel analyses, but how might scientists use these principles in the design of their investigations? We have suggested a golden triangle approach.[41,42] The equilateral triangle in figure 1.4 represents the equal importance of three converging approaches that contribute to an understanding of the neural, hormonal, cellular, and genetic mechanisms underlying social structures and processes.

First, behavioral analyses and assessments are important. According to this approach, the functional consequences of a social phenomenon are decomposed into component representations and processes. These, in turn, may be decomposed into the computations that are likely to be implemented by the brain. What these component processes (or computations) might be will change with advances in theory, methods, and evidence. Evidence from functional or electrical neuroimaging is not necessary in such studies, but it may prove useful either as a source of hypotheses about what these constructs, components, or computations might be or as a means of performing a crucial test between competing hypotheses. Tasks can then be defined that permit the isolation of one or more specific component processes, as verified by behavioral analyses, which in turn permit finer grain analyses of brain function in subsequent research.

Because neuroimaging is noninvasive, it can have an important role to play in the development, testing, and refinement of theories of (and component processes underlying) social phenomena that are difficult to study in nonhuman animals. Correlative evidence from the waking brain using a variety of measurement techniques, therefore, constitutes a second leg of the golden triangle. The brain does not operate exclusively at the spatial level of molecules, cells, nuclei, regions, circuits, or systems, nor does it operate exclusively at the temporal level of milliseconds, seconds, minutes, hours, or days. Any single neuroimaging methodology provides a partial view of brain activity within a very limited range of spatial and temporal levels. Therefore, converging measures that gauge neural events at different temporal and spatial scales can be used to provide a more complete picture of brain function (figure 1.5).[42,43] The neuroimaging studies can then be designed to investigate one or more specific component processes or computations that were isolated in behavioral analyses.

Finally, the third leg of the golden triangle represents experimental evidence from animals and humans. Neuroimaging is a correlative methodology, so random assignment and experimental manipulations including reversible lesions and pharmacological interventions in humans and nonhuman animals are essential to further elucidate the causal role of any given neural structure, circuit, or process in a given task. The term *animal model* can refer to the assay or experimental procedure by which a treatment is produced or measured,[44] but we use it here to refer to investigations in which the participants are nonhuman social animals. We are not the first or the only social species, and the similarities and differences between humans and other social species make comparative, experimental, and mechanistic studies of nonhuman animals an important source of information regarding the role of specific neural, hormonal, and molecular mechanisms in social neuroscience. The protection and ethical treatment of these animals is therefore of paramount importance and is governed by the same ethical principles as is research on human participants.

Each of the legs of the golden triangle has limitations, but yield from the combination of the three is greater than the sum of the parts. Thus, for instance, functional and electrical neuroimaging represents an important part of the methodological armamentarium, but the resulting

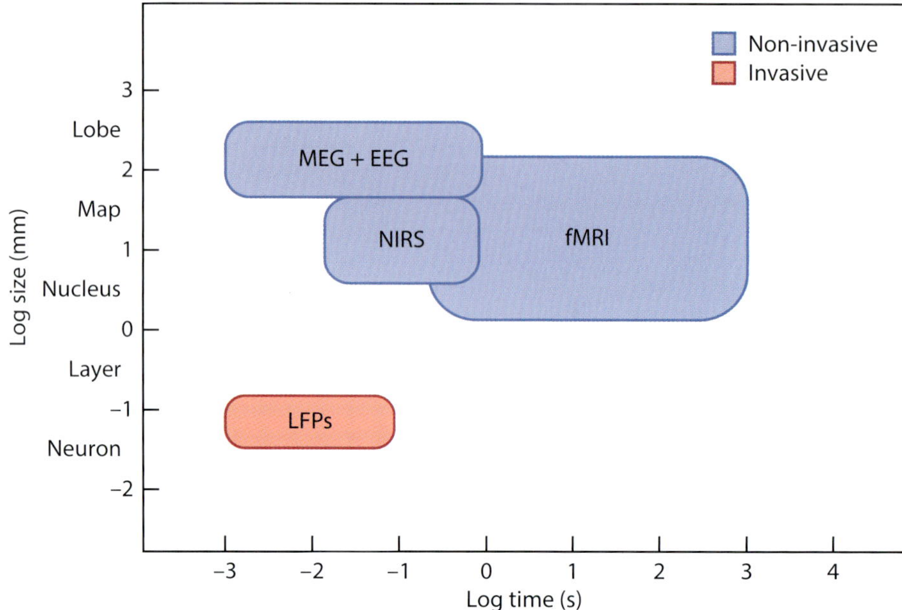

FIGURE 1.5. Illustration of functional neuroimaging and neurophysiological techniques showing comparison of spatiotemporal resolution and penetration depth of neurometabolic optical techniques. The *x*-axis (time in seconds today or size of object/animal/patients) and the *y*-axis (distance from neuron to lobe) are scaled logarithmically. Penetration depths are color-coded from noninvasive (blue) to invasive (red). LFPs = local field potentials, NIRS: near infrared stimulation, fMRI = functional magnetic resonance imaging, MEG + EEG = magneto-encephalography and electroencephalogram. Image is inspired from Van Gerven et al. (2009).[43]

knowledge is more likely to be beneficial when combined with conceptual analyses that decompose complex constructs into component structures, representations, processes, and computations; converging measures that gauge neural events at different temporal and spatial scales; behavioral measures that permit fine-grain analyses of brain–behavior associations; and experimental studies that test the putative role of specific brain structures, circuits, or processes.

1.6 Concluding Remarks

Social species are so characterized because through regular social recognition and interaction, structures (e.g., pair bond, group) are formed that extend beyond any individual member of the species. Social structures and processes differ across species but have evolved hand in hand with neural, hormonal, cellular, and genomic mechanisms because the consequent capacities and behaviors helped these organisms survive, reproduce, and leave a genetic legacy. Social neuroscience is defined as the study of the neural, hormonal, cellular, and genomic mechanisms underlying social structures and processes.

Social neuroscience is built on studies across levels of organization in which variables are measured and/or manipulated to determine the pathways and mechanisms operating within and between each of the levels of organization underlying a phenomenon. Teams of scientists who are investigating brain function in neurological patients, animal models, and healthy individuals are increasingly common. These interdisciplinary collaborations have capitalized on a variety of methods and techniques ranging from behavioral studies, neuroimaging techniques (e.g., magnetic resonance imaging, functional magnetic resonance imaging, high density electroencephalography, optogenetics, receptor autoradiography) to experimental manipulations of neural mechanisms (e.g., transcranial magnetic stimulation, optogenetics, viral vector gene transfer) across scales of neural organization in chimpanzees or healthy humans to cellular and molecular techniques. Even well-traveled techniques such as meta-analyses and electrophysiology have seen upgrades that, for instance, permit investigations of the source and chronoarchitecture of neural structures and processes. The development of experimental manipulations of neural processes in humans through, for instance, the use of transcranial magnetic stimulation and neurotransmitter agonists and antagonists has also helped determine the causal significance of specific neural regions in social cognition, emotion, and behavior. Finally, increases in computational speed and capacities are increasingly simplifying the problem of addressing questions across levels of organization that involve large datasets and/or previously computationally prohibitive simulations or analyses.

Moreover, the twenty-first century presents its own unique array of questions. The development and accessibility of the Internet have transformed major aspects of the social environment, including how and where people study, meet, shop, and interact. Understanding how online interactions are similar to and different from face-to-face social interactions and how they might best be performed to benefit users across the life span is of particular relevance to a world growing more and more dependent upon the Internet and social networking sites for social interactions. And the century is still young. What the short history of social neuroscience has shown is that the number of social neuroscience papers that are appearing in multidisciplinary science journals has grown dramatically since the field was first defined in 1992 (figure 1.6), and the field continues to attract some of the best and

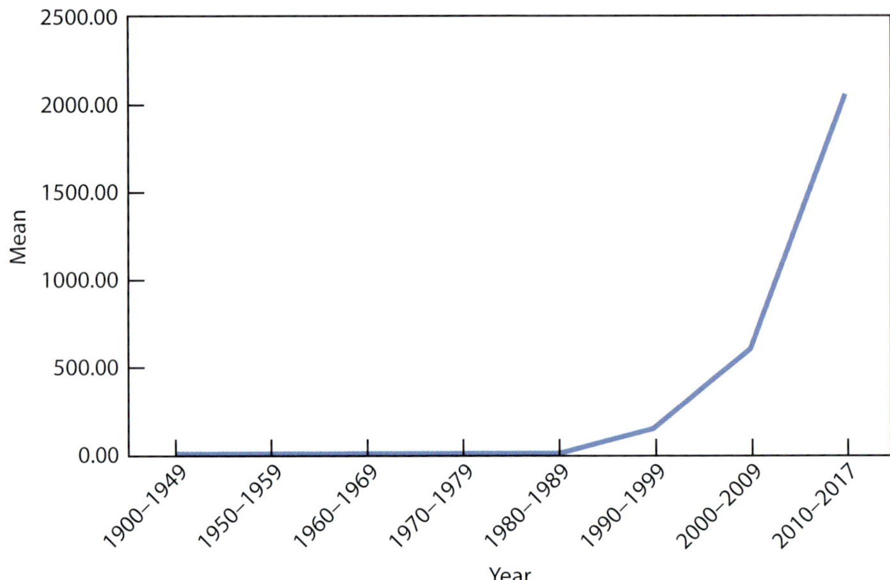

FIGURE 1.6. Mean number of scientific articles published per year on social neuroscience, based on a Web of Science literature search on the topic (social AND neuroscience) OR (social AND brain) for the periods of 1900–1949, 1950–1959, 1960–1969, 1970–1979, 1980–1989, 1990–1999, 2000–2009, and 2010–2017.

brightest young scientists across a wide range of disciplines, with high rates of citation of social neuroscience articles in their dominant disciplinary journals.[45]

The successes that social neuroscience has already met in its first few decades of existence suggest a promising future as it opens a critical avenue for better understanding the neural mechanisms underlying social structures, cognition, interactions, and behavior. A multilevel integration of the social, biological, and cognitive factors underlying behavior should also contribute to the development of new therapeutic interventions to address acute and chronic individual, communicative, and social disorders such as autism, psychopathy, and social phobias. The road ahead is replete with conceptual challenges and methodological issues, but it also promises exciting scientific discoveries. In short, the twenty-first century is an exciting time in which to be a social neuroscientist.

2

SOCIAL CONNECTIONS MATTER

The worst thing you can tell someone who is grieving is "time will heal."

It is not time. . . . It is people. (John T. Cacioppo, 2013)

So imagine there is a condition that makes a person withdraw from others and also makes that person prickly, depressed, and self-centered, and is associated with a 26% increase in the odds of premature mortality. Imagine too that around one in 3 people in America is affected by this condition, and one in 12 is affected severely. The condition is typically reversible, but commonsense solutions are not helpful. Income, education, gender, and ethnicity are not protective, and the condition is contagious. Moreover, the condition reflects the most fundamental feature of social species—the relationships between an individual and conspecifics.[1,2]

There is such a condition—loneliness. And its effects are not attributable to some characterological peculiarity of lonely individuals; they are due to the effects of loneliness on everyday people. Early in its scientific study, loneliness was thought to be an aversive state with no redeeming features, barely different than depression, being alone, introversion, or neuroticism. We now know these commonsense impressions are incorrect. For instance, the discrepancy between an individual's loneliness and the number of connections in a social network is well documented, yet little is known about the placement of loneliness within, or the spread of loneliness through, social networks.

We used network linkage data from the population-based Framingham Heart Study to trace the topography of loneliness in people's social networks and the path through which loneliness spreads through these networks.[3] Results indicate that loneliness occurs in clusters, extends up to three degrees of separation, is disproportionately represented at the periphery of social networks, and spreads through a contagious process.

The Cacioppo Evolutionary Theory of Loneliness (ETL) that we have developed over the past dozen years to generate new hypotheses and to organize the extant data ranging from molecular to sociocultural levels of organization explains the mechanism of action of this contagious process.[4–7] Among the novel predictions from the ETL is that loneliness automatically triggers a set of related pathways that contribute to the observed association between loneliness and premature mortality across the life span. In this chapter, we provide a critical review of the relevant literature on these theoretical pathways.

Before we dive into these different pathways, it is important to understand that feeling connected to others is the glue that makes the creation of stable social structures possible, enabling new behavioral functions such as bonding, communication, empathy, coordination, cooperation, parental caregiving, and social hierarchies. Social contact with conspecifics can be a double-edged sword, however. Moreover, increased social contact potentially offers mutual aid and protection, promotes survival of offspring, enables social learning and reward, permits specialization and a division of labor, enhances immunological and behavioral responses to infectious diseases, and lessens the need for energy expenditure. On the other hand, social contact may increase competition for limited resources including food and mates, exposure to infectious diseases, and risks for oppression and exploitation.[8,9] Because social interactions can range from hospitable to hostile and can change across time, survival depends not only on the ability to detect the presence of conspecifics but to discriminate friend from foe on a continuing basis.

2.1 Salutary Social Connections

There are mental and physical health benefits from the formation and maintenance of stable, salutary connections.[10] In humans, one particularly well-studied pair bond is marriage. These studies have shown that marriage is associated with better physical health and longer life.[10–12] For instance, compared to unmarried individuals, married individuals have fewer physical problems and suffer from fewer long-term illness,[13] have a better survival rate for some illnesses[14] (including some cancers[15–19]) and have a lower overall mortality rate.[20] Moreover, all else constant, the likelihood that married individuals report being happy with life is higher than in those who have never been married or have been previously married.[21]

Marital status per se is neither necessary nor sufficient for these health benefits, however. On average, marriage is associated with a higher household income; a social connection that is mutual, communal, and reciprocal; and a stable, committed relationship that confers mutual aid and protection. The extent that each partner feels connected to and in tune and satisfied with the other partner plays an important role. Accordingly, happy, compared to unhappy, marriages are associated with better health through systems, including cognitive, behavioral, cardiovascular, neuroendocrine, and immunological.[21,22]

For instance, marriage quality and relationship depth are associated with greater life satisfaction and greater systolic and diastolic blood pressure dipping.[23] In contrast, when meaningless, hostile, and toxic connections are sustained, deleterious health effects are observed. Unhappy, compared to happy, marriages are associated with premature morbidity and mortality. Thus, a person's subjective position along the continuum of perceived relationship quality plays a role on the impact of social connections on brain and biology.[10,24–26] The importance of relationship quality and its association with health has been observed across species. For instance, neurobiologist Joan Silk and colleagues showed that the quality of close social bonds has effects on the longevity of female chacma baboons (*Papio hamadryas ursinus*), with females with strong and stable bonds with other female baboons living significantly longer than females who maintain weaker and less stable relationships.[27] These beneficial effects occur through different pathways, e.g., behavioral, cardiovascular, neuroendocrine, and immune system.[21,28–32] The association between health and salutary close relationship resides in the fact that social species all have evolved hand in hand with neural, hormonal, and genetic mechanisms to support them because the consequent social behavior helps these organisms survive.[33]

To investigate the role of stable social connections on brain and biology, we have used a common approach in the neurosciences termed the *subtractive method*. In this method, the effects of the presence of some element in an organism are contrasted with the effects of the absence (or graded absence) of that element. For instance, to investigate the function of a particular gene, one can compare the differences in outcomes on various measures between a mouse that has its full genetic complement and a "knockout" mouse that is genetically identical to the normal mouse except that a target gene of interest has been replaced or disrupted with an artificial piece of deoxyribonucleic acid (DNA). Any differences observed between the normal and the knockout mouse point to the function of the target gene.

Accordingly, to study the effects of social connections on brain and biology, we have investigated differences observed between individuals who have stable social connections and those who do not have such connections. An individual's social connections have both objective and subjective characteristics, however. Among the most fundamental characteristics of social connections are (1) the extent to which an individual is socially isolated (objective social isolation), and (2) the extent to which the individual *perceives* they are socially isolated (perceived social isolation).

In animal studies, animals are typically assigned randomly to socially isolated or to normal (group or paired) housing conditions. People typically exert some choice and control over the extent to which they are objectively socially isolated. Since random assignment to housing conditions is not typically plausible in humans, measurements are typically made of their objective and their perceived social isolation in cross-sectional design (testing a population at a specific time) or longitudinal design (testing a population across several moments in time).[8]

2.2 Measuring Objective and Perceived Social Isolation

Objective isolation in human studies is typically operationalized as being unmarried; having less than biweekly or monthly contact with friends and family; and having no affiliation with voluntary organizations, clubs, or religious groups.[34] For more than three decades, epidemiological studies have noted that objective social isolation is associated with increased risks of morbidity and mortality.[34] The most common explanation for this association is the *social control hypothesis*, which posits that interactions with friends, family, and church congregations incline better health behavior, which in turn decreases risks for morbidity and mortality.[35]

Over the past quarter century, we have taken a somewhat different approach to investigate the association between social isolation and health. In addition to measuring objective social isolation, we have typically also measured an individual's *perceptions* of being socially isolated. The same objective relationship (e.g., sibling, spouse) can be experienced as caring and protective or as uncaring and threatening. Therefore, the brain is the key organ of social connections and processes.[24,36] The measurement of objective and perceived social isolation in investigations shows that objective and perceived social isolation are only weakly related.[28] The number and frequency of contacts with others, for instance, is not as important a predictor of feeling isolated as is how satisfying are a person's social connections.[28,37,38] Consequently, individuals can feel isolated in a marriage, when among friends or family, or in a group or congregation.

Both molecular and behavioral genetic investigations have found perceived social isolation to be in part heritable and in part influenced by a variety of events in the social environment including marital status, homesickness, bereavement, and unrequited love.[24,39–41] The heritable trait of introversion is sometimes confused with perceived social isolation, but these are distinct constructs[4] that do not share the same genetics.[41] Whereas introversion refers to the preference for low levels of social involvement, perceived social isolation refers to the perception that one's social relationships are inadequate in light of their preferences for social connection.[42] Introversion rarely emerges as a strong risk factor for individual outcomes such as broad-based morbidity or mortality.

The human and animal literatures on objective and perceived social isolation developed independently, with an emphasis in the human literature on the potential role of social relationships/isolation on behavioral, neural, hormonal, cellular, and genomic processes and mechanisms,[24] and an emphasis in the animal literature on either the effects of environmental enrichment/isolation on brain plasticity and learning[43,44] or the use of social isolation as a model of behavioral disorders (e.g., depression, anxiety, schizophrenia, aggressive behavior).[45–47] Measures have been developed to assess objective and perceived social isolation in humans and in animals, and these literatures are now coming together to provide new and compelling insights into the biological mechanisms underlying social connections.

Objective isolation in humans has been measured by assigning one point for each of the following: (1) unmarried/not-cohabiting; (2) has less than monthly contact (including face-to-face, telephone, or written/e-mail contact) with one's children; (3) has less than monthly contact with other family members; (4) has less than monthly contact (including face-to-face, telephone, or written/e-mail contact) with friends; and (5) does not participate in organizations such as social clubs or residents groups, religious groups, or committees.[48] Scores range from 0 to 5, with higher scores indicating greater objective social isolation. Objective social isolation in animal studies, in contrast, is typically operationalized by randomly assigning animals either to social normal housing conditions or to physical isolation from conspecifics for an extended period (e.g., two weeks).

Perceived social isolation in humans is known more generally as "loneliness."[49] A person who is *low* in perceived social isolation feels closely connected to, understood by, and safe with those about whom they care. Consequently, they express satisfaction with their social relationships and regard them as close to ideal. In contrast, a person who is *high* in perceived social isolation may have family and friends; but they do not feel truly in tune with anyone, they do not feel they have much in common with anyone, and they do not feel understood by anyone. Consequently, they are not satisfied with their social relationships and regard them as far from ideal. They tend to feel socially isolated even when around others.

Various questionnaire measures of perceived isolation exist, many of which avoid the words "lonely" and "loneliness" because of the response biases engendered by the stigma associated with these terms. The questionnaires instead rely on people responding to statements that have been found to differentiate lonely from nonlonely individuals, such as asking individuals to rate how often they feel "My social relationships are superficial" on a scale ranging from "never" or "rarely" to "frequently"[50,51] (box 2.1).

Although there are aspects of loneliness that may be uniquely human, loneliness represents a generally adaptive predisposition in response to a discrepancy between an animal's preferred and actual social relations that can be found across phylogeny.[25] Experimental and mechanistic studies using animal models provide important information on the effects of loneliness, the mechanisms underlying these effects, and interventions (e.g., pharmacological) for treating loneliness and its harmful effects on health and well-being. Accordingly, animal models for loneliness have been identified based on behavioral (e.g., partner preference) tests that gauge the discrepancy between an animal's preferred and realized social relationships.[8,52–54]

For instance, in the partner preference test, a familiar and an unfamiliar conspecific are tethered in separate chambers, and the test animal is placed in a middle chamber that has ready access to both of the other chambers[55] (figure 2.1). The proportion of time the test animal spends with each conspecific serves as an index of partner preference. When the test animal spends a disproportionate amount of time with one of the two tethered animals, this animal is said to have a strong

BOX 2.1. Brief Loneliness Scale

Directions: The following questions are about how you feel about different aspects of your life. Indicate how often you feel the way described in each of the following statements. Circle one number for each.

Statement	Never	Rarely	Sometimes	Often
1. How often do you feel that you lack companionship?	1	2	3	4
2. How often do you feel left out?	1	2	3	4
3. How often do you feel isolated from others?	1	2	3	4

Note: The brief loneliness scale was adapted from the revised UCLA scale for use in large-scale survey investigations.[217] The score is the sum of the items, with higher scores signifying greater loneliness. Population-based research indicates that the distribution of loneliness scores is positively skewed, indicating that most individuals do not feel lonely at any given moment. The total scale score for the brief loneliness scale ranges from 3 (no loneliness) to 12 (extreme loneliness). A total score of 10 or greater signals an individual who is likely to be dealing with loneliness on a regular basis.

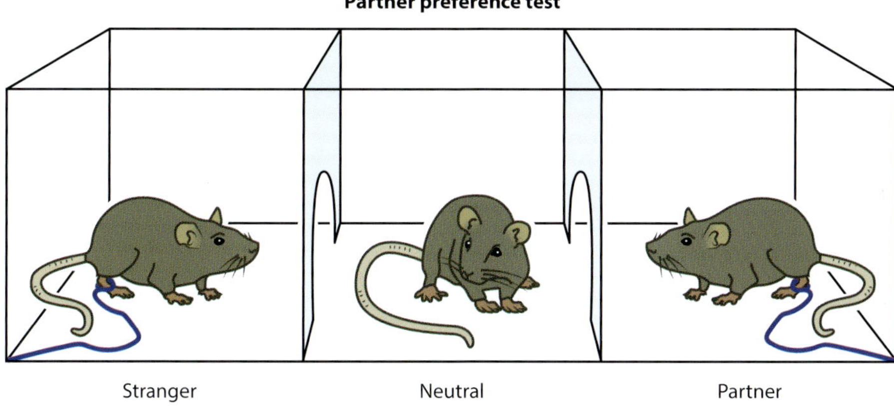

FIGURE 2.1. Partner preference tests were conducted in a three-chambered arena. Stimulus animals (one partner and one stranger) were loosely tethered at each end; the test subject roamed freely for 3 hours. From Ahern, Modi, Burkett, & Young.[55]

partner preference. Note that objective social isolation in the animal literature is the same whether the target animal is experimentally separated from the preferred or the nonpreferred partner, whereas perceived social isolation is greater when the target animal is isolated from the preferred rather than the nonpreferred partner even though the level of objective social isolation is the same for these two conditions.

Given that objective and perceived social isolation can be contrasted with normal social connections in humans and animals, what has been found? We begin by summarizing research on social isolation and mortality. We then survey some of the behavioral, neural, hormonal, cellular, and genomic consequences of perceived social isolation. In chapter 3, we return to this question but focus on brain structures and functions that vary as a function of perceived social isolation. As these studies show, social connections have a surprisingly broad and deep effect on the brain, behavior, and biology.

2.3 The Evolutionary Theory of Loneliness

According to the ETL, the aversiveness that results when an organism perceives it is socially isolated is a biological signal that motivates attention to and repair/replacement of deficiencies in salutary relationships. The neurocognitive and adaptive behavioral effects do not stop there, though. Among the predictions from the ETL is that loneliness automatically triggers a set of related pathways that contribute to the observed association between loneliness and premature mortality across the life span. Our evolutionary theory of loneliness contrasts with traditional theories of loneliness, which conceptualized loneliness as a uniquely human phenomenon.[56–58] Although there may be aspects of loneliness that are uniquely human, there also is continuity intra- and across species.[32,59]

Intraspecies aggression, for instance, represents a significant threat to survival and reproductive success among anthropoid primates, and perhaps especially among humans.[60] The unfettered motivation to form trusting relationships with others in such contexts could therefore prove fatal. ETL[37,5] specifies that the perception of being socially isolated is not only sad, it denotes a dangerous circumstance and promotes *short-term self-preservation by* (1) increasing alertness and implicit vigilance for social threats, (2) increasing the extent to which an individual's response reflects concern for self-interests and personal welfare (i.e., self-centeredness), and (3) triggering the interrelated pathways depicted in figure 2.2.

According to ETL, evolutionary fitness refers to the probability that the line of descent from an individual with a specific trait will remain or increase in the population. Social species are defined by the presence of sufficiently reliable patterns of social interactions that social relationships and structures are identifiable. The social behaviors expressed in these interactions can be classified according to the fitness consequences for the actor and its social partners. From bacteria to

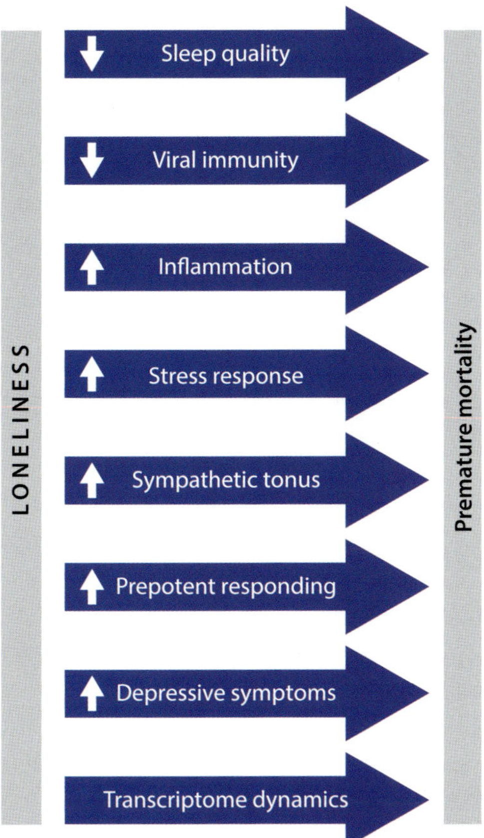

FIGURE 2.2. Illustrative antecedents of loneliness and eight interrelated pathways linking loneliness and premature mortality. Cultural and environmental factors play an important role in the etiology of loneliness. The perception of being socially isolated (lonely) triggers neural changes that initiate eight interrelated pathways whose function is to promote short-term survival. Although historically the activation of these pathways may have promoted short-term survival in the absence of mutual aid and protection, in contemporary societies the chronic activation of these pathways may have deleterious effects on longevity and well-being. Adapted from Cacioppo & Cacioppo.[2]

humans, these social behaviors can be categorized in terms of evolutionary fitness as one of the following: (1) *selfishness*—the actor benefits at a cost to the recipient; (2) *mutual benefit*—both the actor and the recipient benefit; (3) *altruism*—the recipient benefits at a cost to the actor; and (4) *spite* (sometimes termed punitive altruism)—both actor and recipient suffer a loss.[61]

In terms of fitness, perhaps the most common type of behavior across species is selfishness because of the simplicity and general adaptability of behaviors whose benefits (b_a) for the actor exceed the costs (c_a) to the actor (i.e., $b_a > c_a$).* The pain reflex is a simple example of a behavior that has evolved through natural selection because the benefit of the behavior for the actor (b_a; e.g., protection of the physical body from tissue damage) exceeds the cost of the behavior to the actor (c_a). In the case of *social* behaviors categorized as selfish, not only does $b_a > c_a$, but the benefits to the recipient (b_r) is less than the cost to the recipient (c_r) (appendix A).[2] For social behaviors categorized as mutual benefit, the fitness benefits exceed the costs for both actor and recipient ($b_a > c_a$, $b_r > c_r$). Social behaviors that provide mutual benefit therefore also have direct effects on the fitness of the actor and are also favored through natural selection.

Social behaviors that fall under the category of altruism and spite reduce the fitness of the actor, but the same evolutionary processes can select for these behaviors when certain conditions are met.[62,63] According to Hamilton's rule, for instance, altruistic behaviors are favored when the cost to the actor is smaller than the product of the benefit to the other(s) and the relatedness of the other(s) to the actor, where genetic relatedness between the actor and recipient(s) (r_{ar}) describes the genetic similarity between individuals, relative to a reference population. Positive relatedness means that two individuals share more genes than average, and negative relatedness means two individuals share fewer genes than average. A social behavior by an actor for which $r_{ar}b_r > c_a$ is favored by natural selection. Spiteful behaviors are similarly favored by natural selection when the cost to the actor ($c_a > 0$) is smaller than the product of the negative benefit to the recipient ($b_r < 0$) and the negative relatedness of the recipient to the actor ($r_{ar} < 0$) so that $r_{ar}b_r > c_a$.[61,63]

The propositions of the Cacioppo ETL are summarized in appendix A. Briefly, an organism's perception of being socially isolated (i.e., lonely) automatically signals an environment in which the likelihood is low of encountering social behaviors categorized in terms of evolutionary fitness as mutual benefit or altruism, and the likelihood is high of the organism exhibiting behaviors categorized as selfish (appendix A). Interestingly, the seemingly paradoxical effect of loneliness increases the motivation to connect while also increasing self-centeredness and an implicit vigilance for social threats. The proposition that perceived social isolation heightens implicit vigilance for social threats is supported by studies of the neural response to

*Costs and benefits in these computations are not determined by an individual's self-assessment or preferences but rather reflect computations based on evolutionary fitness—a biological construct that represents the probability that the line of descent from an individual with a specific trait will remain or increase in the population.[62,244]

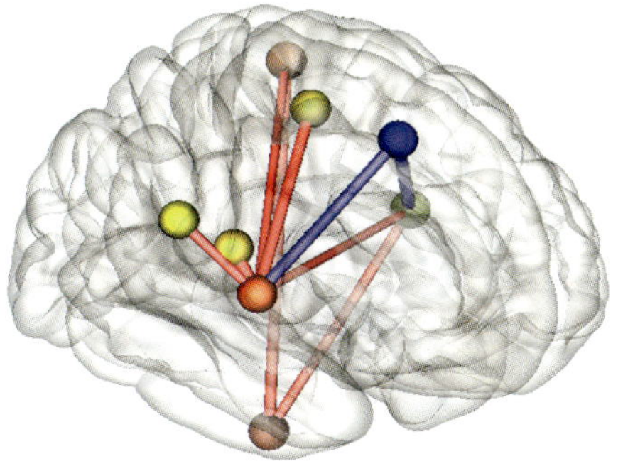
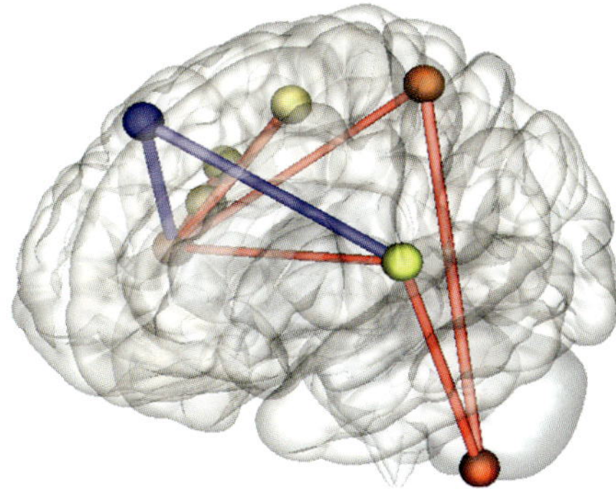

FIGURE 2.3. Analyses of the functional connectivity in the resting brain of 55 health young adult participants (31 females) revealed robust associations between loneliness and increased brain-wide functional connectivity in areas encompassing the right central operculum and right supramarginal gyrus, and these associations were not explained by depressive symptomatology, objective isolation, or demographics. Further analyses revealed that loneliness was associated with increased functional connectivity between several nodes of the cingulo-opercular network (depicted in red), a network known to underlie the maintenance of tonic alertness. In contrast, functional connectivity between loneliness and right middle/superior frontal gyrus was reduced (depicted in blue), a finding associated with diminished executive function in prior literature. Hot or cold colors also correspond to magnitude of the association between network-average functional connectivity and loneliness, controlling for covariates. Adapted from Layden et al.[66]

negative social versus nonsocial pictures,[64] eye-tracking measures of the target of first fixation to social and nonsocial pictures,[65] and functional connectivity of the human brain at rest (figure 2.3).[66] In a recent test of the second proposition, prospective analyses of data from a ten-year longitudinal, population-based study of older adults showed that loneliness increases self-centeredness even after controlling for baseline states of self-centeredness and various covariates.[67] This shift in the fitness consequences of behavior is posited to be evolutionarily old. Consequently, loneliness in the ETL is posited to operate in humans in part through nonconscious processes.

2.4 Pathways and Consequences of Social Isolation

That ETL is unique in a number of its predictions, including the proposition that perceived social isolation (loneliness) triggers the eight interrelated pathways depicted in figure 2.2, is supported by experimental and longitudinal studies in humans and experimental studies

TABLE 2.1. Illustrative human and animal studies of the interrelated pathways associated with increases in loneliness

	Human research	Animal studies
Sleep salubrity	Increased sleep fragmentation daytime dysfunction[73–77,101,150,218–226]	Decreased slow wave sleep and homeostatic rebound[79]
HPA activity	Heightened cortisol awakening response[101,174,226,227] and flatter diurnal cycle[80,172,226,228]	Elevated basal levels of corticosterone[81,82,84,87,130,231] and higher corticosterone levels after a resident-intruder test[83]
Sympathetic tonus	Higher vascular resistance in young adults[97,98] and higher blood pressure and rates of cardiovascular disease in older adults[106,107,109–111,119–121,230–232]	Elevated blood pressure[127–129]
Transcriptome dynamics	Up-regulation of gene expression for inflammatory biology and down-regulation of antiviral gene expression[130,133–135,233]	Up-regulation of gene expression for inflammatory biology and down-regulation of antiviral gene expression[130]
Viral immunity	Decreased viral immunity[101,145–151]	Decreased viral immunity[130]
Inflammatory substrate	Increased inflammation[160,161]	Increased inflammation[162,234,235]
Prepotent responding	Decreased inhibition of prepotent response and increased impulsive responding[143,236–238]	Increased prepotent responding[239]
Depressive symptoms	Increased depressive symptomatology[4,168,169,183,198–203]	Increased depressive behavior[47,54,81–83,208,240–242]

in animal models (table 2.1). Each of the eight pathways represents preparatory tonic physiological adjustments to promote the detection of and response to potential threats, stressors, or pathogens. For instance, across human history a safe sleeping environment has typically meant co-sleeping,[68] and sleep quality predicts better physical and mental health, concentration, and memory.[69–71]

Research has shown that both objective and perceived social isolation serve as independent risk factors for mortality. For instance, a meta-analysis of 70 prospective studies involving more than 3 million human participants who were followed for an average of 7 years revealed that both objective and perceived social isolation increased the odds of mortality about the same as did obesity—around 25–30%.[28]

Objective and perceived isolation affected the odds of a shorter life span through different transduction pathways and mechanisms, however.[8] Objective isolation appears to operate through health behaviors, whereas perceived social isolation operates through a different set of mechanisms. For example, in a US nationally representative sample of 2,101 adults aged 50 years and over from the 2002 to 2008 waves of the Health and Retirement Study, the effect of perceived social isolation at baseline on mortality risk among older adults was determined over the subsequent 6 years.[72] Results indicated that perceived isolation was associated with increased mortality risk over a 6-year period. Importantly, the association between perceived isolation and mortality risk could *not* be explained by objective social isolation or by health behaviors.

The association between perceived social isolation and premature mortality appears instead to reflect the long-term physiological costs of evolutionarily adaptive short-term adjustments that promote short-term survival.[36,5] The evolutionary heritage of humans has shaped the brain to incline individuals toward certain ways of feeling, thinking, and acting. For instance, a variety of biological mechanisms have evolved that capitalize on aversive signals to motivate behaviors that promote short-term survival. Physical pain is an aversive signal that alerts an individual to potential tissue damage and motivates the individual to take care of the physical body. Perceived social isolation takes care of one's social body, which one also needs to survive and prosper. We are a social species.

2.4.1 Theoretical pathways linking loneliness to mortality in the modern world

According to our ETL, loneliness serves as an aversive biological signal to promote the repair or replacement of salutary relationships, and loneliness promotes short-term survival by triggering an interrelated set of behavioral, neural, hormonal, cellular, and molecular adjustments. These interrelated adjustments include (1) increased sleep fragmentation and decreased sleep salubrity, (2) activation of the hypothalamic-pituitary-adrenocortical (HPA) axis, (3) selective sympathetic tonus, (4) altered transcriptome dynamics in leukocytes, (5) decreased viral immunity, (6) increased inflammatory substrate, (7) increased prepotent responding, and (8) increased depressive symptomatology.[1] The theoretical rationale for each of these pathways is described below along with relevant empirical research. Evidence for the causal role of loneliness includes prospective and experimental research in humans and experimental animal studies.

Decreased sleep quality

Simply sleeping with another person does not ensure a safe sleeping environment, however. Sleep quality and salubrity vary inversely as a function of perceived social isolation. For instance, longitudinal and cross-sectional studies of humans show that the perception of being socially isolated (i.e., loneliness) increases sleep fragmentation and decreases sleep salubrity whether or not one is sleeping with another.[73–78] Experimental research using an animal model similarly indicates loneliness decreases sleep salubrity. Adult male mice socially isolated for 5 weeks, compared to pair-housed mice, showed a marked reduction in electroencephalogram (EEG) delta power in non–rapid eye movement sleep (i.e., deep sleep) during baseline conditions and a blunted homeostatic sleep response to acute sleep deprivation.[80]

Heightened activation of the HPA axis

The HPA axis is an important component of the neuroendocrine system that regulates physiological functions including metabolism, digestion, immunity, and energy storage and expenditure; and the physiological preparation for and responses to a perceived harmful event, attack, or threat to survival. Among the major hormones produced in the HPA axis are glucocorticoids (e.g., cortisol in humans, corticosterone in rodents). Increased HPA activation provides metabolic support for sustained responses to threats or stressors, which may enhance short-term survival but at a long-term cost. Research in humans has shown that perceived social isolation is associated with elevated tonic HPA activation (table 2.1).

Psychologist Rebecca Rueggeberg and colleagues[80] reasoned that lonely adults who engaged in what they termed "self-protective coping" (e.g., positive reappraisals, avoiding self-blame) would be buffered from elevated cortisol levels over the course of a day. A two-year follow-up showed that baseline levels of self-protective coping were associated with a reduction in the two-year increases in diurnal cortisol volume found in lonely individuals. This association was *not* found among nonlonely individuals, consistent with the notion that processes in the service of self-preservation, including heightened HPA activation, are more characteristic of individuals who feel socially isolated than those who do not.

Animal studies permit greater experimental control of the various influences of perceived social isolation on HPA activation and show that various species of rodents and nonhuman primates exhibit elevated basal activation of the HPA axis when chronically isolated as an adult from a preferred partner (e.g., a pair bond; table 2.1). For instance, studies in prairie voles show that animals that are chronically isolated from their pair-bonded partner show increased corticosterone levels[81,82] and higher corticosterone levels after a resident-intruder test,[83] whereas prairie voles that are chronically isolated from a conspecific for whom partner preference is low (e.g., same-sex sibling) show no such increase in corticosterone levels.[81,84] Prairie voles separated from their partner—but not when separated from other prairie voles—also show passive coping to stressors and depressive-like behavior. Isolation from a partner leads to decreased excitability of oxytocin neurons, decreased oxytocin messenger ribonucleic acid in the hypothalamus, and decreased oxytocin receptors in the striatum (a dopamine-rich area of the brain). Infusion of oxytocin into the striatum eliminates this depressive behavior, suggesting that perceived social isolation is associated with a withdrawal of oxytocin, which then leads to negative affect and depression.[85]

The quality and the strength of the relationship with the partner is important. This is nicely illustrated in research comparing the stress response between monogamous and polygamous monkeys.[86,87] The monogamous titi monkey is known to form strong mutual pair bonds, whereas the polygynous squirrel monkey (*Saimiri*) does not. Members of both species were housed in heterosexual pairs for several months. However, these species responded differently to social isolation. Following one hour of social isolation from their pair bonds, the normally monogamous titi monkeys (for whom partner preference is high) showed a strong stress response, as characterized by a significant increase in plasma cortisol. On the other hand, the normally polygynous squirrel monkeys (for whom partner preference is relatively low) showed no such response to isolation.[87] The titi monkey is not simply more easily stressed than the squirrel monkey, either. The titi monkeys did not show HPA activation when separated from their infant, whereas the squirrel monkeys showed a significant increase in plasma cortisol when separated from their infant.[86,88–92]

These results are consistent with the notion that it is not the objective presence or absence of a conspecific that determines HPA activation but rather the brain's interpretation of the presence or absence of the conspecific. Paralleling this specific pair-bond effect, adult domesticated dogs (*Canis familiaris*), who show "vocalization and destructiveness immediately after their owner's departure, intense greeting on reunion, and a persistent shadowing to maintain proximity to the owner during other times,"[93] have reduced glucocorticoid levels in the presence of their human caretaker, even when placed in a novel environment, whereas the presence of a long-term familiar (either a same-sex or an opposite-sex) kennel mate does not reduce their stress in a novel environment.[94]

Selectively elevated sympathetic tonus

The sympathetic adrenomedullary system (SAM) is involved in the fight-or-flight response to stressors, and there is evidence that increased broad sympathetic contributions to stress reactivity can increase the risk of the onset or progression of diseases.[94,95] However, according to the ETL, the putative neural, hormonal, and molecular adjustments triggered by loneliness do not represent a generalized fight-or-flight response to an acute stressor, but rather a *tonic preparatory* response that is more selective. For instance, elevated resistance to blood flow through the vascular system (i.e., total peripheral resistance) has served as a marker of threat surveillance in humans,[96] and loneliness in young adults has been associated with higher tonic levels of vascular resistance (but not heart rate) in laboratory studies[97] and during the course of a normal day.[98] More generally, loneliness in humans suggests that it is more closely tied to the tonic activation of the vasculature (hemodynamics) rather than activation of the heart (cardiodynamics).[99]

In addition, fibrinogen is a blood coagulation factor that has been associated with coronary heart disease, and two studies have reported an association between loneliness and elevated fibrinogen levels,[100,101] whereas one reported this association was not significant.[102] No information was provided about the direction or strength of this association by Shankar and colleagues,[102] however, and these authors noted that the levels of reported loneliness were low in this study (e.g., "only 2% of participants reported being lonely all of the time").

The cardiovascular system differs in vulnerabilities across the life span. Blood pressure is a regulated physiological end point, in which control centers in the brain monitor and react to deviations from homeostasis through a negative feedback mechanism, which reverses a deviation from the normotensive set point for blood pressure. Blood pressure is a function of vascular resistance and cardiac output, and when homeostatic mechanisms are robust, blood pressure is maintained within a normal restricted set of values through adjustments to vascular resistance and/or cardiac output.[103] Homeostatic mechanisms undergo wear-and-tear over the course of a life span, however, and their efficacy diminishes. Although elevated vascular resistance and elevated levels of fibrinogen in young adults are risk factors for higher blood pressure later in life, an association between loneliness and blood pressure would be expected to emerge later in life once the homeostatic constraints on blood pressure have weakened.

Loneliness has been associated with elevated basal levels of blood pressure in a number of studies of older adults. In the inaugural scientific investigation of loneliness, Parfitt[104] noted "cardiovascular degeneration" and high blood pressure were associated with loneliness. Four decades later, programmatic investigation by Lynch and colleagues[105–107] provided additional evidence for an association between isolation and chronic cardiovascular conditions such as high blood pressure and cardiovascular disease. Loneliness has been associated with elevated basal levels of blood pressure in a number of studies[97,108–112] but not in all*.[113–115] Increased attention to the diagnosis of high blood pressure and the development of effective treatments may be complicating factors, especially in light of evidence that lonely individuals are more, rather than less, likely to access and use medical services.[116–118]

Cross-sectional and prospective studies have also reported a significant association between loneliness and cardiovascular disease even after controlling for various covariates.[112,119–123] To disentangle potential effects on incidence versus prognosis, Valtorta and colleagues performed a meta-analysis of studies on *new* coronary heart disease (CHD) and/or stroke diagnosis as a function of loneliness or objective social isolation.[124] The meta-analysis included data from 11 CHD studies and 8 stroke studies. No difference was found between the association of CHD incidence with loneliness or objective social isolation, so the meta-analysis was performed collapsing across these measures, which they termed "poor social relationships." Results showed that poor social relationships were associated with a 29% increase in risk of incident CHD and a 32% increase in risk of stroke.[125]

Animal studies of prairie voles indicate that chronic isolation of these typically monogamous animals from a preferred partner induces alterations in cellular functioning in the vasculature (e.g., the release of vascular contracting factors in endothelial cells) that contribute to higher levels of vascular resistance.[126] Studies of rodents[127,128] and baboons[129] have also found that isolation leads to increases in blood pressure. For instance, Coelho et al. performed an experimental study

*Steptoe et al. found loneliness to be related to diastolic blood pressure in response to experimental stressors rather than to basal levels.[102]

of blood pressure in adult male baboons contrasting three social housing conditions: (1) individual housing (social isolation), (2) the standard housing with a social companion, and (3) housing with a social stranger.[129] These conditions made it possible to evaluate the effects of the loss of companionship and mutual protection/assistance, and the effects of social isolation per se. Social isolation per se was not the important factor: solitary housing and housing with an unfamiliar animal were associated with higher blood pressure than housing with a social companion.

In sum, research suggests that the sympathetic nervous system may be affected by or related to loneliness in subtler ways than by triggering a general fight-or-flight stress response. Instead, loneliness is associated with more tonic and specific sympathetic adjustments, such as increased basal sympathetic tonus to vascular and myeloid tissue.[100,130] There is also suggestive evidence that loneliness selectively increases sympathetic activation of myeloid cells to alter myeloid cell population dynamics in a monkey model.[130] The similarities in human and animal literatures on the effects of chronic perceived social isolation on the various mechanisms that could contribute to an increased risk of mortality do not end with the HPA axis or the sympathetic system. The experimental manipulation of perceived isolation in animals is associated with many of the same pathophysiological changes as observed in humans.

Altered transcriptome dynamics

Chromosomes, located in the nucleus of all cells, are made from long strands of deoxyribonucleic acid (DNA) molecules. In humans, there are 46 strands of DNA organized in pairs, constituting 23 chromosomes, with one DNA strand in each pair from one parent. A gene is a short section of DNA, and the genotype represents the full genetic makeup of a cell, and therefore of an individual. The human genotype represents the molecular machinery through which human phenotypes and behavior are expressed, but the environment—including the social environment—operates on these phenotypes over generations through natural selection to shape the genotype (DNA), and within generations across time through the transcription of the genotype to shape the expressed genotype.

Genes exert effects because DNA molecules serve as templates to construct ribonucleic acid (RNA) copies, a process that is known as transcription. RNA, in turn, codes for a sequence of amino acids that together form the proteins (e.g., hormones and neurotransmitters) that regulate processes in the brain and body.[131] The transcriptome refers to the entire set of RNA molecules in one cell. Unlike the genotype, which represents a genetic blueprint expressed in the form of a set of DNA molecules that are the same in all cells in the body, the transcriptome represents gene expression, which can differ across cells and across time. *Transcriptome dynamics* refer to changes in the transcriptome, for instance, as a function of the environment, whether perceived (e.g., loneliness) or real (e.g., exposure to toxins).[132,133] Investigations of transcriptome dynamics, therefore, are focused on gene functioning rather than genetic structure.

In an early investigation, we discovered that the transcriptome dynamics of leukocytes differed between older adults high versus low in loneliness, with individuals high in loneliness showing differences in the expression of hundreds of genes including the up-regulation of proinflammatory genes and down-regulation of genes involved in glucocorticoid receptor signaling and interferon responses (i.e., viral immunity).[133] We replicated this association in a more comprehensive investigation of older adults,[134] and the association has since been replicated in several other studies.[130,135,136]

The pattern of threat-related or stress-related changes in gene expression, which has been termed the Conserved Transcriptional Response to Adversity (CTRA), has potential evolutionary significance. For instance, the white blood cells in people who feel socially isolated, compared to those who do not feel isolated, show the activation of CTRA, which involves changes in gene expression that lead to decreased inflammatory control, decreased antiviral defense, and increased glucocorticoid insensitivity[133,134] (box 2.2)—responses that have putatively evolved through natural selection to promote short-term survival when an organism perceives itself to be socially isolated.

Social behaviors characterized by mutual benefit have historically involved close or frequent contact between conspecifics, which increases the likelihood of exposure to a viral infection. In such contexts, a bias in immunity toward antiviral readiness would be adaptive. However, when there is a shift in social behaviors from mutual benefit (or altruism) to selfishness (or spite)—as when an individual feels lonely—the adaptive state of readiness for the immune system shifts to an up-regulation of proinflammatory gene expression and a down-regulation of antiviral responses to better deal with bodily injury and bacterial infection from hostile human contact or increased predatory vulnerability due to separation from the social group.[132,134] These changes may have promoted short-term survival across human history, but in industrialized societies where people are living closer together and longer than ever before, chronic inflammation has become

BOX 2.2. Gene Regulation by the HPA Axis

Several studies now suggest that perceived social isolation is associated with glucocorticoid resistance and a complementary increase in proinflammatory gene expression that may contribute to some of the adverse health outcomes associated with perceived social isolation.[8,131] In this box, we briefly describe the mechanism:

Activation of the HPA axis releases glucocorticoids that circulate through the bloodstream to reach virtually every cell type in the body.

Glucocorticoids (GCs) refer to a class of steroid hormones that bind to a particular receptor on the surface of cells and inside cells called the glucocorticoid receptor. GCs regulate a wide array of physiologic processes by simultaneously altering the transcription of hundreds of genes. GC molecules are small and diffuse across cell membranes and into the cytoplasm, where they can bind to intracellular glucocorticoid receptors (GRs). These GRs then travel into the nucleus of the cell, where they can bind to genes that contain specific DNA sequences called glucocorticoid response elements (GRE).

Some anti-inflammatory effects of glucocorticoids are mediated by transcriptional induction of molecules that inhibit immune responses. GR molecules can also inhibit the transcription of specific genes, either by binding to their DNA sequences in locations that block access by other stimulatory molecules, or by binding to stimulatory molecules in the cytoplasm and blocking their translocation to the nucleus. For example, many anti-inflammatory effects of glucocorticoids are mediated by glucocorticoid receptor antagonism of the proinflammatory transcription factors NF-κB (nuclear factor kappa-light-chain-enhancer of activated B cells) and AP-1 (activator protein 1). GR transcriptional repression also mediates the negative feedback loop in the hypothalamus that prevents accumulation of excessive GC levels. The combination of strong transcriptional activation of some gene sets and transcriptional repression of other gene sets allows one specific hormonal signal to influence a diverse array of biological processes in a wide range of different cell types.

These dynamics can result in a state of "glucocorticoid resistance" in which normal or high levels of HPA activity have little or no effect on cellular function because the glucocorticoid receptor fails to translate the hormonal stimulus into a gene transcriptional response.

associated with various diseases including cancers and cardiovascular disease.

To investigate the potential causal role of loneliness, cross-lagged panel models were calculated in our population-based longitudinal Chicago Health, Aging, and Social Relations Study of older adults.[137] Results indicated that increases in loneliness led to an up-regulation of the expression of genes underlying inflammation and a down-regulation of the expression of genes that defend against viral infections that were measured one year later (the CTRA). Reciprocal effects were also observed, with the CTRA predicting higher loneliness one year later—putatively through the affective and behavioral effects of proinflammatory cytokines on the brain.[130] These results were specific to loneliness and could not be explained by demographic factors or various other factors such as depressive symptomatology or social support.

Recently, textual analyses of natural language have been used as indicators of gene regulation in the human immune system.[138] Of the self-report measures that were included in this study (e.g., stress, anxiety, depression, loneliness), only loneliness was significantly related to CTRA gene expression. In addition, total language output and patterns of word use covaried with CTRA gene expression. Specifically, CTRA gene expression was related to a low prevalence of third person plural pronouns (e.g., they, them, their) and a high prevalence of adverbs (e.g., really, very, certainly), impersonal pronouns (e.g., it), and third person singular pronouns (e.g., he, she, him, her). The authors suggested the language structure associated with the CTRA may reflect greater CNS arousal and a relatively inward orientation toward a social world perceived implicitly to be threatening:

> Given the observed relationship between personal expression and gene expression, patterns of natural language use may provide a useful behavioral indicator of nonconsciously evaluated well-being (implicit safety vs. threat) that is distinct from conscious affective experience.[138]

We have also investigated the cellular mechanism underlying the association between loneliness and the CTRA in rhesus macaques behaviorally classified previously as high in loneliness or controls.[53] Results from the rhesus macaque model revealed a selective expansion of the monocyte pool that was limited to a subset of monocytes representing a less mature phenotype, which are more effective in producing a reactive oxygen species in response to bacteria but are also inflammation-primed, interferon-impaired, and glucocorticoid insensitive.[130] This selective change in the circulating monocyte population served as the primary mechanism underlying the differences observed in transcriptome dynamics as a function of loneliness.

Finally, and consistent with the pathways posited by the ETL, the macaques that had been classified behaviorally as high in loneliness showed heightened sensitivity to social threat, elevated activity of the sympathetic nervous system, lower levels of interferons under basal conditions, reduced glucocorticoid target gene expression, reduced cellular sensitivity to circulating glucocorticoids, and an impaired response to a viral infection.[130]

Cole, Levine, et al., who replicated the positive association between loneliness and the CTRA, also reported a negative association between subjective well-being (eudaimonia) and the CTRA in a small subsample of older adults from the HRS.[136] In a joint analysis, the association between subjective well-being and the CTRA vitiated the association between loneliness and the CTRA. Longitudinal analyses have shown that loneliness and subjective well-being are reciprocally influential,[139,140] and together the results suggest that interventions designed to enhance well-being may diminish the effects of loneliness on the implicit surveillance for social threats and the associated physiological preparation for social stressors.

Together, these studies support a mechanistic model in which chronic loneliness predicts a sympathetically mediated increase in the release of immature monocytes from the bone marrow, a down-regulation of glucocorticoid receptor sensitivity and antiviral gene expression, and an up-regulation of inflammatory gene expression. However, these studies do not address possible differences in transcriptome expression in the brain. To address this gap in knowledge, genome-wide RNA levels have been measured recently in postmortem nucleus accumbens from donors for whom stable measures of loneliness were available.[141] The nucleus accumbens was selected because of its previous association with social reward and differential activation in response to positive social, in contrast to nonsocial, stimuli.[142] Results suggested that loneliness was associated with differentially expressed transcripts previously associated with behavioral processes, neurological disease, psychological disorders, cancer, organismal injury, and skeletal and muscular disorders. The authors noted that the highest up-regulated gene in the nucleus accumbens of lonely individuals was Cocaine and Amphetamine Regulated Transcript Protein (CART), and they speculated that the dopamine-countering effects of high levels of CART may contribute to reduced nucleus accumbens activation in lonely individuals, making positive social interactions feel less rewarding and possibly contributing to feelings of loneliness.[141] Previous research has shown that loneliness is associated with less positive social interactions[37,98,143] and less activation of brain regions involved in reward in response to positive social images of strangers than positive nonsocial images.[142,144]

Decreased viral immunity

Perceived social isolation is also associated with a reduction in viral immunity in humans (table 2.1). Empirical work in humans suggesting that loneliness is associated with diminished viral immunity dates back more than three decades. The first such study we found reported that elderly adults high, in contrast to low, in loneliness are characterized by lower levels of immunoglobulin (i.e., antibodies IgG, IgA, IgM).[145] Subsequent research showed an association between loneliness and cellular immunocompetency (e.g., lower levels of natural killer cell activity) in psychiatric patients[146] and medical students.[147] Kiecolt-Glaser and colleagues also found that medical students high, relative to low, in loneliness showed larger increases (from basal levels measured one month earlier) in antibody titer levels to Epstein-Barr virus (EBV) in response to the stressors of medical exams, indicative of decreased viral immunity.[148,149]

The association between loneliness and vaccine response has also been investigated. In an illustrative study, loneliness was measured four times daily for the two days preceding influenza immunization, 11 days following immunization, and biweekly over the following 14 weeks.[150] The average of these measures served as a measure of the extent to which a student felt socially isolated (lonely). Results indicated that perceived social isolation was related to variation in the antibody response for the A/New Caledonia vaccination measured one month and four months after the immunization. Social isolation was associated with the lowest antibody response, and greater psychological stress.

The evolutionary explanation for the transcriptome dynamics identified in leukocytes extends to viral immunity, and the transcriptome changes associated with loneliness in humans and rhesus monkeys suggest that an important aspect of the molecular substrate for an association between loneliness and reduced viral immunity is in place. To investigate the potential *functional* significance of these transcriptome changes, the expression of Type I and II interferons was assessed in an additional sample of macaques before and at 2 weeks and 10 weeks following experimental infection with the simian immunodeficiency virus (SIV).[130] Measures at baseline again showed that lonely, compared to control, macaques showed lower levels of interferon gene expression. Two weeks after the experimental infection (peak of acute viral replication), interferon gene expression was significantly elevated and did not differ as a function of loneliness. However, 10 weeks after the experimental infection (after establishment of a long-term viral replication set-point), lonely macaques showed lower levels of interferon gene expression than the control animals. The lonely animals also showed poorer suppression of SIV gene expression between the postinfection measurement periods as well as an elevated SIV viral load and reduced anti-SIV immunoglobulin G (IgG) antibody titers at ten weeks. These results are in accord with the notion that loneliness is associated with a reduction in viral immunity and underscore the importance of the timing of the immune response in studies of loneliness and viral immunity.

Finally, and paralleling the monkey study, perceived social isolation in human immunodeficiency virus (HIV)–infected men has been associated with higher levels of herpesvirus 6 antibody titer levels,[151] and lower CD4+ (cluster of differentiation 4) helper cells,[152] suggesting that perceived social isolation in HIV+ men may be associated with higher odds for viral progression.

Increased inflammatory substrate

Inflammation is an important component of immune functioning that is responsible for an arsenal of weapons to kill germs and combat physical, microbial, autoimmune, and metabolic insults to the body. The proinflammatory response to a trauma or pathogen, for example, is part of an adaptive immune response to remove pathogens and dead or dying cells and restore homeostasis. According to the ETL, such a response is advantageous evolutionarily, especially under conditions such as perceived social isolation (loneliness), which in the absence of mutual aid and protection may have been associated with an increased likelihood of exposure to germs through cuts and abrasions.

The functional utility of the modulation of the proinflammatory response by loneliness is illustrated in animal research on wound healing. In an experimental model of wound healing that involves an inflammatory component, social isolation in the monogamous *Peromyscus californicus* mouse *facilitated* wound healing, whereas social isolation in the polygynous *Peromyscus leucopus* did not affect healing relative to the group-housed comparison condition*.[153] The shift toward an inflammatory substrate under conditions of perceived social isolation may have contributed to the likelihood of short-term self-preservation.

In the modern age, chronic proinflammatory conditions may produce short-term benefits, but they contribute to a host of chronic diseases, including stroke, obesity, diabetes, cardiovascular disease, Alzheimer's

*Social isolation from a partner in a monogamous murine species produces behavioral and neurological evidence of loneliness, whereas social isolation from a partner in a polygynous murine species does not produce evidence of loneliness.[33,60,245]

Disease, and a number of cancers.[154–157] The transcriptome changes in leukocytes associated with loneliness also suggest that loneliness in humans is associated with an increased inflammatory substrate (see above), but several studies have failed to find a significant association between loneliness and indirect markers of tonic levels of inflammation (e.g., C-reactive protein).[100,102,135,158]

However, the changes in inflammatory biology suggested by the transcriptome differences in circulating leukocytes may be better reflected in the synthesis of proinflammatory cytokines rather than indirect circulating markers of inflammation. In a study bearing on this notion, Steptoe and colleagues investigated the association between loneliness and inflammatory responses to a laboratory stressor in middle-aged adults from the Whitehall cohort. Interleukin-6 (IL-6), interleukin-1 receptor antagonist (IL-1Ra), and the chemokine monocyte chemotactic protein (MCP-1) served as inflammatory markers.[159] Results showed that loneliness in women, but not men, was significantly associated with elevated levels of MCP-1 at baseline and throughout the task, and with the intensity of the IL-6 and IL-1Ra response to the psychological stressor.

Subsequent studies have found an association between loneliness and the inflammatory response to acute experimental stressors in both men and women. For instance, in study 1, lonely healthy adult men and women showed higher levels of the proinflammatory cytokines IL-6 and tumor necrosis factor-α (TNF-α) in response to laboratory stressors than their nonlonely counterparts, and in study 2, lonely posttreatment breast-cancer survivors exhibited greater synthesis of the proinflammatory cytokines IL-6 and interleukin-1β (IL-1β), and nonsignificantly greater synthesis of TNF-α, than their nonlonely counterparts.[160] In a subsequent investigation, participants were categorized based on a composite measure of loneliness, anxious attachment, fear of negative evaluation, and rejection sensitivity. Results showed that participants high, relative to low, on this composite measure showed stronger IL-6 and TNF-α responses to an inflammatory challenge (endotoxin).[161]

Experimental studies in animals also indicate a positive association between loneliness and *inflammation*. For instance, socially isolated mice, relative to socially housed mice, show elevated circulating levels of the proinflammatory cytokine IL-6 following an experimental stroke[162] or cardiac arrest.[163] In addition, microglial cells are a type of glial cells found throughout the central nervous system. The phasic activation of microglial cells (microgliosis) is involved in the production of local inflammatory responses, removal of dead cells, and regeneration following brain injury, whereas prolonged activation leads to prolonged neuroinflammation, neurodegeneration, and diminished recovery.[164] Social isolation in the murine model produces larger increases in microglial activation and increased neuronal damage following brain injury.[162,164] As Karelina and DeVries note, "These data provide compelling evidence for a role for neuroinflammation as a mechanism by which social interaction influences health" (p. 73).[164]

In sum, there is clear experimental evidence for an association between loneliness and inflammation in animal models, and experimental animal research shows that social isolation, in contrast to group housing, produces an elevated inflammatory response that facilitates the healing of bodily wounds in monogamous rodents but not in polygynous rodents—that is, in species in which social isolation produces behavioral evidence for the condition of loneliness. Inflammation, like immunity more generally, is a multifarious process that is influenced by a host of factors that are more difficult to control in human than animal studies, however. The evidence in human studies suggests that the operation of these uncontrolled influences may be more problematic in studies of loneliness and inflammation based on circulating markers of chronic inflammation (e.g., C-reactive protein) than on the gene expression in leukocytes contributing to the synthesis of proinflammatory cytokines, and inflammatory responses to an acute stressor.

Increased prepotent responding

Perceived social isolation increases the likelihood of prepotent responding (table 2.1). A prepotent response is a behavior that has priority over other response tendencies due to maturational primacy or practice or repetition with positive reinforcement. Impulsive responses are among the responses that are prepotent, so by increasing prepotent responses, perceived social isolation may also decrease impulse control. Prepotent responses exist because they have generally proven to be adaptive for an organism even though, in a specific instance, responses lower in the response hierarchy (and that require self-control to express) may be superior. Learning, for instance, often involves overcoming a prepotent response to learn a new, more adaptive form of responding, whereas performance is best when it relies on prepotent responses such as a sequence of behaviors that has been well-learned. The likelihood of the expression of a prepotent response increases under conditions of high activation and uncertainty because it has the highest effective habit strength. Accordingly, prepotent responses are often adaptive in conditions in which short-term self-preservation is a priority, such as when an organism perceives itself to be socially isolated.

One of the most obvious symptoms of perceived social isolation is that it makes people feel despondent and dejected. Numerous studies have reported significant correlations between perceived social isolation and depressive symptomatology,[165,166] and for decades many clinicians believed that perceived isolation was simply an aspect of depression with no distinct concept worthy of study.[167] However, psychometric analyses have shown that these states are stochastically and functionally separable,[4,168] and coheritability analyses in genome-wide association studies have confirmed that perceived social isolation and depression are distinct phenotypes.[41] Longitudinal research[169] and animal studies[25] have further shown that perceived social isolation increases depressive symptomatology above and beyond what can be explained by initial levels of depressive symptomatology. Even these effects of perceived social isolation may have been adaptive across human history. For instance, the auditory and expressive signals of sadness and depression function, in part, as a safe call to others for aid and protection, and the associated lethargy reduces the likelihood of conflict in a potentially hostile social environment.

In sum, human and animal research indicates that perceived social isolation triggers a set of interrelated behavioral, neural, hormonal, cellular, and molecular responses that promote short-term self-preservation, even though these responses may have long-term costs for health and well-being. The fact that many of these responses have a deep phylogenetic basis means that many—such as the implicit hypervigilance for social threats and decreased impulse control—can be orchestrated by the human brain without a person's intention or awareness.

Increased depressive symptomatology

Prior research has clearly shown that depression can have adverse health effects.[170–173] Among the research we have reviewed thus far are studies showing that the association between loneliness and various pathways is not simply mediated by depressive symptomatology.[77,111,134,174,175] However, an early evolutionary analysis of depressed states presaged the ETL predictions. This analysis proposed that depressive symptomatology evolved to minimize risk of social exclusion or harm when their social value in the interaction is less than their social burden.[176] The ETL builds on this early work to argue that the effects of loneliness on depressive states and behaviors may prove deleterious in the long term, but these effects increased short-term fitness.

For instance, the depressed state that results from loneliness lessens the likelihood of attempts to force one's way back into an important group from which one feels isolated and increases the likelihood that an individual will exhibit facial displays, postural displays, and acoustic signals that may serve as a call for others to come to its aid to provide companionship, protection, and support.[5,6] In this section, we review evidence that loneliness increases depressive symptomatology, representing yet another pathway through which chronic or repeated loneliness may contribute to premature mortality.

The most common clinical focus on loneliness has been on its association with poor mental health, with an emphasis on depressive symptomatology. Many studies have reported significant correlations between loneliness and depressive symptomatology,[50,57,179–196] and for decades many clinicians believed that loneliness was simply an aspect of depression with no distinct conceptual features worthy of study.[167,197]

Importantly, longitudinal research indicates that loneliness and depression are separable, loneliness predicts increases in depressive symptomatology above and beyond what can be explained by initial levels of depressive symptomatology, and the prospective associations between loneliness and depressive symptomatology are reciprocal.[168,169,183,198–203] In addition, experimental manipulations of loneliness have been found to produce higher negative mood, anxiety, anger, and depressive symptomatology,[4] and, as we noted, coheritability analyses in genome-wide association studies show that loneliness and depression are distinct phenotypes.[41]

In sum, the cumulative research suggests that loneliness contributes to depressive symptomatology, which in turn may have increased short-term survival but in the modern era has long-term costs in terms of mental and physical health. As such, the effect of loneliness on depressive symptomatology represents yet another important and powerful pathway through which loneliness may contribute to premature mortality.

2.5 Perceived Social Isolation and the Brain

Selected components of the neural network underlying the interrelated adjustments are illustrated in figure 2.4. Additional details about the brain network sustaining perceived social isolation are discussed in chapter 3.

In brief, several parts of the brain are involved in the way you feel connected to or disconnected from other people. In the case of loneliness, key areas of the social brain are hypo-activated, while others (that

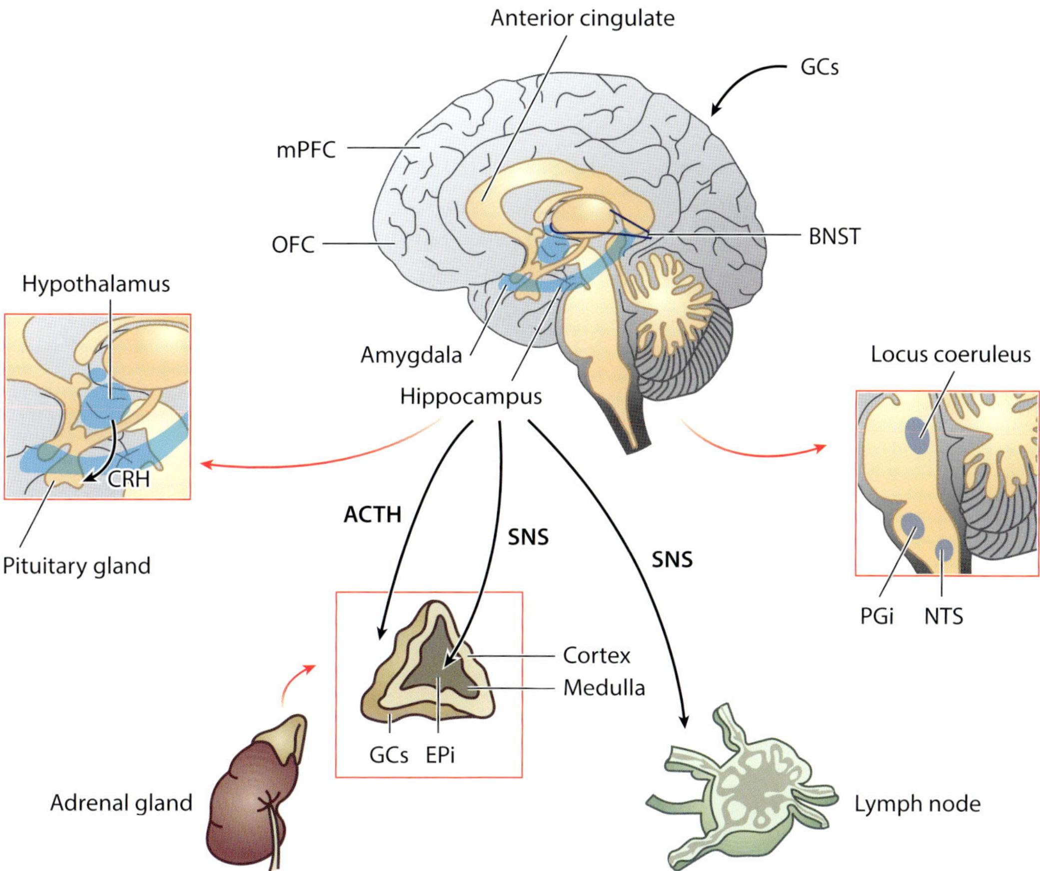

FIGURE 2.4. Schematic of illustrative neural components of the network underlying the neural adjustments to loneliness. The orbitofrontal cortex (OFC) and medial prefrontal cortex (mPFC) are involved in perceived social isolation (loneliness) and project to posterior regions such as the BNST, which orchestrates tonic adjustments in the HPA axis, the sympathetic adrenomedullary (SAM) axis, and the innervation of the vascular, lymph node, and myeloid tissue by the sympathetic nervous system (SNS). The HPA axis controls circulating levels of glucocorticoids (GCs) through a cascade that starts with signals from the prefrontal cortex and limbic regions (e.g., amygdala, BNST) to the paraventricular nucleus of the hypothalamus, which secretes corticotropin-releasing hormone (CRH) into the hypophyseal portal circulatory system. This then stimulates the anterior pituitary to release adrenocorticotropic hormone (ACTH). ACTH travels through the blood to the adrenal cortex where it acts on melanocortin Type 2 receptors (MCR2s) to stimulate the secretion of glucocorticoid hormones (cortisol in humans and most mammals, corticosterone in rodents) into circulation. Adapted from Cacioppo et al.[8]

are not supposed to be activated during social connection) are hyper-activated. For instance, brain areas that are involved in pair-bonding and feelings of closeness tend to be deactivated when you feel lonely. On the other hand, brain areas that are involved in your surveilling your surroundings for potential social threats are hyper-activated.

The prefrontal cortex (PFC), for instance, is involved in the perception of social isolation (i.e., loneliness), and the amygdala and bed nucleus of the stria terminalis (BNST) to which the prefrontal cortex projects are involved in orchestrating preparatory tonic physiological adjustments for potential threats and stressors. The central and medial nuclei of the amygdala and the BNST are connected by cells throughout the stria terminalis, and both the amygdala and the BNST project to hypothalamic and brainstem areas that mediate autonomic, neuroendocrine, and behavioral responses to aversive or threatening stimuli.[8] The amygdala is especially important for rapid-onset, short-duration behaviors that occur in response to specific threats, whereas the BNST mediates slower-onset, longer-lasting responses that frequently accompany sustained threats (or the surveillance for threats) and that may persist even after threat termination.[204]

In a series of investigations of perceived social isolation in mice, neuroscientist Gillian Matthews and colleagues found synaptic changes in dopamine neurons in the dorsal raphe nucleus (a brain area important for social connection and reward) following a 24-hour period of social isolation.[205] Specifically, synaptic inputs onto dopamine neurons in the dorsal raphe nucleus were potentiated following social isolation. These changes appeared to be specific to the dopamine neurons in the dorsal raphe nucleus, as no differences were found in dopaminergic neurons of another brain area involved in social connection and reward, the ventral tegmental area. Moreover, the synaptic changes in the dorsal raphe nucleus were not observed following movement of the animal into a new cage but were observed following social isolation in a new cage, suggesting that these synaptic changes were driven by the experience of social isolation rather than by a nonspecific salient environmental change.

To determine whether acute social isolation influenced the naturally occurring activity in the dopamine neurons in the dorsal raphe nucleus, the activity of these neurons was measured when the target mouse made first contact with a novel juvenile mouse that had been introduced into the home cage of a target mouse that had been group-housed or socially isolated for 24 hours. Neuronal activity in the target mouse increased significantly more when it had previously been isolated than when it had been group-housed. These studies, together with experiments in which activity in the dopamine neurons in the dorsal raphe nucleus was experimentally manipulated, suggest that dopamine neurons in the dorsal raphe nucleus play an important role in motivating reconnection following an acute period of social isolation.

The optical stimulation of dopaminergic neurons in the dorsal raphe nucleus led to the release of dopamine in two brain areas that are two major players in the HPA axis: the amygdala and the BNST, with a greater dopamine release in the BNST than in the central amygdala. These results suggest a mechanism in which perceived social isolation affects basal HPA functioning and behavior via inputs to the BNST (and to a lesser extent the amygdala), and that exposure to conspecifics modulates activity in the BNST, HPA functioning, and behavior through neurons in the dorsal raphe nucleus that modulate activity in the amygdala and BNST.[9]

Animal models also indicate that perceived social isolation diminishes neurogenesis and nerve growth factors[52] and increases inflammation[162] in the brain. For instance, social isolation decreases myelination in the PFC of mice.[206,207] Moreover, the ongoing myelination that occurs in the adult PFC represents a form of myelin *plasticity* to adapt brain structures and functions to environmental demands.[207] To test reversibility (i.e., plasticity), mice previously isolated for eight weeks were group-housed for four weeks. As predicted by the adult myelin plasticity model in which myelination plasticity serves to adapt brain function to environmental demands, the myelin transcripts in PFC and social behavior returned to control levels in the social reintegration group.[207]

Research has identified additional molecular mechanisms through which social isolation may impact neurogenesis in parts of the adult brain. For instance, social isolation has been shown to reduce levels of nerve growth factors including brain-derived neurotropic factor (BDNF); cellular transcription factors that modulate the production of proteins in the brain (e.g., cyclic adenosine monophosphate response element-binding protein); and the endogenous neurosteroid, allopregnanolone.[cf. 52] Moreover, the exogenous administration of allopregnanolone (or its precursors) reduces the effects of perceived social isolation on HPA activity, BDNF expression, and depressive behavior.[208,209]

These effects of loneliness on the brain may have real-world consequences, as well. In a large prospective study of individuals at risk for Alzheimer's disease, neurologists David Bennett, Rob Wilson, and colleagues showed that loneliness at baseline predicted greater cognitive declines in multiple cogntive domains and increased risk for dementia even after controlliong for objective social isolation, education, gender, age, and other health-related factors.[210] Analyses further showed that loneliness predicted late-life dementia even after controlling for objective social isolation and standard neuropathological measures derived from brain autopsies.

Although the mechanism underlying the association between loneliness and cognitive decline has not yet been identified, there are two lines of evidence that point to neuroinflammation as possibly playing a role. First, studies in mice have found that social isolation decreases central anti-inflammatory responses and survival rate, and increases neuroinflammation, swelling, and cell death following the experimental induction of stroke in mice.[162,211,212] Second, research in humans has shown loneliness to be related to increased gene expression of proinflammatory NF-κB (nuclear factor kappa-light-chain-enhancer of activated B cells) transcripts[133,213] and inflammation.[101,159,160]

We have focused in this section on the effects of the perceived absence of salutary social connections, but the purpose was to investigate the influence of salutary connections on the brain. The normal state for social animals is to have stable, salutary connections, which are maintained in part because these relationships have intrinsic value. Neuroimaging research has shown, for instance, that the decision to maintain a social relationship produces greater activation in regions of the brain associated with pleasure, reinforcement, and value (striatum, ventromedial prefrontal cortex, septo-hypothalamic)[214] than the decision to maintain a nonsocial relationship. Additional benefits of such relationships are discussed in chapter 8.

2.6 Concluding Remarks

The effects of perceived social isolation on the brain and biology of humans and social animals are surprisingly broad and deep. Every living thing inherits systems of physiological carrots and sticks that direct its behavior. Social connections can engage components of a reward network, and the perceived absence or diminution of salutary social connections can engage components of a defensive network.[24] Moreover, the perception of isolation from others—of being on the social perimeter—is not only unhappy, it signals danger across phylogeny. Fish have evolved to swim to the middle of the group when predators approach,[215] mice when housed in social isolation rather than in pairs show sleep disruptions and reduced slow wave sleep,[79] and prairie voles when isolated from their partner and subsequently placed in an open field show less exploratory behavior and more predator evasion.[216] For instance, fish on the edge of a school are more likely to be attacked by predatory fish, not because they are the slowest or weakest, but because it is easier to isolate and prey upon those that are isolated from others.[215]

Given the danger involved in a social animal being absent or isolated from salutatory social connections, the brain has evolved to monitor the status of an organism's social body just as it monitors the status of an organism's physical body. Early in our history as a species, we survived and prospered by banding together—in dyads, in families, in tribes—to provide mutual protection and assistance. The signal of perceived social isolation—triggered by a discrepancy between an individual's preferred and actual social relations—is similarly part of a biological warning system that has evolved to warn us of threats or damage to our social connections, which as a member of a social species increases our likelihood of surviving, reproducing, and leaving a genetic legacy. Perceived isolation motivates individuals to repair, renew, or replace the social connections needed to insure survival and to promote social trust, cohesiveness, and collective action while also increasing the focus on self-preservation in a social context that is perceived as potentially dangerous. These goals are worthwhile only if the individual survives a social environment in which the brain perceives it is socially isolated, however. Thus, many of the behavioral and physiological consequences of perceived social isolation found in human and nonhuman social species (table 2.1) are designed to give primacy to short-term self-preservation rather than to reestablish trust in and connections with others, who may be more inclined to be hostile than hospitable. The fact that many of these responses have a deep phylogenetic basis means that many—such as the implicit hypervigilance for social threats and the decreased impulse control—can be orchestrated by the human brain without a person's intention or awareness.

3

THE SOCIAL BRAIN

The coordinated movement of a flock of birds, a herd of animals, a school of fish, a pod of whales, or a classical ballet group of dancers can be a marvel to behold. Each consists of a collection of individual organisms whose movement within the collective is controlled not by a single mind but by the brain within each individual animal. The flawlessness and fluidity of the collective action across time and space belie the complexity of the orchestration of movements and adjustments required by each animal within a dynamic social context. The calculus of social perception and social interactions is possible because of the development within social species of neural and hormonal mechanisms underlying complex cognitive capacities that have been sculpted over generations by the specific environmental challenges by a species. These developments differ by species, so in this chapter we emphasize the neural developments that contributed to the ascendancy of *Homo sapiens* to the top of the food chain. There is little fossil record of brains across millennia, however; so paleontological, archeological, and comparative investigations have provided critical evidence bearing on the emergence of the human brain.[1,2]

Human evolution was marked by a variety of milestones (e.g., see box 3.1). One of the earliest hominids, the genus *Ardipithecus*, evolved in Africa about 6 million years ago, living primarily in trees like other primates. However, hominids from this group took the first steps toward walking upright, which helped them survive in the forests and grasslands in which they lived. Approximately 4 million years ago, the genus *Australopithecus* evolved in Africa. Open grasslands and wooded areas remained their primary habitats, and they still climbed trees, but they walked upright most of the time. Around 3 million years ago, the genus *Paranthropus* evolved in Africa. This group of early hominids had large teeth and powerful jaws, enabling them to consume a wider variety of foods.

BOX 3.1. Major Milestones in Human Evolution

Bipedal walking appeared about 6 million years ago in an early human known as *Sahelanthropus*. Approximately 4 million years ago, the bodies of early humans evolved in ways that allowed for sustained upright walking but still permitted the climbing of trees to take advantage of nearby open areas and dense woods. Approximately 2.6 million years ago, tool use appeared, and more than 2 million years ago tools were used to pound, crush, and access new foods. Approximately 800,000 years ago, humans began to use fire as a tool, including for cooking, warmth, safety, and socializing. Dramatic climate changes occurred from 800,000–200,000 years ago—a period that also saw a rapid expansion of human brain size and complexity, which enabled humans to interact with each other and their surroundings in new and creative ways to survive this period. About 12,000 years ago, humans began to transform the Earth's natural landscapes by deliberately influencing the growth and breeding of various plants and animals. Farming and herding (e.g., planting rather than eating the seeds) depend on a sophisticated sense of time and the capacity to plan and delay immediate gratification—capacities that are served by a well-developed prefrontal cortex.[38]

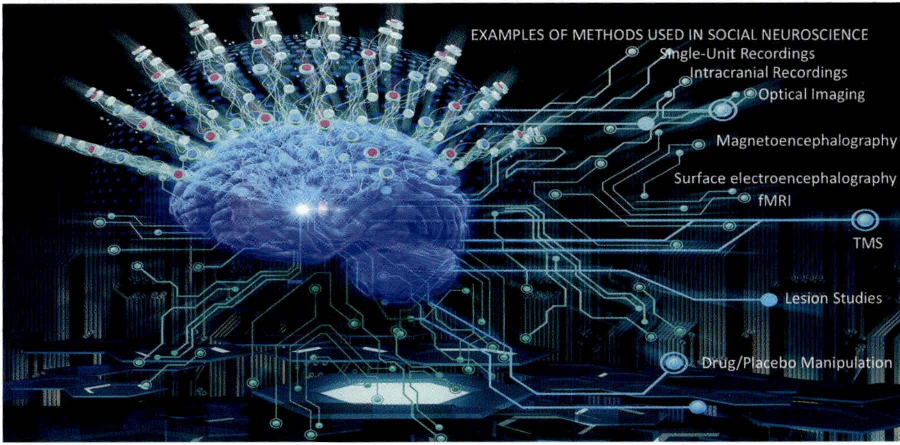

FIGURE 3.1. Methods in social neuroscience. Graphical illustration of neuroimaging techniques that may be used in social neuroscience. Excerpted from http://humanorigins.si.edu/evidence/human-evolution-timeline-interactive. iStock.com/Devrimb.

Excerpted from http://humanorigins.si.edu/evidence/human-evolution-timeline-interactive.

3.1 Methods for the Study of the Social Brain

With more than 85 billion brain cells working together in malleable networks to produce our mind, consciousness, and behavior, the scientific investigation of the human brain represents one of the most complex and exciting scientific frontiers in the twenty-first century.[3] Since Angelo Mosso's discovery of the "human circulation balance" in the nineteenth century,[4] significant neuroimaging developments and refinement have been made, for instance, in terms of neuroimaging power (e.g., from 1T to 3T or 7T for fMRI; from 32 electrodes to 64 and then to 128 electrodes or 256 electrodes for surface EEG),[3,5] computational capacities and analytic tools, and statistical approaches (multi-kernel density analyses, multi-voxel pattern analyses, network modeling of brain connectivity, graph theoretical analyses).[6–8] In addition to traditional physiological measures (e.g., facial electromyography, impedance cardiography and electrocardiography, eye-tracking, electrodermal activity),[9] contemporary neuroimaging techniques (such as positron emission tomography, PET; functional magnetic resonance imaging [fMRI]; electroencephalogram [EEG] and event-related potentials [ERPs], magneto-encephalography [MEG], or transcranial magnetic stimulations [TMS]; figure 3.1) offer unprecedented access to the human brain during normal waking states.

Also of importance in social neuroscience are the advances made in genetics and molecular biology. For instance, a growing body of research demonstrates that the social environment can modulate gene expression, thereby influencing neural and neuroendocrine functioning. Methods and models are also being developed in social neuroscience to bridge the gap between animal and human research. In line with the fact that most of our human social behavior arises from neurobiological and psychological mechanisms shared with other social species, studies are being performed across phylogeny to understand the neural, hormonal, chemical, and genetic bases of social behavior.[3,10] Such interdisciplinary investigations across social species (and across cultures within social species) are becoming more common in the field. Because of space limitations, however, we limit our focus here on the neuroimaging technique most used by social neuroscientists investigating the human brain—that is, fMRI.[11] Moreover, our discussion of methods is designed to promote the understanding and interpretation of neuroimaging studies in the literature rather than to provide a guide on how to collect or analyze data using these methods.

3.1.1 Neuroimaging Methods

With high spatial (mm) and intermediate-to-low temporal (seconds) resolution, fMRI has attracted considerable attention as a method for investigating the functional neuroanatomical bases of psychological

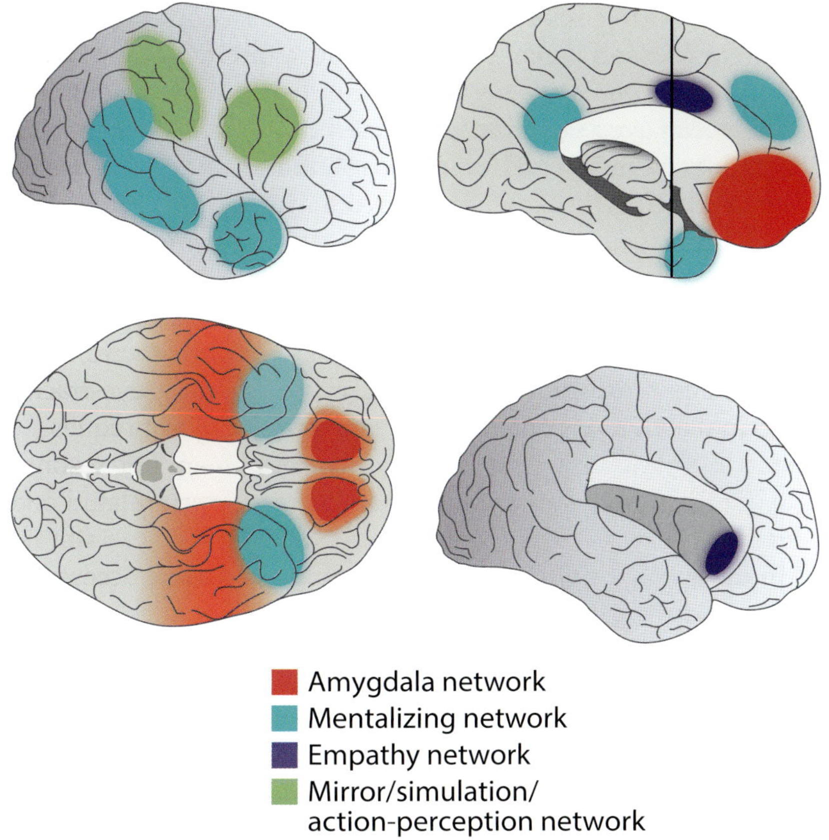

FIGURE 3.2. Four core networks involved in social cognition: amygdala network, mentalizing network, empathy network, and mirror/simulation/action observation network. Adapted from Kennedy & Adolphs.[12]

states and processes. Since its introduction in the 1990s, fMRI has attracted a large number of scientists and led to the discovery of several brain areas involved in various social functions and behaviors. From hundreds of fMRI studies on social cognition and behaviors, a general social brain matrix can be drawn (figure 3.2).[11,12]

Kennedy and Adolphs identified four core neural networks in the social brain:[12]

1. the "empathy network" that is engaged when individuals experience emotions vicariously (or remotely) from observing others (or from imagining the pain others are going through or the pain they might have gone through);
2. the "mentalizing network" that is activated when one thinks actively about one's own internal states and also when thinking about others;

3. the mirror (or simulation) network that is recruited when individuals observe/imagine/predict actions and intentions of others; and
4. the "social perception" network (or amygdala network) that is activated with a wide range of social behaviors from emotional processing to detection of socially salient stimuli to social affiliative behaviors and decision making in response to social threat to social pain.

This depiction is illustrative of the social brain matrix rather than exhaustive, of course, as, for instance, social (e.g., face, gaze) perception involves the fusiform gyrus, middle temporal gyrus, superior temporal sulcus, and amygdala,[13–15] not to mention the neural circuitry underlying human speech perception and production, attachment and love, and social rejection.

Functional MRI provides a wealth of data on how these various neural networks may be selectively activated or co-activated by specific tasks that have been designed (typically through behavioral studies) to isolate one or more information-processing operations underlying a social behavior. There are a few considerations that may be helpful when reading, designing, or interpreting fMRI research.[5,16,17]

Keep things simple, but not one bit simpler. The current fMRI model of the hemodynamic response posits that a transient increase in neuronal activity within a brain region leads to a relative deoxygenation of the blood flowing near these cells followed quickly by increased blood flow to the region. As a result, blood near a region of local neural activity can have a higher ratio of oxygenated to deoxygenated hemoglobin than blood in locally inactive areas. The BOLD fMRI provides a measure of these dynamic adjustments and, by inference, the transient changes in neuronal activity in the proximal brain tissues.[18] This makes it possible to investigate the likelihood of one or more neural regions showing changes in activation (increases or decreases in activity level) as a function of experimental tasks that are thought to vary one or more specific psychological states or processes.

No one doubts that the operations underlying behavior emanate from the brain. Early fMRI studies concluding that a particular psychological state or process resulted in brain activation left many wondering what the alternative hypothesis might be. It does not follow necessarily that a particular behavioral function, or even a single psychological operation, is achieved through the operation of a single neural region. More generally, the functional localization of component social processes is not simply (or, at least given the current state of knowledge, even mostly) a search for centers.[5] The angular gyrus, for instance, is a brain area that has been shown to be activated not only in one social function, but also in a broad variety of social functions, such as bodily-self representation, embodied cognition, love, self-esteem, semantic processing, reading, number processing, memory retrieval, attention, spatial cognition, left-right orientation, reasoning, metaphors, and abstraction.[19] This example highlights the importance of investigating neural circuitry/networks, rather than a neural center, when mapping the social brain. Other regions, including the amygdala, medial prefrontal cortex, and fusiform gyrus, are similarly associated with a variety of putatively different processes. Whether the simple computation being performed by each region has not yet been accurately described or the net effect of the activation of a given region differs as a function of its coupling with one or more other brain regions has yet to be determined. For instance, think of how different sodium chloride (a nutrient) and hydrogen chloride (an acid) are, even though the element chlorine is common to both. At this juncture, it is perhaps judicious to keep in mind Einstein's take on Occam's razor, which is to make things as simple as possible, but not one bit simpler. To avoid oversimplified explanations that lead to false conclusions, when designing or interpreting the results of neuroimaging studies a one-to-one mapping between a specific brain region and a specific social state or process should be subjected to strong empirical tests rather than assumed.

Causal inferences are justified from neuroimaging methods only when all plausible alternative interpretations have been eliminated. An image of activated regions of the brain associated with a specific component process may make it appear as if the neural basis for that process has been discovered, but is it sufficient given the correlational nature of current neuroimaging methods? To illustrate the issue, consider a simple physical metaphor in which Φ represents a physical mechanism, a heater (analogous to a neural mechanism in the brain), and ψ represents an invisible but measurable state, the temperature inside a house (analogous to a social state or process). Although the heater and the temperature are conceptually distinct, the operation of the heater represents a physical basis for the temperature in the house. In this case, $\psi = f(\Phi)$. A bottom-up approach—that is, $P(\psi/\Phi)$—makes clear certain details about the relationship between ψ and Φ, whereas a top-down approach—that is, $P(\Phi/\psi)$—clarifies others.[20] For instance, when the activity of the heater is manipulated (i.e., Φ is stimulated or lesioned), a change in the temperature in the house (ψ) is observed. The fact that manipulating the activity of the heater produces a change in the temperature in the house can be expressed as $P(\psi/\Phi) = \text{E}$, where E represents an effect size that differs from zero. Note that the $P(\psi/\Phi)$ need not equal 1 for Φ to be a physical substrate of ψ. This is because in our illustration there are other physical mechanisms that can affect the temperature in the house (ψ), such as the outside temperature (Φ') and the amount of direct sunlight inside the house (Φ''). That is, there is a lack of complete isomorphism specifiable, at least initially, between the functional dimension (ψ) and a physical basis (Φ).

Now consider the indicator light on a thermostat that illuminates when the heater is operating. In this case, the indicator light represents a physical element that would show the same covariation with the temperature in the house as the operation of the heater as long as a top-down (e.g., functional brain imaging) approach was used. If the complementary bottom-up approach were used, it would become obvious that disconnecting (lesioning) the heater can have effects on the temperature in the house whereas disconnecting (or directly activating) the indicator light has none. This simple example should makes it clear that a region of differential brain activation that corresponds to a specific information-processing operation does not necessarily mean that this brain region is the neural substrate for the information-processing operation.

To be able to draw causal interpretation (rather than correlational assumptions) about the link between biology and behaviors often requires other methods, such as lesion studies, transcranial magnetic stimulation, and pharmacological interventions (e.g., ligands, drugs) in human or nonhuman animals to elucidate the causal role of any given neural structure, circuit, or process in a given task. Any single neuroimaging methodology provides a partial view of brain activity within a very limited range of spatial and temporal levels, and it is the confluence of methods that advances our understanding of the neural mechanisms underlying social and cognitive behaviors.

Consider moderator variables. Social process and behavior are important and interesting in part because they are so complex. Social phenomena, ranging from aggression to discrimination, are multiply determined. The multiply determined nature of these phenomena calls for the parsing of big research questions into smaller, tractable series of research questions that ultimately constitute systematic and meticulous programs of research. Where to parse a phenomenon may not be obvious without empirical evidence, however. Therefore, the generalizability problem, or the absence of dependence on an originally unmeasured variable, represents a recurring issue in research on human social behavior.

Let's return to the analogy of the heater and temperature described above to illustrate the issue. Let's further assume that a contrast method was used to identify the link between the heater and home temperature. Specifically, the temperature of the test room was found to be higher at noon than at 4 a.m. on a winter day in Chicago, and the contrast between the images of the living room taken with a thermo-imaging camera at 4 a.m. and noon showed the heater to appear in this contrast image. A colleague in southern Florida replicates the procedures precisely, also finding that the temperature of the test room was higher at noon than at 4 a.m. in southern Florida. However, the contrast image failed to reveal any evidence that the heater was involved at all; instead the curtains covering the large living room windows appeared in the contrast. If room temperature were *only* a function of a heater in a room, then such a result would call into question the replicability of the original observations. However, the multiply determined nature of room temperature and the differences across contexts in the operation of one or more of these determinants raise the real possibility that both findings are replicable but neither provides a comprehensive account of the phenomenon. Treating such discrepancies as a theoretical question rather than simply noting there were "methodological differences" should foster the development of testable hypotheses and, ultimately, more comprehensive theories in social neuroscience.

Science is cumulative. In addition, several factors (such as sample size) can influence the data. For instance, Button et al. provide a tutorial on how and why neuroimaging studies with small sample size reduce the likelihood of detecting a true effect (due to low statistical power), increase the likelihood that the effect size of a true effect is overestimated (due to the use of $p < .05$ to identify when an effect has been "detected" and the larger sampling error associated with smaller sample sizes), and increase the likelihood that a statistically significant effect is not truly different from zero (due to differences in the base rates for tests of true and untrue effects).[21]

During the past five years, several narrative reviews have tried to address this issue by accumulating evidence from small-sample-size fMRI studies and grouping them together. Unfortunately, narrative reviews are qualitative rather than quantitative. To better address this question and identify the clusters that are statistically activated beyond chance level, one needs to perform quantitative meta-analyses.[7,8,22] There are several meta-analysis tools (e.g., activation likelihood estimate [ALE], multilevel kernel density analysis [MKDA]) one can use to test whether peak activations occur randomly throughout the brain or whether they occur statistically above chance in response to some patterns related to an experimental condition.

Given the small sample size in most neuroimaging research, small but theoretically important effects are likely to go undetected (due to low statistical power), thereby providing at best an incomplete and at worst a misleading depiction of underlying neural mechanisms. Given the cost of fMRI, simply increasing sample size may be a challenge. If the effect sizes and confidence intervals for all regions that reached some minimal threshold of effect size (e.g., $d = 0.1$) were provided in supplementary materials, quantitative reviews and the cumulative nature of science might be fostered.

Consider the time course within and between brains. Most of the advances in our understanding of the neural basis of the social brain have been based on studies in which people have been considered as isolated, often static entities. For instance, hundreds of studies on action observation and on the understanding of intentions performed by others have examined the kinematics of volunteers (stimuli) performing an action toward an object without the intention to interact with the observer/participant. Although such studies have successfully identified the role of particular functions of the social brain and are reviewed in papers using meta-analysis related to different aspects of social cognition, they do not inform us about real-time interpersonal interactions, nor, given the temporal resolution of fMRI, do they provide rich information about the timing of the activation of the associated neural regions that have been identified.

The neural mechanisms underlying shared representations have also tended to focus on single individuals performing tasks that vary aspects of these representations. Although important, the focus on the individual as the unit of analysis may not capture aspects of shared representations that manifest during normal social interactions.[23–25] Neuroimaging methods such as hyperscanning and statistical methods such as multilevel modeling make it possible to investigate the effects of the social aggregate (e.g., dyad) on interacting brains in addition to the effects of the individuals within the social aggregate. Hyperscanning was introduced by Montague and colleagues,[26] who performed an fMRI study using two scanners and pairs of individuals competing against each other in a simple game designed to measure the effect of deception in a competitive context. Montague's method for hyperscanning uses the Internet to allow investigators to synchronize and control two or more scanners to dissect the effects of individual factors, social (relational/interactional) factors, and potential interactions among these factors on regional brain activation.

The hyperscanning approach permits investigation of unique aspects of the social brain, but it raises some methodological issues to consider, as well. First, the recording parameters and operating characteristics among the linked scanners should be as identical as possible. Even at the same field strength, different scanners may have different gains, different gradient strengths, head coil sensitivities, and shimming protocols that may influence the data. Another issue is the accuracy and stability of the synchronization of the scanners across the Internet. Finally, multivariate statistical analyses for dependent, multilevel data structures provide useful estimates of the influence of factors at multiple levels of organization.

Since Montague's demonstration of hyperscanning, other hyperscanning approaches have been developed, including functional near-infrared imaging hyperscanning and EEG hyperscanning. EEG hyperscanning, for instance, provides not only information about where changes in brain activity are observed, but also when they occur. EEG hyperscanning typically uses different devices located in the same laboratory with similar sampling rates, calibration and amplifiers, which simplifies concerns about the synchronization of the different acquisition machines.

3.1.2. Epigenetic Processes and Gene Expressions of the Social Brain

Two other approaches also of interest in social neuroscience are (1) the study of the modifications of gene expression (transcriptomics), and (2) the study of the alterations of gene activity that can be transmitted to the next cell generation but that occur without changing the genetic code of the DNA (epigenetics)*.[27] In the first approach, labeled transcriptomics, researchers need to have access to RNA, as extracted from blood, to study gene expression. In the second approach, labeled epigenetics (or epigenomics), researchers focus on DNA and study the fundamental differences that make us unique. Several epigenetic processes, such as DNA methylation or chromatin modification, can be studied, but DNA methylation is the most studied in neuropsychiatry and neuroscience, as DNA methylation is considered a critical factor influencing gene expression.[27,28] Methylation involves the direct chemical modification of the DNA by binding of a methyl molecule to a cytosine basic unit, which then leaves an "epigenetic mark"[7] and impairs gene transcription, with the assumption that patients or persons

*The Greek preposition "epi" (meaning "what goes beyond") is used in the latter approach as these changes occur on top of the genetic code.[27,28]

with a social disorders (such as autism) would show a greater degree of methylation than a control group. Such studies are, as yet, only nascent in social neuroscience.[27,29]

A popular design in gene expression studies entails that a group of individuals with a specific psychiatric or medical condition is compared to a control group. For instance, in the case of a transcriptomic study of perceived social isolation (loneliness), a group of people with very high scores on a loneliness scale is compared to a group with very low scores on that same measure. Researchers then look for genes whose expression is up-regulated (i.e., relatively overexpressed in lonely individuals) or down-regulated (i.e., relatively overexpressed in nonlonely individuals).[30] The complete pattern of up-regulated and down-regulated genes then represents the specific gene expression profile of loneliness. Once such a profile is obtained, researchers check the function of each of the genes involved in a gene database to better understand which biological systems are involved. In 2007, a germinal study performed by Cacioppo and colleagues compared six lonely to eight nonlonely individuals, and revealed the upregulation of expression in 78 genes and down-regulation in 131 genes, including up-regulation of genes bearing response elements for proinflammatory NF-κB/Rel transcription factors, and down-regulation of genes bearing anti-inflammatory glucocorticoid response elements, mature B lymphocyte function, and type I interferon response.[31] The authors interpreted the impaired transcription of glucocorticoid response genes and increased activity of proinflammatory transcription control pathways as a possible functional genomic explanation for elevated risk of inflammatory disease in individuals who experience chronically high levels of subjective social isolation.[32] The overall pattern of results of this discovery was confirmed in a larger replication sample on older adults that compared 25 chronically lonely individuals to 68 controls. Together these results help us to understand why lonely people show heightened vulnerability to a broad variety of diseases (thought to emerge through excessive nonspecific immune activity) and impaired reactions to viral infections (thought to be linked to insufficient specific immune activity).[10] Further studies need to be done in this field to better understand the transciptomic and epigenetic of the social brain.

In sum, significant technological advances over the past few decades have led to the development of new methods and theories of the social brain. These developments have both transformed the nature and amount of data available on brain structure and function at various scales, and expanded the breadth of theories of the social brain. With such advancements come specialized methodological, analytical, and conceptual expertise that may benefit from the cumulative expertise of an interdisciplinary team, but also risks increasing subdisciplinary specializations that may make it difficult for nonexperts in the field to understand or evaluate research designs and interpretations.

3.2 Evolution of the Social Brain

As we mentioned in chapter 1, the genus *Homo* emerged approximately 2.5 million years ago in Africa. Species in this genus developed stone tools (approximately 2 million years ago) and harnessed fire (approximately 800,000 years ago), which not only led to a change in their diet but also led these humans to gather around campfires to find safety from predators, to socialize, and to share food, information, comfort, and warmth. Between 800,000 and 200,000 years ago, human brain size increased rapidly. This was also a period of dramatic fluctuations in climate. Larger, more complex brains made it possible for humans to interact with each other and with their changing environment in new and more flexible ways. In turn, as behavior became more complex, the amount of folding (convolutions) of the cerebral cortex increased (figure 3.3)—making the amount of folding and connections between brain areas a better indicator of complex intellectual abilities than the size of the brain.

Throughout this period, humans occupied a modest position in the food chain. Humans are not particularly strong, fast, or stealthy relative to other creatures. They lost their canine teeth long ago, and they do not possess the defenses from predation provided by camouflage, natural armor, or flight. As historian Yuval Harari[33] noted, "humans who lived a million years ago, despite their big brains and sharp tools, dwelt in constant fear of predators, rarely hunted large game, and subsisted mainly by gathering plants, scooping up insects, stalking small animals, and eating the carrion left behind by other more powerful carnivores" (p. 11).

3.2.1 The Emergence of *Homo sapiens*

Homo sapiens, who were once thought to have evolved fairly quickly in East Africa about 200,000 years ago, now appear to have evolved over a more gradual period dating back more than 300,000 ago across Africa.[34] Like other hominids, *Homo sapiens* spent most of this period as marginal creatures. About 70,000 years ago, they spread from East Africa to the Arabian Peninsula and Eurasia, where they were confronted by other *Homo* species, most notably the Neanderthals. Although

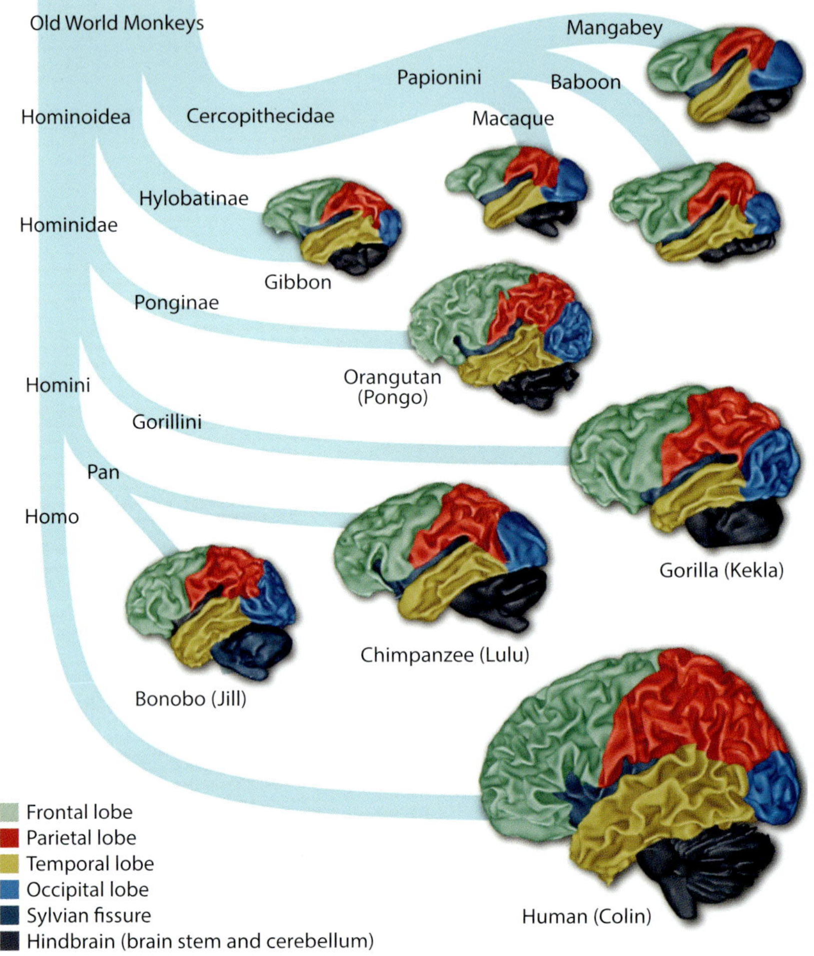

FIGURE 3.3. Visualization of evolution of brains in primates based on the surface of the cerebral cortex. Each lobe is represented by a specific color: frontal lobe (green), parietal lobe (red), temporal lobe (yellow), occipital lobe (blue). The cerebellum is also represented here (dark blue). Copyright © 2009 Dahnke@http://dbm.neuro.uni-jena.de.

Neanderthals were larger, stronger, and had superior vision and similar total brain volumes, the organization of the brains of Neanderthals and *Homo sapiens* differed, with greater brain volume in Neanderthals devoted to vision and physical robusticity, and greater brain volume in *Homo sapiens* devoted to the collection of capacities and skills that are important to living in groups (i.e., social cognition).[35] Consequently, *Homo sapiens* are thought to have been capable of more complex cognitive operations such as intention understanding, deception, counterfactual reasoning, and social learning, which permitted them to learn more quickly from successes and failures and to develop more effective strategies for dealing with challenges and conflicts. By around 13,000 years ago, *Homo sapiens* were the sole remaining human species, the result in part from genocide and in lesser part from interbreeding.[33,36]

Changes in social and cultural structures and processes contributed to their ascendancy.[37,38] *Homo sapiens* domesticated plants and animals and began forming permanent settlements approximately 12,000 years ago (the Agricultural Revolution). The behaviors required for successful farming and herding, such as planting rather than eating seeds, depend on a sophisticated sense of time and the capacity to plan and delay immediate gratification—capacities that are served by a well-developed prefrontal cortex.[39] Social groups were based largely on kinship until kingdoms and empires began to emerge some 4,000 to 5,000 years ago. The corresponding social behaviors, such as social recognition, communication, and coordination across large groups of people, placed yet additional demands that favored the development of rapid and sophisticated computational capacities applicable in social contexts. Even though there are no neural cell types that are unique to the human brain, there are changes in the morphology and profusion of both excitatory and inhibitory neurons, which potentiate integrative connectivity, as well as glia cells, whose functions include support and protection for neurons.[40] For instance, the axonal projections that connect the temporal and parietal lobes with frontal neocortical areas (i.e., the arcuate fasciculus) play an important role in human language (figure 3.4). These axonal projections are more rudimentary in chimpanzees and absent in macaques (figure 3.4). Moreover, damage to these connective structures in humans diminishes linguistic capacities. Similar species differences are evident in the primary axonal tract connecting lateral frontal with lateral parietal neocortical regions, a connective pathway that underlies frontoparietal functions including spatial attention to the actions of others, social learning, and tool use.[40]

Beginning around 500 years ago, there was a rapid rise in the accumulation of knowledge based on empirical observations and logical reasoning (the Scientific Revolution). The accumulation of knowledge stimulated innovation, and about 200 years ago economies based on predominantly agrarian activities began to give way to those based on machinery, factories, and mass production (the Industrial Revolution). Human and cultural activities stimulated by the Scientific Revolution continued to expand and, within the past half century, technology, information, and analytics have emerged as driving forces in economic development and expansion (the Information Revolution).

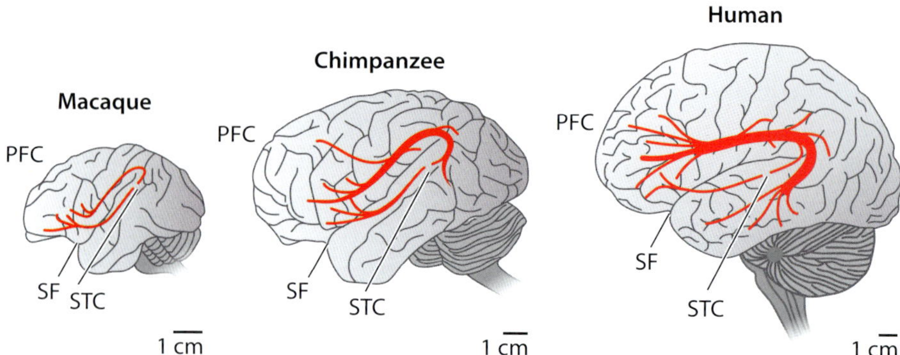

FIGURE 3.4. Language-related pathways are strongly lateralized and modified in humans. Compare the arcuate fasciculus projections in humans and in nonhuman primates. In humans, projections extend far into the medial and inferior temporal gyri, whereas projections to the chimpanzee temporal lobe are less extensive into the inferior temporal gyrus. In macaques, the projection into the inferior temporal gyrus appears to be completely absent. PFC, prefrontal cortex; SF, Sylvian fissure; STC, superior temporal cortex. From Sousa et al.[40]

Importantly, capacities for exploiting the potential benefits for individuals in a social context and for coping with the challenges of social life also created selective pressures that favored the evolution of larger, more interconnected, and more complex brains in anthropoid primates, which in turn permitted more sophisticated social, cognitive, and cultural processes.[41,42]

As discussed in chapter 1, primates have exceptionally complex and large brains for body size compared to other vertebrates.[43] Explanations for the evolution of larger, more complex brains in primates initially emphasized developmental constraints, ecological problem solving such as the demands of foraging and navigation, and diet.[44–46] Over the past few decades, these explanations have been largely replaced by the notion that the forces of natural selection have favored larger brains and more complex cognitive capacities to cope more effectively with the demands of social life (figure 3.5).[47]

While some basic functions have remained common among great apes, humans, and other species[48–52] (see box 3.2 for an example), the general idea is that some brain regions (figure 3.6) are not necessarily conserved or regressive, but may even be enhanced in recent human evolution.[53] For instance, anthropologist Katerina Semendeferi and her team performed a series of experiments[53–56] and showed that, compared with other great apes, the volume of brain regions involved in the processing of emotion, such as the hippocampus, the lateral nucleus

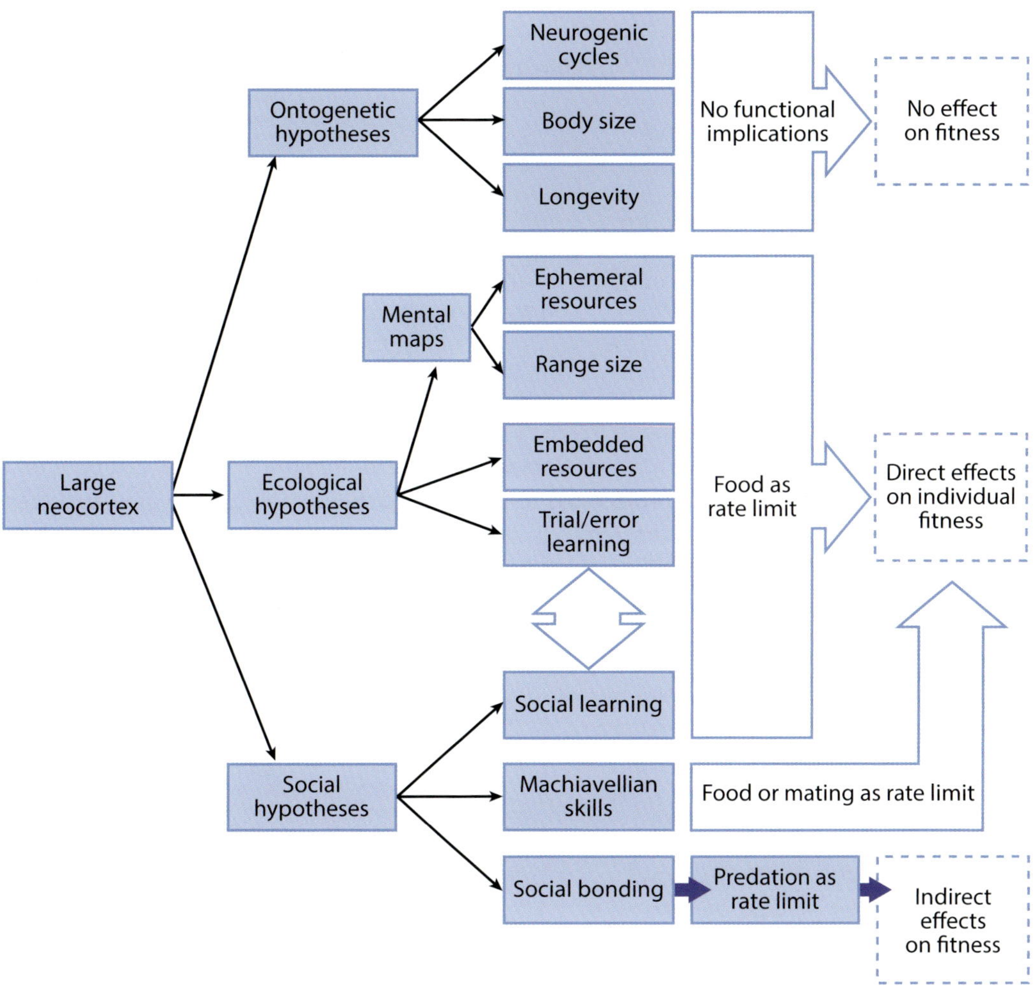

FIGURE 3.5. Alternative hypotheses outlined by Robin Dunbar[47] for the evolution of large brains in primates. Hypotheses differ in whether their central claim is about ontogeny, ecological, or social processes (left). Emphasis is on whether primates view food or predation as the rate-limiting process in population dynamics, and on whether they view the fitness benefits from large brains as being direct or indirect. Republished with permission of Oxford University Press from Dunbar R. I., "Evolutionary basis of the social brain" in: Decety J, Cacioppo JT, eds. *The Oxford Handbook of Social Neuroscience*. (pp. 28–38). Copyright © 2011 by Oxford University Press. Permission conveyed through Copyright Clearance Center, Inc.

of the amygdala, and the orbital frontal cortex, are, respectively, 50, 37, and 11% greater in humans than predicted for an ape.[53]

When Semendeferi and her team compared the relative size of the frontal cortices in living specimens of several primate species, including all extant great apes, using magnetic resonance imaging, they found

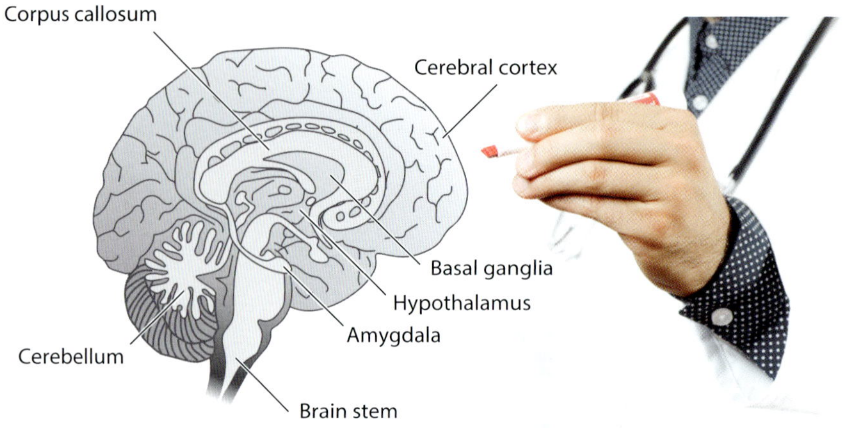

FIGURE 3.6. Schematic representation of brain areas involved in emotion.

BOX 3.2. Evolution of Laughter from Great Apes to Humans

Laughter is the shortest distance between two people.
—Victor Borge, Danish comedian

Evolutionary biologists and neuroscientists believe that laughter has evolved from nonhuman displays. Psychologist Marina Ross and colleagues tested this hypothesis and provided evidence of a common phylogenetic origin of tickle-induced vocalizations from five species: (1) orangutans, (2) gorillas, (3) chimpanzees, (4) bonobos, and (5) humans. It remains unknown whether laughter and squeaks emerged prior to or in the common ancestor of great apes and humans.

It is also unclear why we cannot tickle ourselves and laugh. One interpretation comes from neuroscientist Sarah-Jayne Blakemore and her colleagues—using fMRI, they found that the activation of brain areas detecting our movement as soon as we move and sending signals to other parts of the brain involved in laughter allows for the subject to quickly predict the intent of the movement and inhibit the laugh.

that although the absolute differences in size of the frontal cortex between humans and other primates were large, the frontal cortex in great apes and humans occupied a similar proportion of the cortex of both cerebral hemispheres.[55] Based on these results and the fact that differences were observed between humans and apes in the spatial organization of neurons within certain regions of the frontal pole,[54] it

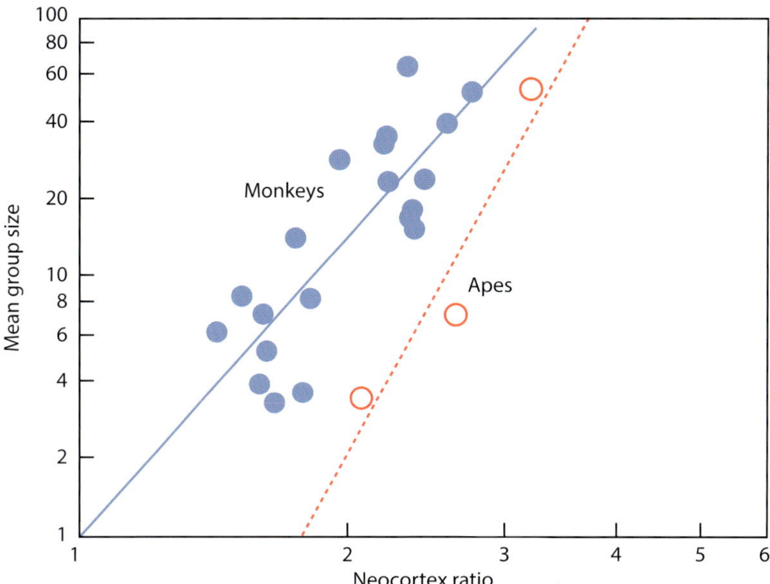

FIGURE 3.7. In anthropoid primates, Dunbar and colleagues showed that the size of the mean social group increases with relative neocortex volume. Adapted from Dunbar and Shultz.[57]

is suggested that the special cognitive abilities attributed to a frontal advantage may be due to differences in individual cortical areas and to a richer interconnectivity, none of which required an increase in the overall relative size of the frontal lobe during hominid evolution (see also figure 3.7).[55] In humans, the frontal lobe is involved in different components of social cognition.[11,12] As demonstrated in the human social brain matrix (figure 3.2),[11] different parts of the frontal lobes subserve a key role in each of the four functional brain networks of the social brain.

3.2.2 The Social Brain Hypothesis

There are variations in name and details,[37,47] but the general notion underlying the social brain hypothesis is that ecological problems are solved socially, and their complexity is generally less than the social environment. Accordingly, neural mechanisms that enhance social cohesion and social problem solving were major drivers in the evolution of the human brain.[57] A spider monkey's brain is perfectly adequate for the challenges of finding enough to eat while not stepping on a snake. It is also adequate for getting by in a small troop with rigid social rules. But greater behavioral latitude—such as shifting alliances to meet selfish goals—means greater social complexity. The more demanding mental challenges come from sorting out friend from foe when both are capable of sophisticated deception, negotiating power structures that include shifting rivalries and alliances based on complex motivations, using language to communicate (as well as to manipulate others), juggling long- and short-term mating relationships not rigidly constrained by the female's ovulatory cycle, and grappling with ever-changing cultural evolution.

Social recognition of and bonds between conspecifics contribute to the creation of stable social structures, enabling or enhancing behavioral functions such as communication at a distance, coordination, cooperation, parental caregiving, empathy, dominance hierarchies, and knowledge transmission through culture. Sociality also carries costs (e.g., competition for food and mates, increased risk of pathogen transmission), and social interactions can range from hospitable to hostile, trustworthy to exploitive, and protective to murderous. Moreover, the nature of social interactions and relationships can change, sometimes abruptly, across time. Accordingly, simple stimulus-response reflexes and fixed action patterns are insufficient for survival in anthropoid primates. The cognitive demands on individuals in a complex social group include social recognition; learning by social observation; recognizing the shifting status of friends and foes; anticipating and coordinating efforts between two or more individuals; using language to communicate, reason, teach, and deceive others; forming, monitoring, and orchestrating relationships, ranging from pair bonds and families to friends, bands, and coalitions; navigating complex social hierarchies, social norms, and cultural developments; subjugating self-interests to the interests of the pair bond or social group in exchange for the possibility of long-term benefits; recruiting support to sanction individuals who violate group norms; and doing all this across time frames that stretch from a person's distant past to multiple possible futures.[41,58]

The greater cognitive demands faced by individuals living in stable social groups placed a premium on evolutionary and cultural developments that resulted in individuals becoming smarter.[37,42] Moreover, much of the period associated with neocortical expansion in humans was characterized by warfare and/or dramatic climate changes, each of which places a premium on mutual aid and protection within families or groups and on the capacities of anticipation, deception, and exploitation between conflicting families or groups. Consequently, selective pressures favored evolutionary and cultural developments that improved one's likelihood of effectively negotiating the dynamic social environment to promote one's genetic legacy.

For instance, social and cultural learning permits individuals to leverage the experiences of others for their benefit at greatly reduced costs. Extensive postnatal brain growth and differentiation enables the development of larger, more complex brains but at the cost of a longer period of immaturity. In primates, this extended period of immaturity is typically spent with parents, caretakers, and generally tolerant adults, thereby providing more opportunities for supervised individual learning and for social and cultural learning.[37]

Anthropologist Robin Dunbar and colleagues have summarized evidence that in anthropoid primates the ratio of the neocortex volume to the volume of the rest of the brain correlates with various indices of social complexity, including group size, grooming clique size, number of females in the group, the prevalence of social play, and the frequency of social learning, tactical deception, and temporary coalitions.[57] This quantitative association is not a robust characteristic in other vertebrates, although larger relative brain size is associated with pair-bonded monogamy. Dunbar[59] suggests that at some early point in their evolutionary history, anthropoid primates used the kinds of cognitive capacities used for pair-bonded relationships by vertebrates to create relationships between individuals who are not reproductive partners:

> The relationships of anthropoid primates involve a form of "bondedness" that is only found elsewhere in reproductive pairbonds.[57] (p. 1346)

This permits an expansion of social behaviors, but it also brings with it demands for more sophisticated forms of social cognition, for instance, to maximize the benefits of social behaviors directed at them that fall under the categories of mutual benefit and altruism, and to minimize the costs associated with social behaviors directed at them that fall under the categories of selfishness and spite (see chapter 1).

Social monogamy is illustrative of a social structure that has evolved in different species as a result of evolutionary developments in the brain, whose influence on behavior, in turn, has contributed to further evolutionary developments in the brain. Social monogamy, or pair-bonding, is much more common in birds (90% of species) than mammals (less than 5% of species), and within mammalian species the primate species account for more than a quarter of the species that are characterized by social monogamy.[60] In birds, biparental investments in incubation, feeding, and caregiving constrain the adults to remain a pair. In mammals, in contrast, female internal gestation and lactation typically result in high female parental investment while males continue to search for additional reproductive opportunities. Why, then, has social monogamy evolved particularly in primate species?

Anthropologist Christopher Opie and colleagues considered three possible explanations for the evolution of social monogamy in primates: (1) the parental care hypothesis, which specifies that the cost of raising offspring is so high that females must rely on the help of others; (2) the mate guarding hypothesis, which specifies that males may form a pair bond, particularly when females occupy small but discrete ranges, to guard the female from rival males seeking to mate with her; and (3) the infanticide risk hypothesis, which specifies that social monogamy arises when the risks of infanticide are high and resident males can provide protection against infanticidal males. Data from primate species were used to test for correlated evolution between social monogamy and a marker for each hypothesis (parental care, female ranging patterns, and male infanticide, respectively). The results suggested that the presence of infanticide increased the probability of a shift toward social monogamy, and social monogamy increased the probability of a shift to greater parental care and discrete ranges.[60] The shift to greater parental care not only decreased the likelihood of infanticide, it also permitted longer periods of dependency of offspring, which in turn permitted additional brain development and expansion after birth.[61]

3.3 Development of the Social Brain in Infancy

It was once thought that infants required only their materialistic needs to be addressed, but evidence now clearly shows that how an infant relates to others is important both for healthy development and for optimal functioning.[62] Human infants are born into a world filled with other people and are completely dependent on care and protection by others for their survival. Infants are not born blank slates but instead enter this world with rudimentary cognitive capacities and predispositions for relating socially to others. For instance, Smith and others, who recorded videos of what infants are looking at during their awake hours, showed that newborns preferentially attend to social stimuli including faces, voices, and biological motion.[63–66] Most precisely, for the first six months of life infants mostly look at faces. After that, hands take over, and they mostly look at their own hands or those of others manipulating toys and objects.[64–66] The idea that Smith and colleagues suggest, based on their statistical analysis of videos, is that infants' visual experience shapes the nature of the brain pathways that develop in the neurotypical brain. Similar mechanisms have been suggested for other modalities than vision.

Language acquisition by infants exposed to speech material for a brief period in the laboratory has been explained in terms of computational learning theory, in which a predictive function is identified based on the speech sounds to which the infants are exposed. The concept of *neural commitment* was advanced to describe how the brain develops to achieve this function. According to this principle, the neural architecture underlying the detection of phonetic and prosodic patterns of speech is established in infancy to maximize processing for the language experienced by the infant. Once established, this neural architecture facilitates the learning of patterns that conform, and it impedes the learning of new patterns.

However, developmental psychologist Patricia Kuhl[67] has found that social interactions play an important role in complex natural language learning. Infants learn phonetic units early in life, and their ability to discriminate foreign-language phonetic units declines rapidly between 6 and 12 months. Kuhl and colleagues investigated factors that influenced this decline. Nine-month-old American infants were assigned randomly to one of four sets of language materials across 12 sessions: (1) native Mandarin Chinese speakers, (2) videotapes of native Mandarin Chinese speakers, (3) audiotapes of native Mandarin speakers, and (4) native language (English) speakers. Results indicated that exposure to native Mandarin speakers reversed the decline in the discrimination of foreign-language (Mandarin) phonetic units, but only when this exposure involved interpersonal interaction. Exposure to videotapes or audiotapes of the same Mandarin material produced the same decline in phonetic discrimination as did exposure to English only.[68] Kuhl[67,69] suggested that language evolved to address a need for communication across conspecifics, and therefore social factors "gate" the computational processes underlying speech learning to ensure speech learning would be limited to signals from humans rather than other sources, thereby protecting infants from meaningless calculations.

Over the first year, infants develop cortical mechanisms that potentiate social capacities and establish a foundation for the emergence of social cognition, interactions, and relationships.[63,67] Developmental psychologist Tobias Grossmann[63] has described six such neurocognitive developments, three of which typically emerge within the first six months of life and three of which typically emerge within the second six months (see box 3.3).[70–73] Grossman's model is in line with prior models, such as Baron-Cohen's model,[74,75] proposed on developmental data from humans and animals, as well as data from subjects with social cognition disorders.[74–76]

The first neurocognitive function refers to developments in the prefrontal cortex (PFC), especially the medial PFC, which are associated

BOX 3.3. Human Brain Development: A Gray Matter

Neuroscientists have shown that scanning children's brains year after year reveals drastic changes in various parts of the brain. From the first weeks after conception to adolescence, the gray matter becomes less dense as the brain matures, and the brain fissures, gyri, and sulci develop week after week postconception.[70] The total of gray matter volume increases with age, followed by a maturation (a thinning of the gray matter). Paradoxically, the thinning of the gray matter corresponds to increased cognitive abilities, improved neural organization, and increased white matter, which allows brain cells to better communicate.

with a sensitivity to another person's action (e.g., eye contact, infant-directed smiles, contingency) or communication that signals the action or communication is directed at the infant (i.e., self-relevance). This function is one of the key components needed for mentalizing, which corresponds to the ability to read and understand the mental states of others solely from body language (e.g., from the eyes).[74–76] We further address the progression from social perception to social cognition in chapters 4 and 6. The second refers to developments in the medial PFC and regions such as the ventral striatum that are associated with a sensitivity to another person's action (e.g., eye gaze) or communication regarding an external object or event that signals the action or communication is shared with the infant (i.e., joint engagement). The third refers to developments in the inferior frontal and premotor cortex that are associated with a sensitivity to aspects of another person's actions (e.g., action kinematics) that signals the next step or goal of that action (i.e., action prediction or intention understanding). The development of self-relevance contributes to the detection of joint engagement, and perceptions of joint engagement and predictability contribute to the coordination of action, cooperation, and competition. All three processes are practiced during social play.

The second triad of neurocognitive developments is centered in the temporal cortex.[63] Developments particularly in the inferior temporal cortex are associated with a sensitivity to features of another person (e.g., facial morphology, skin pigmentation) that mark the category or group (e.g., gender) to which the person belongs (i.e., categorization). Second, developments particularly in the right superior temporal cortex are associated with a sensitivity to any observable features of another person or their actions that can change over time (e.g., facial expressions) and signal the state of the person (e.g., happy, angry) that index the person's likely future behavior (i.e., discrimination). Finally, developments particularly in the multisensory regions within the superior temporal cortex are associated with a sensitivity to any information about another person or the person's action (e.g., temporal co-occurrence) that promotes the fusion of input across modalities and channels (i.e., integration). The emphasis in categorization is on stable features of a person, whereas the emphasis in discrimination is on changeable features or actions of a person. Integration, in turn, builds on these functions to draw associations between information from different sources and to use this information to form predictions regarding the person's actions, intentions, or goals.

3.4 Development of the Social Brain with Salutary Relationships

Humans have evolved to be predisposed toward social connection and salutary relationships.[77] A salutary relationship is characterized by past and expected future benefits from interactions with a person; perceived mutuality, which portends a sustained relationship, lessening the risks of commitment and behavioral investments; and an emotional connection with another individual. Commitment, which reflects a focus on long-term goals at the expense of short-term desires and temptations, engages the region of the brain involved in executive functioning and impulse inhibition, the dorsolateral prefrontal cortex.[78] The neuropeptide oxytocin also plays a role in pair-bonding and commitment (see box 3.4).[79–86] As we saw in chapter 2, the same group size in humans, even the same objective relationship (e.g., sibling, spouse), can be experienced as caring and protective or as callous and threatening. Additionally, we reviewed evidence that the loss of salutary social relationships (e.g., perceived social isolation or loneliness) triggers a set of processes that promotes short-term self-preservation but that, when extended over time, has deleterious effects on health and well-being.

In a neuroimaging study, participants selected articles from the *New York Times* to read personally or to share with others.[87] Three sets of brain regions were investigated. To investigate if selecting and sharing information activated regions associated with inherent value, activity was measured in regions identified in meta-analyses as engaged by subjective valuation—the ventral striatum and ventromedial prefrontal cortex (VMPFC).[88] To investigate whether self-related processing was associated with sharing information, activity was measured in regions identified in meta-analyses as engaged by self-related processing—the medial prefrontal cortex and posterior cingulate cortex.[89] Finally, to investigate whether thinking about the mental states of others was

BOX 3.4. Oxytocin and Pair-Bonding

Oxytocin, a peptide hormone and neuropeptide, plays a role in selective social bonding toward familiar, significant partners (e.g., the mother-infant bond) across social species from monogamous prairie voles to dogs to humans. When healthy men who are in a monogamous relationship receive oxytocin via a nasal spray, they prefer to keep a greater distance from an unknown attractive female than men who are single.[85]

associated with sharing information, activity was measured in regions identified in meta-analyses as engaged by social cognition—the VMPFC, middle medial prefrontal cortex, dorsomedial prefrontal cortex (DMPFC), precuneus, temporoparietal junction (TPJ), and right superior temporal sulcus (STS).[90] Results revealed that decisions to select and share articles to read, in contrast to decisions to select articles to read but not share, were associated with increased activity in all three sets of regions, consistent with sharing information engaging self-related and social cognition and having intrinsic value.[87]

The brain is energetically expensive and has evolved generally to minimize these costs. For instance, beyond standard developmental processes, specific brain regions have been hypothesized to grow larger primarily when needed to meet functional demands.[91] It follows from this notion that regional neuroanatomical adjustments should occur contingent on the demands of social versus isolated living conditions. Consistent with this reasoning, experimental studies of social isolation on brain size indicate that the effects are not uniform across the brain but instead are most evident in brain regions that reflect differences in the functional demands of solitary versus social living for that particular species. In chapter 1, we saw that the desert locust transforms from a solitary to a social state, with the solitary state producing an approximately 30% reduction in locust brain size (which comprises a midbrain flanked by paired optic lobes).[92,93] Despite the desert locust in its asocial phase having a smaller brain overall, it has disproportionally large primary visual and olfactory neuropils, putatively due to the increased individual predation risk and the need for the solitary locust to detect visual stimuli at a greater distance.[92] By contrast, the desert locust in its social phase has a larger midbrain to optic lobe ratio, and within both the visual and olfactory systems, higher multimodal integration centers are disproportionately larger than the primary sensory neuropils.[92,93] Thus, across the social and asocial phase, the size of specific brain regions is modulated on environmental demands.[92]

Does social isolation affect the morphology of the brain in other animals? There are important inconsistencies in this animal literature, but several robust results have emerged: (1) social isolation has significant effects on brain structure and processes in social animals across the life span, (2) these effects are not uniform across the brain or across species but instead tend to be most evident in brain regions that permit adaptation to differences in the functional demands of solitary versus social living for a particular species, and (3) the effects are not simply weakened recapitulations of what is observed as a result of isolation during development.[94] For instance, studies in fish, birds, and mammals indicate that enriched social environments and complex social interactions (such as those observed during the breeding season) enhance cell proliferation and neurogenesis in the brain, notably in the regions critical for social interaction, memory, and communication; social isolation, in contrast, reduces brain cell proliferation in various taxa, such as birds and prairie voles.[95] Moreover, rodents isolated from a preferred companion, relative to nonisolated rodents, show regional brain changes in several areas that are involved in the processing of social information, memory, sensorimotor integration, and spatial information processing, such as the prefrontal cortex, occipital cortex, and hippocampus.[94]

The effects of isolation are not limited to neuronal cell proliferation. Myelin is an electrically insulating fatty layer surrounding the axons of some nerve cells that is critical for proper neural functioning. Research on myelination in the prefrontal cortex of mice shows that isolation during a brief critical period early in life produces a permanent decrease in the myelination of the prefrontal cortex,[96] whereas isolation in adulthood produces a reversible decrease in the myelination of the prefrontal cortex.[97] These results indicate that there are important differences in the initial myelination of neurons during critical periods in the developing prefrontal cortex and in the ongoing myelination that occurs in the prefrontal cortex in response to environmental demands. Specifically, once the critical period for myelination in the PFC has passed, the process of myelination is characterized by a *plasticity* to adapt brain structures and functions to environmental (including social environmental) demands.

Perceived social isolation, or loneliness, has also been associated with morphological changes in the human brain. A study by neuroscientist Ryota Kanai and colleagues of the association between loneliness and brain size indicated that loneliness was correlated negatively with gray matter density in the left posterior superior temporal sulcus (pSTS), an area involved in biological motion and social perception.[98] Previous research had shown that the smaller the size of a participant's online social network, the smaller the pSTS, middle temporal gyrus, and entorhinal cortex (brain regions involved in social perception and associative memory).[99] Kanai and colleagues therefore also examined the extent to which the association between loneliness and the size of the pSTS could be explained by social network size (an index of objective social isolation), empathy, or anxiety. Results showed that factoring out these variables did not change the correlation between loneliness and pSTS size. Instead, they found that loneliness and pSTS size were related to poorer performance on gaze perception, and gaze perception performance mediated the association between loneliness and pSTS. Previous research has shown that loneliness is related to differences in social *perception* rather than social contact,[94,100] so the results are consistent with the notion that brain morphology in adult humans can vary as a function of the demands of their social environment.[98]

The studies of Kanai and colleagues focused on cortical gray matter in the human brain, which consists primarily of neuronal cell bodies, dendrites, and associated glia cells. White matter in the brain, in contrast, consists primarily of bundles of myelinated axons that connect various gray matter areas (i.e., nuclei). The white matter tracts, in essence, represent the highways underlying communication among the nuclei in the brain. If the increased integrative connectivity in the human brain evolved in response to the demands of the social environment, then one might expect loneliness to be associated with reductions in regional white matter density. Consistent with this reasoning, loneliness has been found to be associated with lower regional white matter density in areas involved in self- and social cognition. These areas include the same region identified in studies of gray matter density (left pSTS) as well as the inferior parietal lobule, anterior insula, posterior temporal parietal junction, dorsomedial prefrontal cortex, and rostrolateral prefrontal cortex in both male and female young adults. There was no brain area in which white matter density was greater as a function of loneliness.[101]

Loneliness has also been found to be associated with changes in the brain at rest. Such changes can be investigated by measuring functional magnetic resonance imaging (fMRI) activity during a baseline period during which time participants have no task to perform. The brain continues to function, of course, and one can investigate the extent to which loneliness is associated with the *functional* connectivity in the resting human brain. Functional connectivity is not a measure of gray matter or white matter density, but instead it is a measure of the extent to which the activity in two or more nuclei in the brain are correlated when the brain is at rest. The notion is that brain regions that are frequently activated in a coordinated fashion in everyday life are especially likely to show correlated activity at rest. Analyses of functional connectivity indicated that loneliness is associated with stronger functional connectivity between several nodes in the cingulo-opercular network, a network known to underlie the maintenance of tonic alertness,[102] and weaker connectivity between nodes in the self-regulatory network involved in the inhibition of impulsive responses. These results are in accord with the evolutionary model of loneliness and behavioral data, in which perceived social isolation (loneliness) automatically triggers an increased implicit vigilance for social threats and a decrease in the inhibition of impulsive responses (i.e., an increase in prepotent responses).[103,104]

Several fMRI studies have examined how loneliness influences regional brain activation in response to a task.[105] In the first such fMRI study, lonely and nonlonely participants were exposed to unpleasant social images (figure 3.8). To control for the effects of unpleasantness per se, participants were also exposed to unpleasant nonsocial images (i.e., objects) that were matched on various dimensions (e.g., visual complexity, rated unpleasantness).[106] Activation of the visual cortex has been found previously to vary with attention to a visual stimulus. In this study, loneliness was related to stronger activation in visual cortex in response to unpleasant social images, in contrast to pleasant nonsocial images. Despite the greater attention apparently afforded to negative social stimuli by lonely participants, they showed weaker activation in the temporoparietal junction to unpleasant social images—a region that has been found previously to be activated in theory of mind tasks and in tasks in which individuals take the perspective of another.[106] These results are in accord with behavioral studies showing that loneliness is associated with higher levels of self-centeredness[107] and lower levels of empathy for others.[108]

Participants in this fMRI study of loneliness were also exposed to pleasant social images and pleasant nonsocial images to investigate the effects of loneliness on appetitive social information processing in the brain.[106] The ventral striatum, a key component of the mesolimbic dopamine system, is rich in dopaminergic neurons and is critical in reward processing and learning. The ventral striatum is activated by primary rewards such as stimulant drugs, abstinence-induced cravings for primary rewards, and secondary rewards such as money.[94] Evidence that social reward also activates the ventral striatum has begun to accumulate in studies of social cooperation,[109] social comparison,[110] punitive altruism,[111] and passionate romantic love[112,113] (see chapter 8). In

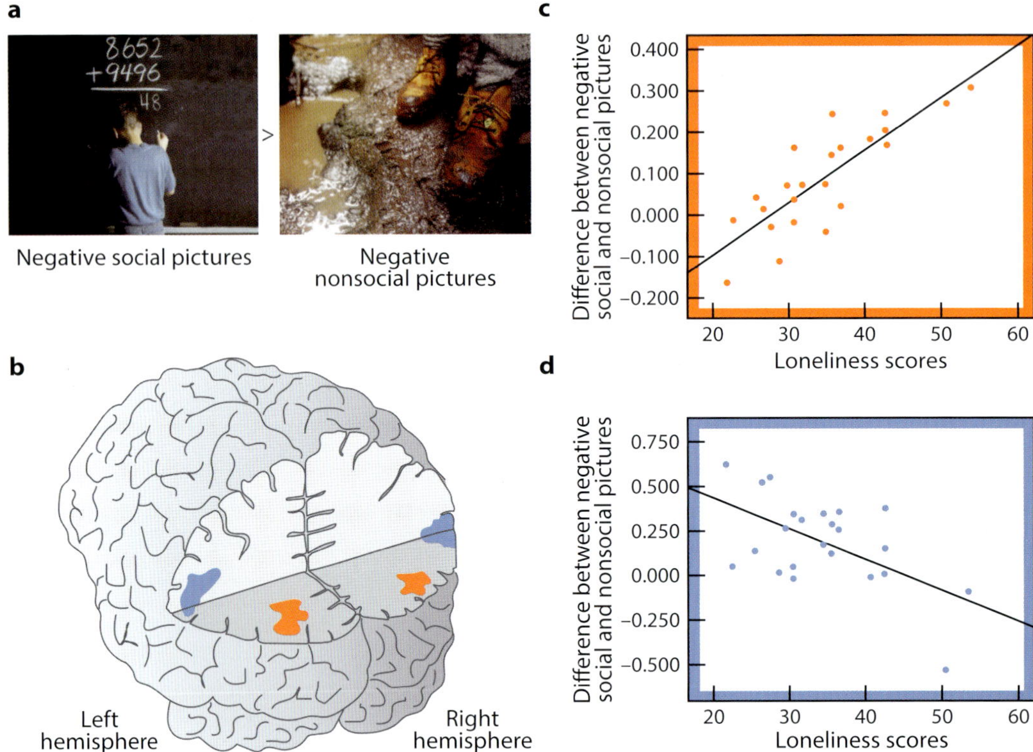

FIGURE 3.8. Neuroimaging of social disconnection. A. Example of stimuli used in fMRI experiments on loneliness. Brain activity is recorded and compared in response to negative social minus nonsocial pictures. B. fMRI results illustrating the differences in the comparison between the brain activity obtained from lonely versus nonlonely participants while they were looking at negative social pictures compared to the brain activity from the same participants while they were looking at negative nonsocial pictures. Results are shown on an average human brain. The visual cortex (in orange) shows more activation for negative social pictures than for negative nonsocial pictures in lonely people, in contrast to nonlonely. The temporoparietal junction (which is involved in empathy and perspective taking, shown in blue) shows less activation to negative social pictures compared to negative nonsocial pictures in lonely people. C. and D. The scatter plots depict the significant positive relationship between loneliness scores and activation of the right visual cortex in response to negative (C) or positive (D) social pictures compared to negative nonsocial pictures, suggesting that the more participants reported feeling lonely, the more the right visual cortex was activated.[106] From Cacioppo and Cacioppo (2013).[117]

an fMRI study, participants were exposed to pictures of strangers and to pictures of objects matched to be equally pleasant, arousing, luminous, and visually complex. Results confirmed that pleasant pictures of strangers and pleasant pictures of objects activate the ventral striatum. In addition, nonlonely individuals showed greater activation in the ventral striatum to pleasant pictures of strangers compared to matched pictures of objects, whereas lonely individuals show the reverse response.[106] These neuroimaging results mirrored behavioral results in a previous study, showing that lonely and nonlonely undergraduates were equally likely to experience positive interactions but these interactions were rated as less pleasant by lonely than nonlonely students.[114]

In a follow-up fMRI study, participants were asked to send two digital photographs of "a person they could go to for help or for comfort, such as a family member, a close friend or significant other."[115] Loneliness was measured before the scan, and during the scan participants viewed images of the person to whom they felt close, and images of a gender, race, and age-matched stranger. Results showed that loneliness was negatively related to the activation in the ventral striatum in response to the pictures of strangers, replicating the previous fMRI study.[106] In addition, participants showed greater activation in the ventral striatum during exposures to the pictures of the familiar person to whom they felt close, compared to exposures to pictures of the stranger. Finally, this result was shown more strongly by individuals who were high rather than low in loneliness, consistent with the rewarding value of salutary relationships, particularly when such relationships are scarce.

The emphasis in the research on loneliness is on the perception by the brain of the availability of preferred relationships, from which mutual benefits or altruism can be expected. The greater the discrepancy between the relationships they seek and those they perceive, the greater the loneliness. Using fMRI, psychologist Todd Heatherton and colleagues investigated the effects of expecting loneliness in one's future life on the neural response to positive or negative social scenes.[116] Following a personality survey, participants received feedback that was putatively based on their answers to the survey. In reality, participants were randomly assigned to one of two conditions. Half of the participants received feedback that led them to believe their future lives would be isolated and lonely (future life of high loneliness/exclusion), whereas the other half received feedback that led them to believe that their lives would be filled with long-lasting, stable relationships (future life of low loneliness/inclusion). Participants then were scanned while viewing pictorial stimuli that varied in valence and sociality.

A potentially important difference between this study and the prior neuroimaging studies of loneliness is that it was not the level of loneliness or perceived isolation at the moment that was manipulated, it was the participants' expectations for their future social life, which, in turn, was designed to influence how participants thought about the pictures they were shown during the scan. The differences in brain activation between the two experimental conditions were focused on an area previously shown to be activated when a person thinks about the mental states of others—the dorsomedial prefrontal cortex. For instance, when participants viewed negative social pictures, the activity in the dorsomedial prefrontal cortex was greater for participants in the low-loneliness/inclusion condition than in the high-loneliness/exclusion condition. In addition, the level of activity in the dorsomedial prefrontal cortex was similar in participants in the low-loneliness/inclusion condition whether they viewed positive, neutral, or negative social scenes. In a future life filled with social interactions, it is important to anticipate the mental states and behavioral predispositions of others whether they are enjoying themselves or in need of help. For participants in the high-loneliness/exclusion condition, in contrast, the level of activity in the dorsomedial prefrontal cortex was least active in response to negative social scenes, where self-preservation is served by focusing on one's own rather than a stranger's interests and welfare.

3.5 Concluding Remarks

The genus *Homo* emerged approximately 2.8 million years ago in Africa. Species in this genus developed stone tools (approximately 2 million years ago) and harnessed fire (approximately 800,000 years ago), which not only led to a change in their diet but also led these humans to gather around campfires to find safety from predators, to socialize, and to share food, information, comfort, and warmth. Larger, more complex brains made it possible for humans to interact with each other and with their changing environment in new and more flexible ways.

Evidence now suggests that *Homo sapiens* emerged approximately 300,000 years ago in Africa and spread from East Africa to the Arabian Peninsula and Eurasia around 70,000 years ago. Our species was not the first or only humans to roam the earth, and we were not at the top of the list in terms of sensory acuity, strength, or agility. However, the migration of *Homo sapiens* led to conflict with the stronger Neanderthals, who ultimately lost the battle for survival. Only 12,000 years ago, *Homo sapiens* began farming, herding, and forming permanent settlements, and only about 4,000–5,000 years ago human settlements began to be shaped by kingdoms and empires. There was a rapid rise in the accumulation of knowledge based on empirical observations and logical reasoning beginning around 500 years ago, and these developments led to the Industrial Revolution of 1760–1850 and more recently to the

Digital Revolution that is transforming how people work, play, search, congregate, shop, study, learn, and communicate.

At each step, social behaviors and stable social relationships became more central to successful survival and reproductive success, and the social environment became increasingly complex. Individual variations in the brain (e.g., size, encephalization) underlying cognitive capacities and social behaviors that provided an individual with a competitive advantage in an increasingly important and complex world (e.g., empathy, social learning, understanding the intentions of others, deceiving others regarding one's own true intentions, deferral of gratification, inhibition of impulsivity) were favored by natural selection. According to the social brain hypothesis, ecological problems are solved socially, and their complexity is generally less than the complexity of the social environment. Therefore, mechanisms that enhance social cohesion and social problem solving were major drivers in the evolution of the human brain.

The evolution of more complex, computationally more powerful brains also enabled the development and utilization of larger, more complex social structures and cultures. Both evolutionary developments of the brain and the development of human cultures have contributed to an expansion of human capacities. For instance, evolutionary changes that enabled extensive postnatal brain growth and differentiation enabled the development of larger, more complex brains but at the cost of a longer period of immaturity. This extended period of immaturity is typically spent with parents, caretakers, and tolerant adults who provide more opportunities for supervised individual learning and for social and cultural learning, which acts to leverage the experiences of others for their benefit at greatly reduced costs. Developmental studies of the human brain suggest that newborns have capacities that favor the care and affection of their caregivers, and that ontogeny recapitulates phylogeny, for instance, with frontal lobe development unfolding across the period of human immaturity.

Finally, the human brain has evolved to prefer salutary social relationships, and where one falls along the continuum of having adequate access to salutary relationships has implications for brain development, morphological adjustments in adults, and patterns of activation at rest and in response to social events. Studies comparing individuals who feel socially isolated and those who do not have identified morphological differences in specific brain regions related to self- and social cognition. Differences have also been observed in the patterns of regional brain activation at rest and in response to social stimuli, with the specific nature of these changes reflecting differences in functional responses to the social environment.

4

CONNECTING FORCES

The metaphor for the brain in social neuroscience is not a solitary computer, but a smart phone—a computationally powerful, mobile, and broadband-connected computing device. This metaphor underscores the connective forces between brains that have evolved in primate social species. We take for granted that we are immersed in a world of invisible *physical* forces whose presence we recognize because of their effects. Gravitational forces hold us close to the earth, the electromagnetic energy emanating from the sun not only turns night into day but bathes us in potentially harmful radiation, the tilt and position of the earth in its journey around the sun influences our seasons, and rising global temperatures are increasing the frequency of extreme weather events.

We are also immersed in a world of invisible forces that operate through our brains to connect us to others. Among these forces are identification, imitation, emotional contagion, empathy, and deliberate thought about the thoughts, feelings, intentions, and goals of others. These forces can vary by the context, the nature of the relationship we have with others, and our goals and expectations for an interaction. For instance, we are likely to reason differently about the intentions of an opponent whose serve we are awaiting in a tennis match and the intentions of a defendant in a court trial, whose guilt or innocence may depend on the conclusions we draw about intentions based on the evidence presented in a court trial. Less obvious, perhaps, is that processes such as imitation and contagion also can vary, often without volitional input, in ways that can impact our interactions and relationships. Understanding the mental state and needs of another person contributes to effective caregiving of infants, mutual aid and protection among adults, and defenses against an antagonist. The alignment of the goals and intentions across individuals can also promote feelings and empathy, salutary relationships, and cooperation more generally.[1] We begin with the social forces of imitation and identification to illustrate connections at a distance.

4.1 Emotional Contagion, Affective Perspective Taking, and Empathy

Understanding rapidly and effortlessly the emotional states of others is advantageous for social interactions ranging from caretaking to combat. We begin by discussing three such processes: emotional contagion, affective perspective taking, and empathy. Characteristics of each of these processes are summarized in table 4.1.[2]

Emotional contagion refers to the reproduction of an emotional state in an observer—for instance, through an automatic mimicking (unintentional rapid imitation) of somatovisceral responses including facial expressions, vocalizations, and postures.[3–6] Automatic mimicry can be both partly implicit and consciously controlled (such as motor mimicry that can be controlled by the motor muscles) or it can also rely on an unconscious signaling system that is controlled by the autonomic nervous system (figure 4.1).[5]

In emotional contagion typically, observers fail to recognize that their emotional states resulted from observing another person. Emotional contagion occurs effortlessly in infants, with the underlying mimicry and somatosensorimotor resonance (a phenomenon that is described when the perception of an individual's actions and sensory experiences produces brain activity that is similar to that which would be observed if the person were to perform the same actions) between the infant and others subserved in part by the human mirror neuron system.[7] Individuals within a social context who attract greater attention are more likely to trigger emotional contagion, as well. Investigations using instrumentation that permit the detection of micromovements in expressive facial muscles (i.e., facial electromyography) have shown that observers often show motor mimicry to the emotional expressions they see, and that these microexpressions can be so subtle that they produce no visually perceptible changes in facial expression.[8,9] In addition to mimicking facial expressions, observers also tend to mimic postures and vocal utterances, including utterance duration, speech rate, and latencies of response. These processes are rapid, spontaneous, and subserved by neural circuitry including the amygdala and the anterior insula.[10,11]

TABLE 4.1. Characteristics of core emotional processes

State	Characterization	Illustrative components
Automatic mimicry	The rapid and unintentional imitation of speech and movements, gestures, facial expressions, and eye gaze	Unintentional matching of the movements of a model that may facilitate emotional contagion
Imitation	An intentional copying of the appearance, behavior, or pursuit of the goal of another individual	Imitation is observed in nonhuman primates,[32] but in contrast to apes, we excel at imitating novel, opaque, or intransitive responses
Emotional contagion	The unintentional reproduction of an emotionally aroused state in (and unbeknownst to) an observer through the mimicking of the observed emotional responses	Aligns the emotional state of an observer with that of an observed individual, but is blind to the specific needs of the observed individual
Somatosensorimotor resonance	The perception of an individual's actions and sensory experiences produces brain activity that is similar to that that would be observed if the person, who is observing, were to perform the same actions)	Resonance facilitates anticipation of another's actions and sensory experiences
Affective perspective taking	Inferences drawn about the emotional state of another person based on the imagined perspective of the other and the perception of observable cues (e.g., emotional expressions)	Identifies the source of the emotional cues and fosters an understanding of an observed individual's emotional state, but it does not align the emotional states of these individuals
Emotional empathy	An emotional response triggered by the perception of another's condition or emotional state that is similar to what the other person is expressing or would be expected to feel in the given situation, thereby aligning the emotional states between people	Aligns the emotional states of an observer with that of an observed individual, attributes the affective state to the other person, and maintains the executive control necessary to attend and respond to the emotional state and needs of the other person
Cognitive empathy	Conscious drive to recognize and understand an individual's emotional state from their perspective	Consciously aligns the cognitive emotional states of an observer with that of an observed individual

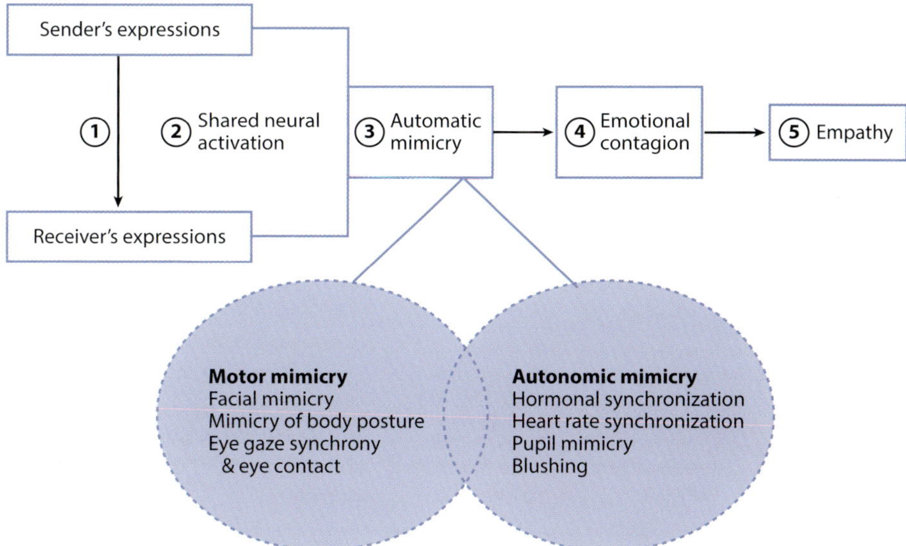

FIGURE 4.1. Schematic representation of the development of empathy. 1. First a sender expresses emotions in her/his/its body language (e.g., facial expressions, eye-gaze, body postures) and in her/his/its physiological response (e.g., heart rate) that a receiver may perceive. 2. Perceiving other's emotions activates the observer's brain network that she/he/it activates when they experience that same emotional state. 3. In turn, that shared neural representation activates somatic and autonomic responses resulting in motor mimicry and autonomic mimicry. 4. Then emotional contagion occurs. 5. Feelings of empathy become conscious. From Prochazkova & Kret.[5]

Affective perspective taking refers to an individual apprehending the emotional state of another person based on a consideration of the perceptual input, perspective, and emotional response of the other. Affective perspective taking fosters an understanding of another individual's emotional state, but it does not align the emotional states of these individuals. Emotional contagion, in contrast, fosters the alignment of the emotional states between individuals, but it is blind to the specific needs of the source of the contagion.

Empathy combines features of emotional contagion and affective perspective taking. Accordingly, empathy can be subdivided in emotional empathy and cognitive empathy (see table 4.1). Specifically, emotional empathy entails an emotional response triggered by the perception of another's condition or emotional state that is similar to what the other person is expressing (or would be expected to feel in the given situation), thereby aligning the emotional states between these individuals. Cognitive empathy corresponds to the ability to understand one's

BOX 4.1. Discriminating Self from Other

Neuroimaging studies in humans suggest that thinking about one's own traits or goals is associated with activity in the VMPFC, whereas thinking about another person's traits or goals is associated with activity in the DMPFC. There is growing evidence that activity in the VMPFC also varies as a function of perceived personal significance.[80] In addition, regions of the right TPJ and inferior parietal cortex—areas involved in the integration of sensory and somatosensory input and social perception—tend to be active when discriminating between one's own actions and the actions of others.

The cognitive representation of the self and the representation of others are separable, but no network of regions has been identified for which the self is necessary and sufficient to activate.[81] The representation of the self serves many different functions, including creating the perception of agency and effectance, distinguishing self from other, organizing and guiding one's perceptions of and interactions with the physical and social environment, and creating a sense of meaning and coherence.[81,82] Accordingly, the brain regions associated with self-related processes vary as a function of the goal or task. Self-attribution, for instance, is associated with activation in the VMPFC, whereas self-regulation is associated with activity in regions such as the dorsolateral prefrontal cortex.[75] In addition, our sense of self, which we typically experience as continuous and unitary, is shaped by information from others. For instance, a characteristic of the self is more likely to become salient to the extent that this characteristic distinguishes the person within a social group.[83] If there is only one player on a basketball team who is short, this distinctive characteristic is likely to spontaneously become a salient and accessible aspect of the self even if this person is taller than an average person. Moreover, our opinions, abilities, status, traits, values—even our physical features, health, and well-being—are influenced in part through social comparisons.[84] Children may know their height in feet and inches, but they may not think of themselves as short or tall until they compare themselves to others their same age. The flexibility in the neural mechanisms associated with the self therefore appears to serve the breadth and complexity of the physical and social stimuli to which the brain responds across a life span.

emotional state from their perspective. In addition, empathy identifies the affective response as originating in the other person and includes the executive control necessary to attend and respond to the emotional state of the other person (see box 4.1). If the executive functioning falters, empathy devolves into distress, compromising the ability of the observer to provide aid or protection to the person in need.

Empathy is thought to have evolved initially from increasingly complex forms of parental care in mammals to improve the odds of offspring survival, but empathy now serves additional functions. For instance, empathy promotes affective communication and understanding not only of a person's needs but of the actions that can be taken to best address the person's needs. In the appropriate context (e.g., when goals are shared), empathy can promote affiliation and prosocial behavior, whereas in other contexts (e.g., zero-sum situations) empathy can produce the feeling of counter-empathy (e.g., schadenfreude) and exploitation of another person's weakness.[12,13]

Functional magnetic resonance imaging (fMRI) studies of the neural correlates of empathy have identified activation in a distributed set of regions including the insula, amygdala, superior temporal sulcus, ventromedial prefrontal cortex (VMPFC), dorsomedial prefrontal cortex (DMPFC), orbitofrontal cortex, and anterior cingulate cortex.[14–16] However, fMRI results are correlational. Lesion studies complement neuroimaging studies by providing causal evidence for the involvement of specific brain regions in empathy. Most such studies have involved lesions or diseases such as dementia or traumatic brain injury, which are associated with broad or diffuse damage to the brain.

However, neurologist Argye Hillis identified a set of studies of patients with lesions that were localized to specific brain regions. Figure 4.2 illustrates the neural regions that were involved in empathy, emotional contagion, and affective perspective taking.[2] This work suggests that empathy involves the amygdala, temporal pole, anterior insula, and anterior cingulate cortex. The right orbitofrontal cortex and inferior frontal cortex were identified as critical for emotional contagion, whereas the right medial prefrontal cortex was identified as critical for perspective taking. This work not only advances our understanding of the widely distributed and recursive network of regions that are involved in these social processes, but it also helps physicians, patients, and families understand and deal with diseases and injuries to the brain that involve one or more of these regions.

Timing also matters in terms of the engagement of brain regions in the social brain. Using high-density electrical neuroimaging, neuroscientists Stephanie Cacioppo and Jean Decety investigated the spatiotemporal dynamics of intention understanding of emotional actions (and indirectly empathy) in the human brain.[17] In their study, high density EEG was recorded while participants viewed short video clips depicting individuals being hurt either intentionally or accidentally. Using high-density electrical neuroimaging allows scientists to identify spatiotemporal patterns of communication between regions that contrast analyses (such as fMRI) with low temporal resolution may not detect.

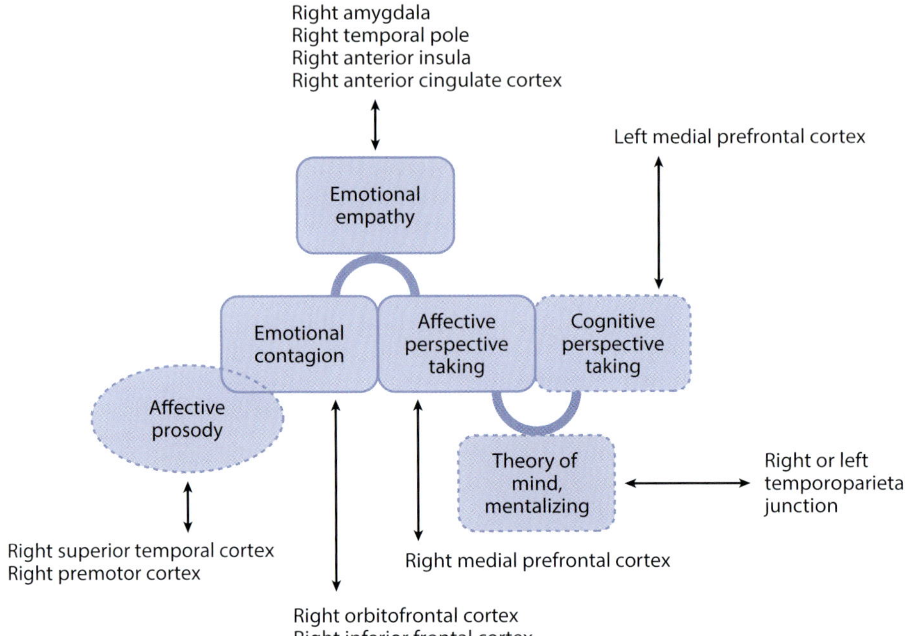

FIGURE 4.2. Schematic representation of the proposed cognitive and neural mechanisms of emotional empathy (in solid borders) and cognitive empathy/perspective taking (in dashed borders). List of brain areas hypothesized to play a critical role in emotional empathy, emotional contagion, and both cognitive and affective perspective taking are also listed here.[2] Fig. 1 from Hillis AE. Inability to empathize: Brain lesions that disrupt sharing and understanding another's emotions. *Brain*. 2014;137(4):981–97. Reprinted by permission of Oxford University Press.

When the participants viewed an individual being hurt intentionally, their brains showed a very rapid response localized to a specific brain network known to sustain computational systems underlying affective states, cognition, and motivational processes (the posterior superior temporal sulcus [pSTS], amygdala, and VMPFC; see figure 4.3).[17] Interestingly, the right pSTS distinguished intentional vs. accidental actions as fast as 62 milliseconds following the onset of the stimulus—faster than the blink of an eye. The amygdala/temporal pole was activated next at around 122 milliseconds following the onset of the stimulus, followed by the activation of the VMPFC around 182 milliseconds after the stimulus onset (see figure 4.3).

In contrast, observation of *accidental* harm to another person activated the pSTS but did not evoke a response in the amygdala or the VMPFC. These data support the notion that the rapid emotional processing and perception of intentionality are the first inputs to

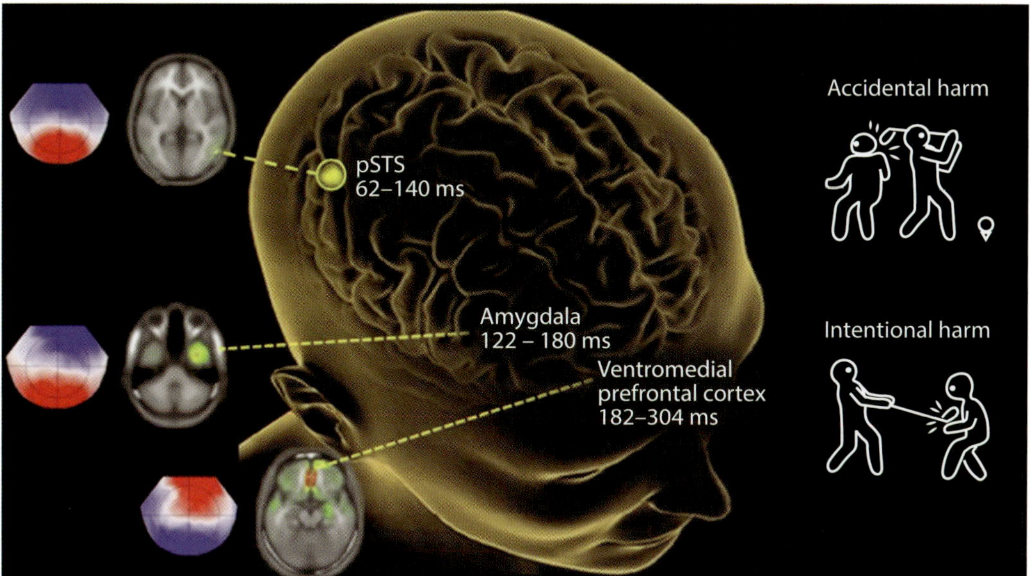

FIGURE 4.3. Schematic representation of temporal brain dynamics involved in empathy and understanding of intentional (vs. accidental) harmful actions. Brain areas that are part of the social brain matrix are activated very fast (within 300 ms post-stimulus onset) when the subject observes someone who performs an intentional harm. Adapted from Decety & Cacioppo.[17]

empathy and moral computations, rather than slower mentalizing processes.[17]

Performing moral computations involves more than just the empathizing and the mentalizing brain networks for Stanley and Adolphs (see chapter 3).[18] According to Jorge Moll and colleagues,[19–21] moral cognition emerges from the integration of contextual social knowledge (that is mainly stored in the PFC), social semantic knowledge (mainly stored in the anterior and posterior temporal cortex), and motivational and basic emotional states (that are subserved by the corticolimbic amygdala network; figure 4.4).[20] In line with the identification of these networks, neuroimaging studies of psychopaths show reduction in gray matter in the prefrontal cortex, and abnormal brain activation in limbic regions and the prefrontal and temporal lobes.[22–24]

Animal studies could potentially provide even more detailed information about the neural mechanisms underlying social processes such as empathy, but doubts have existed that nonprimate mammals are capable of expressing empathy. An early study showed that a rodent who was injected with a painful substance showed more writhing when it could visually observe a cagemate writhing in pain following

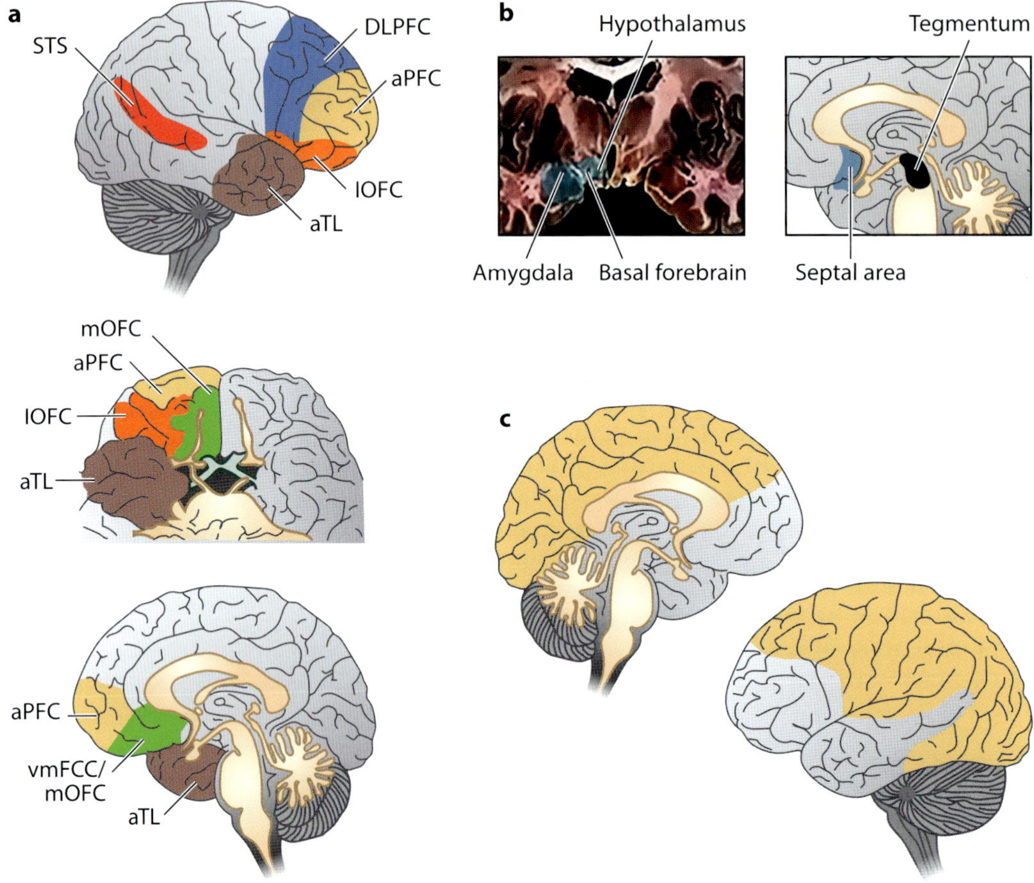

FIGURE 4.4. Cortical and subcortical areas involved in moral cognition. Performing moral computations involves more than just the empathizing and the mentalizing brain networks. Brain regions that have not been consistently associated with moral cognition and behavior in patient studies include the parietal and occipital lobes, large areas of the frontal and temporal lobes, the brain stem, basal ganglia, and additional subcortical structures. Adapted from Moll et al.[20]

the same injection than when (1) the rodent could visually observe an unfamiliar rodent writhing in pain following the same injection and display of pain, and (2) the rodent could visually observe the cagemate when the cagemate was not injected and therefore did not show pain. These results suggested that rodents could express emotional contagion,[25,26] but empathy remained a possible alternative interpretation.

In contrast to emotional contagion, an empathic observer recognizes that the observed individual is in distress and is motivated to help. To test whether rodents show empathetically motivated helping behavior, rodents were placed in an arena where a cagemate was trapped in a restrainer.[27] In the absence of reinforcement, observations of the rodents across sessions revealed that the rodents learned to free their cagemates, and thereafter, the rodents quickly freed their cagemates when placed in the arena. When rodents were placed in an arena with two restrainers—one trapping their cagemate and another containing chocolate—the rodents opened both restrainers and shared the chocolates in over half the trials.

Was the helping behavior by the rodents motivated by empathy or was it motivated to reduce the distress aroused by observing a cagemate in distress? The latter is called negative affect relief, which has been shown previously to be rewarding to the helper in situations that could appear to reflect empathy.[28] Follow-up research has provided additional evidence that the helping behavior is motivated by empathy.[29] In a follow-up study of rodents using the same paradigm as above, the helping behavior by the observer rodent was blocked when affective arousal in the rodent was blocked pharmacologically, suggesting that an emotional response to the trapped cagemate was instrumental in motivating the helping behavior. Importantly, and contrary to the negative affect relief hypothesis, observer rodents who showed the highest stress (i.e., corticosterone) responses to the trapped cagemate were less, rather than more, likely to help the cagemate. These data suggest that observer rodents were motivated by an other-oriented emotional response such as empathy rather than simply by the stress of seeing their cagemate's suffering.

The growing evidence that nonprimate mammals express empathy suggests that other-oriented emotional responses and their consequent social behaviors evolved long ago. The development of animal models of complex social states such as empathy also opens the opportunity to investigate these social states and component processes using far more powerful experimental techniques and measures than are available in humans.

4.2 Imitation and Identification

Imitation refers to an intentional copying of the appearance, behavior, or pursuit of the goal of another individual. Imitation can be contrasted with mimicry, which represents an unintentional matching of the movements of a model.[30,31] Imitation is observed in nonhuman primates,[32] but in contrast to apes, we excel at imitating novel, opaque, or intransitive responses.[33] Given the importance of observing another person's actions in imitation, there is considerable overlap between the neural correlates of imitation and the neural correlates of action observation, including the pSTS, inferior parietal lobule, inferior frontal gyrus, and premotor cortex.[34–36]

Psychologists Christine Legare and Mark Nielsen proposed that imitation functions to promote social learning and cumulative cultural transmission. Imitation and innovation (e.g., using old tools in new ways), they suggest, work in tandem to promote the cultural learning necessary to ensure cultural and individual adaptation to novel and changing challenges.[37] For instance, children use imitation to learn both instrumental skills and social conventions. When first learning observationally how to do something, young children often overimitate, that is, they replicate causally irrelevant as well relevant actions for achieving a goal. This has been termed the "copy-when-uncertain" strategy for social learning in which achieving the goal has a higher priority than behavioral efficiency.[37]

Why might the copy-when-uncertain strategy have evolved? Many of the rituals and social conventions (e.g., norms, practices) that children must learn to become competent members of their cultural community lack obvious causal connections between specific actions and goals. Legare and Nielson suggest that the copy-when-uncertain strategy prepares children to acquire relevant social conventions so that they, too, might become contributing members of their group:

> As specialized cultural learners, children are well-equipped to engage in socially motivated, high fidelity imitation, a potential indicator of group affiliation through conformity . . . high fidelity imitation may have evolved social functions, notably encoding normative behavior, affiliation within groups, and detecting ostracism from ingroups.[37] (p. 690)

The likelihood of imitation is increased by the process of *identification*, which refers to perceiving one's own qualities (desired or real) in another person, the emotional attachment to that person that comes from seeing the other person as an extension of oneself or an idealized self, and the assimilation of the desired characteristics or qualities of the other person to oneself. Identification promotes social connection and social learning in part by focusing social attention and cognition on others who may serve as models.

Thinking about oneself or others produces partially overlapping regions of brain activation, with thinking about one's own traits or goals typically associated with activation in the VMPFC and thinking about another person's traits or goals associated with activation in the

DMPFC.[38–40] Thinking about another person with whom one has a strong identification, however, is associated with a pattern of activation that is more similar to what is seen when thinking about oneself (i.e., VMPFC) than when thinking about strangers.[41]

4.3 Mentalizing

Mentalizing refers to the implicit or explicit processes underlying the endeavor to understand or predict the mental states and behavior of others. As described in chapter 3, mentalizing is a function that is sustained by a specific brain network that differs from other networks involved in understanding emotions or actions of others.[18] Although the capacity for mentalizing is more expansive in humans than in other animals, animal research has contributed significantly to our understanding of these capacities and their underlying neural mechanisms. Monkeys, for instance, recognize conspecifics and understand the meaning of their social interactions and those they observe between others. Research with monkeys has identified a putative precursor to human mentalizing—a dedicated neural network dedicated for processing social interactions (box 4.2).

In this section, we describe three perspectives on mentalizing and the neural mechanisms associated with each. The first is the *egocentrism perspective,* which is based on using one's own mental state as a proxy for the mental state of others. The second is the *simulation perspective,* which covers a family of theories that share the notion that understanding the mental states of others results directly from neural simulations of the observed or imagined actions of others. The third perspective on mentalizing rests on our capacity to recognize that others have mental states that differ from one's own and on reasoning to determine the likely mental states of others. This *theory of mind perspective* is based on the notion that we think about the mind like naïve scientists, developing a theory of another's mind based on observations and relationships between observations to explain our prior beliefs and observations about the person and to predict the person's future behavior.

4.3.1 Egocentrism Perspective

A common way of thinking about the mental states of others is through egocentric reasoning—thinking about the world and others with whom we interact from a self-centered point of view. Egocentrism in social reasoning emerges early ontogenetically. For instance, a young child

BOX 4.2. Evidence of a Dedicated Brain Network in the Macaque Monkey for the Processing of Social Interactions

Many neural mechanisms involved in social interactions, ranging from olfaction and vision to navigation and foraging, underlie capacities that are domain general. As outlined in chapter 3, however, social cognition is a core cognitive component in primates, and neural mechanisms have evolved that are selected for social information processing.[85,86] Animal research has contributed to the elucidation of these domain-general and domain-specific mechanisms. For instance, interactions can reveal hidden features of objects (e.g., mass) and agents (e.g., goals). In a recent neuroimaging study of macaque monkeys, neuroscientists Julia Sliwa and Winrich Freiwald identified a dedicated network for processing social interactions that putatively represents a forerunner of the capacity for human mentalizing.[76] During scanning, monkeys were exposed to real-world videos depicting (1) social interactions between monkeys, (2) physical interactions between objects, (3) monkeys engaged in independent goal-directed behaviors, (4) objects moving independently, (5) nonacting monkeys, and (6) stationary objects.

Videos of monkeys, whether stationary, moving, or interacting, activated areas of the brain that are associated specifically with facial stimuli and body parts (see chapter 5). Videos of interactions between monkeys and interactions between objects activated the mirror neuron system, suggesting that this system may contribute to understanding interactions in the physical as well as social environment. Importantly, a network of regions in the macaque brain activated exclusively by videos of social interactions included the medial prefrontal cortex, DMPFC, anterior cingulate cortex, ventrolateral prefrontal cortex, inferior parietal lobule, and orbitofrontal cortex. This neural network showed deactivation in response to all other videos. Sliwa and Freiwald noted that this neural network shares functional and anatomical characteristics of the human theory of mind mentalizing network.[76]

may assume a parent wants candy because the child wants candy. However, even adults may rely heavily on their own egocentric perspective, especially when a person is pressured to respond quickly, the person is anxious or under cognitive load, the person is in a high-powered role, or the person is mentalizing about someone perceived to be similar.[42]

Although far from perfect, egocentric mentalizing historically may have provided a reasonable approximation of what another person is thinking because individuals shared so many similarities in terms of their prior experiences and current challenges. Moreover, although egocentrism in contemporary society may not provide particularly accurate depictions of what another person is thinking, these inaccuracies may not be readily apparent to the perceiver. This is because what we believe about another person influences how we act toward that person which, in turn, can cause the person to act toward us in a fashion that confirms our initial expectations. For instance, if we think someone has hostile intentions toward us, we are likely to act in a hostile and defensive fashion toward that individual, which may then elicit defensive and hostile behavior from the other person. This a form of self-fulfilling prophecy called *behavioral confirmation*[43] that can create the impression that egocentric reasoning about the mental states of others is more accurate than it is.

As noted above, thinking about one's own mental states is associated with heightened activation in the VMPFC, whereas thinking about another person's mental states is associated with heightened activation in the DMPFC.[38,39,44] Not surprisingly, therefore, neuroimaging research on egocentrism suggests that egocentric mentalizing is also associated with activation in the VMPFC.[41]

4.3.2 Simulation Perspective

According to the simulation perspective, we directly comprehend the mental states of others by internally replicating the observed actions and state of a person without explicit reasoning about what the other person is thinking.[36,45,46] During simulation, one does not really need to know what the other's internal mental state is to anticipate their actions. This is the fundamental distinction between "simulation theory" and "theory theory." Interestingly, both theories are represented by different brain networks (see chapter 3, Stanley and Adolphs's simulation network vs. theory theory/mentalizing network).

These simulations can operate at various levels of abstraction. At one end of the spectrum, simulation can involve matching the actions we observe with internal models (e.g., efferent copy, motor mimicry) of the same action. At the other end of the spectrum, mental simulations operate on more abstract representations, for instance, forming a mental impression of what another person is thinking in novel situations.

Interest in simulation theories of mentalizing burgeoned following the discovery of mirror neurons by neurophysiologist Giacomo Rizzolatti.[47] Neurons are the basic building blocks of the human brain. Among these neurons is a subset that fires when an individual performs an action and when the individual observes someone else perform the same action at a distance (see figure 4.5).[47,48] That is, the neurons

FIGURE 4.5. Illustration of human-animal imitation. iStock.com/Nevena1987.

respond to the actions by another person as if the observer was viewing oneself in a mirror. Mirror neurons not only fire when an action is observed, but also when the intention of the action is clear but the complete action cannot be observed.

There is growing evidence that mirror neurons exist in humans as well as nonhuman primates, specifically that regions of the inferior frontal gyrus (e.g., pars opercularis) and the rostral part of the inferior parietal lobule in humans include mirror neurons. The discovery that neurons in your brain might be doing the same thing as comparable neurons in the brain of someone you are watching has led to a variety of theories and debates about the role of mirror neurons in social perception, cognition, and behavior.

Across evolutionary history, primates battled predators and other primates for survival, reproduction, and valued resources. To the extent that one individual could rapidly anticipate the intention and actions of another through simple sensorimotor observation—whether competing for food, position, or survival—the individual would be more likely to emerge victorious. Mirror neurons putatively serve this function.

Although hand-to-hand combat can still be seen today, we are more likely to see athletes engaged in far less mortal battle on playing fields and courts. The way in which athletes read and anticipate the actions of their opponent during fast-ball sports, such as a tennis, is a challenging and complex process that is a remarkable feat in itself. A tennis player's ability to predict an opponent's intentions quickly and accurately is particularly important during the return of serves, during which the time required to plan and initiate a response typically exceeds the flight time for the ball. For instance, the speed of a professional tennis serve can reach 130 miles per hour, which reaches the receiver in less than 4 tenths of a second. To return a serve directed to

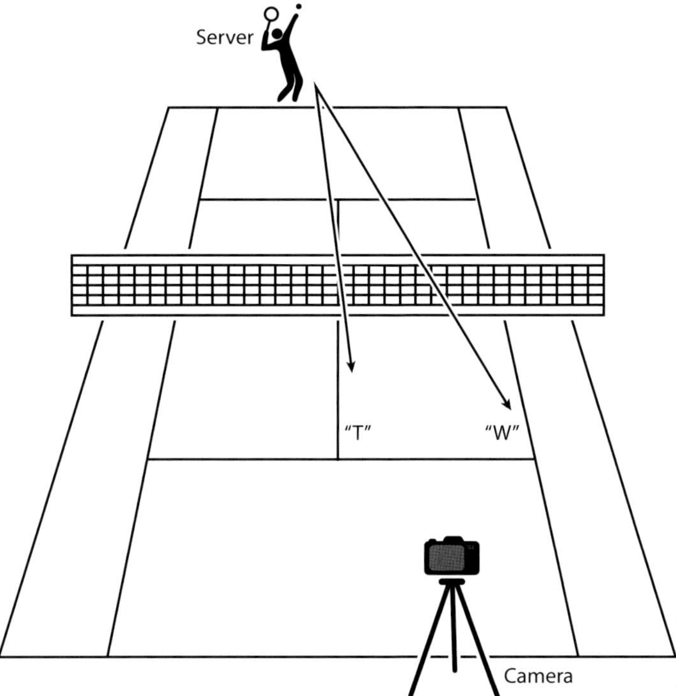

FIGURE 4.6. Schematic representation of two conditions presented during the Cacioppo et al.'s tennis study. Participants had to predict whether their opponent would serve on the wide side of the tennis court (W) or to the center of the court (T). Based on Cacioppo et al.[49]

the center of the T or to the wide side of the service box (T or W; see figure 4.6), the receiver must make a decision regarding the server's intended target based, at least in part, on information identified prior to the server striking the ball. This ability is so critical in professional tennis that at the US Open in 2013, ESPN tennis commentator (and former professional tennis star) John McEnroe said, "If there is something you don't want to be on a tennis court, it is predictable."

Through simulations performed automatically by the mirror nervous system, the more an observed action is congruent with a neurocognitive template from a person's previous intentions and actions, the easier it is for the person to read the intention and action of the person who is being observed. Expert tennis players have received many thousands of serves and have well-developed neurocognitive (i.e., sensorimotor) templates for recognizing and responding to tennis serves. According to simulation theory, when an expert tennis player waiting to receive a serve is watching an opponent toss and strike the ball, activity in the mirror neuron system quickly and spontaneously leads to an understanding of the server's intention, for instance, to serve to the T or wide side of the court.

To investigate this hypothesis, neuroscientist Stephanie Cacioppo and colleagues imaged the brain of expert tennis players while they were lying in a magnetic resonance imaging scanner watching film clips of another expert tennis player serving to either the T or to the wide side of the service box (see figure 4.6).[49] The level of skill of expert tennis players is impressive. When the digital videos were overlapped to contrast an expert's serve to the T versus to the wide side of the box, the differences were so subtle and so rapid that they were rarely noticed. As McEnroe's commentary for the ESPN audience suggested, expert tennis players do their best to mask their intentions.

Typically, a tennis player decides where to serve before the ball-toss, but an expert can execute the same blistering serve whether the instruction of where to serve (T or wide) is shouted during bouncing the ball prior to the serve, or following the ball-toss just as the ball approaches its peak. The server's *intention* differs across these two instructional conditions, however. When the server knows the location of the serve when bouncing the ball, the ball-toss and serve are executed with the intention of striking a specific target on the court (T or wide). In contrast, when the server learns of the target for the serve only as the ball approaches its peak of the toss, all of the actions up to that point are performed without an intended target for the serve.

Digital videos were made of an expert tennis player serving in each of these two instructional conditions, with the video ending just prior to the racquet striking the ball. Overlapping the digital video of the expert's serves in these two instructional conditions confirmed that to a normal observer they looked the same. Participants in the study watched the videos of the serves while lying in the scanner. Following each video, participants were asked to predict the target location of the serve (T or wide). All participants (perceivers) in the study were tennis players, but their level of expertise varied.

When the server in the video did not know the target of the serve until the ball had been tossed—and, hence, simulations of the observed serve were the same until moments before the ball strike—the receivers lying in the scanner were no better than chance in predicting the target of the serve, regardless of their level of expertise. However, when the server in the video knew the intended target prior to the ball-toss, receivers could predict the location better than chance, and the greater their expertise, the better they were at predicting where within the service box the ball was being served. Moreover, and as predicted by simulation theory, the accurate prediction of the serve was characterized by activation of cortical areas within the human mirror neuron system and subcortical areas involved in motor memory. That is, the activity within regions of the mirror neuron system was associated with the very rapid and accurate identification of the server's intentions.

The human brain is an energetically expensive organ, so its activity is selective. For instance, the mirror neuron system is more likely to be activated when the person being observed is relevant to a person's goals, such as a tennis opponent preparing to serve in contrast to an usher who is seating patrons in the grandstand. The intentions of the opponent are important to detect if the interaction is to end well, whereas the intentions of the usher are less critical to the outcome of the task at hand.

The simulation perspective holds that the understanding of the mental states of others is not limited to simulations of observed behavior or activation of the mirror neuron system. Simulations can also be of more abstract events, such as the sensorimotor actions that would have occurred if an action by another had been performed. Consistent with this reasoning, neuroimaging studies show that simulations are associated with heightened activity in the action observation network, which includes areas in the mirror neuron system (inferior parietal lobule, inferior frontal gyrus) and areas involved in the performance of actions (e.g., pSTS, dorsal premotor cortex, ventral premotor cortex) (see figure 4.7).

Although expertise facilitates intention understanding among athletes,[50,51] one does not have to be an athlete to receive the mental assist from neural simulations. Consider a relationship with a significant other. The more connected you feel to a person, the more likely neural simulations will be engaged when you observe the person perform a relevant action. As a consequence, you would be more likely to anticipate and predict motor intentions faster (and better) than if a stranger were to perform that same action[52–54] (see chapter 8). Because a key to a successful interpersonal interaction rests in part on our ability to understand the mental states and intentions of others (an ability that occurs often automatically), neural simulations may promote consonance in salutary relationships as well as survival in belligerent relationships.

Simulations are not the only process that can be invoked to understand the mental states of others. For example, in the fMRI study of tennis players described above, the more expertise participants had in

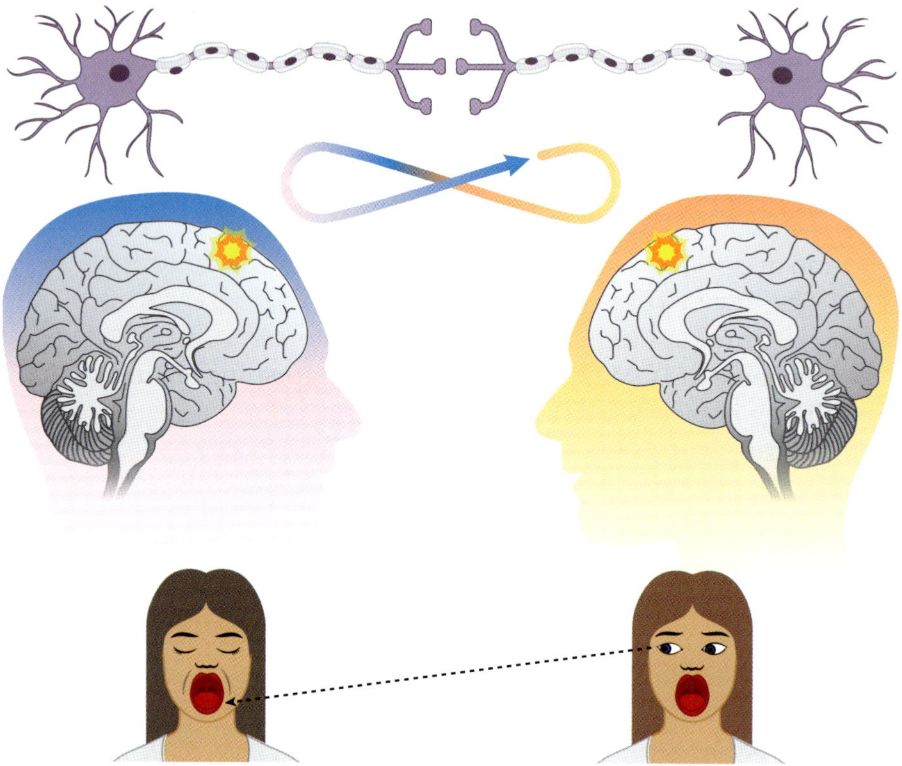

FIGURE 4.7. Schematic representation of the human mirror neuron system, which includes the inferior frontal gyrus and inferior parietal lobule, including the anterior intraparietal sulcus which is associated with goal-directed actions. iStock.com/normaals.

tennis, the more accurate were their predictions regarding the target of the serve, the more brain regions associated with the mirror neuron system were activated, and the more the predictions appeared to be based on an intuitive understanding of the likely target location of the serve. For participants with less expertise, however, predictions were near chance levels. These participants appeared to treat the task as a problem to be solved through observation and deliberate reasoning. Rather than activation in the mirror neuron system, these participants showed activation in brain regions involved when people try to understand the mental states of others by slower, more deliberate and effortful reasoning.[49] These results reinforce prior studies on the dynamics of intention understanding.[55,56] We turn next to this third perspective on mentalizing.

4.3.3 The Theory of Mind Perspective

The theory of mind (TOM) perspective emphasizes that we think about the mind of others like naïve scientists, developing a theory of the other person's mind to permit anticipation and prediction of the person's mental state and behavior. The information that contributes to such a theory includes observations of the person's behavior, knowledge we learn about the other person's behavior (and the behavior of similar others, e.g., those who occupy the same social role), the person's description of or explanation for their mental states and behavior, and so forth. The theory of mind regarding a specific person, in turn, can guide our thinking, emotions, and behavior. Although simulations may contribute to TOM mentalizing,[57] a distinguishing feature of TOM mentalizing is the role of deliberate and effortful reasoning in understanding the mental states and intentions of others.

There are numerous instances in which people consider and integrate a wide range of data to understand and predict others—for instance, diplomats representing two perennial enemies who are negotiating a possible peace treaty, jurors who are deliberating on the guilt or innocence of a defendant, and faculty who are considering applicants for admission to a doctoral program. In each case, the individuals draw from multiple sources of information in a deliberate attempt to understand and predict the goals, state, and behavior of another person. The TOM mentalizing network serves an adaptive function in these social contexts.

Early in life, young children use their own mental states to infer the mental states of others. If two young children see an adult place a toy beneath a red box, they each believe the other shares their belief that the toy is beneath the red box. If one of the children leaves the room and the remaining child sees the adult move the toy from beneath the red box to place it beneath a blue box, the child relies too heavily on their own knowledge during mentalizing and believes the other child also knows that the toy is now hidden beneath the blue box. That is, they engage in egocentric projection when mentalizing.[58]

During development, children learn that their own experiences, beliefs, and perspectives are not always the same as those of others. They also develop the executive functions needed to inhibit their own perspective when considering another person's mental state, and to hold relevant information in working memory. Children as early as two years of age have been found to show components of TOM mentalizing.[59]

The TOM mentalizing network is so called because it refers to brain regions whose activity is modulated when thinking about the thoughts or feelings of others to form a theory of the mental states (mind) of the others. The activation of the TOM mentalizing network, therefore, extends beyond situations in which you see another person in action, and it generally requires more time and effort to reach an understanding than direct motor simulations.

A variety of tasks have been used to investigate the neural correlates of TOM mentalizing. The variations in the tasks can produce activity in regions that are not central to TOM mentalizing. These task-specific regions of activation can be identified when quantitatively summarizing regions of activation across these tasks. Doing so reveals that activity in the DMPFC and the right and left regions of the temporoparietal junction (TPJ) are activated across tasks.[60,61] Activity in the DMPFC has been found previously to be involved in processing socially or emotionally relevant information about others, particularly others who the perceiver regards as socially relevant or close,[62–64] and activity in the TPJ has been found to be involved in the processing of the mental perspective of others.[65,66]

Although the theory of mind path has some advantages, it can be a treacherous path that also leads to errors, especially under time pressure. The tennis study described above is an example. As the receiver in a tennis match with an expert, you do not have much time to think about and predict where the server intends to serve once the ball is tossed in the air. If you try to figure out why the server tossed the ball in this way or that way (e.g., to disguise her intentions, to show off, or to hide the fact she is tired), you may still be deep in thought as the tennis ball flies past you. Indeed, the tennis players in the fMRI study we described above who showed evidence of using theory of mind to predict the tennis serve performed at chance levels.

Psychologist Eugene Caruso showed the timing of social stimuli influences social apprehension, as well. Participants viewed either normal speed or slow-motion replays of scenes in which the action of one person resulted in harm to another.[67] When participants viewed the action in real time, the harmful action was judged to be unintentional. However, when participants viewed the action over an artificially long period of time, there was time for TOM mentalizing processes to produce the false impression that the actor had time to premeditate, resulting in more intentionality (and culpability) being attributed to the actor.

In most instances, however, the TOM mentalizing network improves one's understanding others. As would be expected, social circumstances that demand a person be responsive to another's state of mind increase the engagement of the TOM mentalizing network. For instance, activity within this network is elevated in individuals low, in contrast to high, in social status, presumably because they have more to gain by accurately understanding and predicting the other's mental states and behaviors.[68] Moreover, forming trait inferences about others spontaneously or deliberately activates the TOM mentalizing network, and doing so deliberately activates this network to a greater degree than doing so spontaneously.[69,70]

4.3.4 Multiple Pathways

There is growing evidence that each of these three perspectives—egocentrism, simulation, and TOM mentalizing—describes mechanisms that have evolved in the human brain for understanding and predicting the mental states and behaviors of others. These mechanisms differ in terms of the capacities, time, and effort they require, and therefore in their adaptive value across individuals in specific social contexts. Accordingly, these mechanisms tend to operate in different situations and in different individuals in the same situation (e.g., as a function of their role or expertise).

Economist Shinsuke Suzuki and colleagues provided computational evidence for the influence of multiple mechanisms for understanding the state or intentions of other brains when an individual seeks to build a consensus.[71] Egocentric mentalizing—the overweighting of one's own thoughts, feelings, and preferences—was especially apparent early in efforts to reach group consensus. Consensus could be reached quickly when everyone shared the same egocentric perspective. However, when individuals were confronted by the expression of different views or preferences by others, evidence emerged for the operation of TOM mentalizing—the view that minds of others are a puzzle to be solved. For instance, as the difficulty in reaching a consensus grew, individuals for whom reaching a consensus was more important than maintaining their preferred option were increasingly likely to yield to the options thought to be held by those who were unyielding.[71]

Some of the participants in the study were also scanned during consensus formation. Results show: (1) egocentric mentalizing was associated with elevated activity in the VMPFC (a region involved in self-referent processing and egocentric mentalizing); (2) the prior expressed view of the majority was associated with elevated activity in the right pSTS and the adjacent area, the TPJ—regions involved in self-representation, mentalizing, and learning bodily signals for predicting others' behavior; and (3) the inferred commitment by others to a particular option (their estimated "stickiness") was associated with activity in the intraparietal sulcus adjacent to the inferior parietal lobule. Interestingly, activity in the intraparietal sulcus was observed in a nonsocial control task as well as the consensus-building task, which makes it unlikely to reflect mirror processes. Instead, the activation of the intraparietal sulcus has been observed during domain-general tasks such as evidence accumulation and encoding the probability of events, processes that were engaged when participants computed how committed others were to their positions.[71]

4.4 Social Learning

Natural selection influences behavior across generations, but natural selection has also produced a genetic constitution for learning, which influences behavior over much shorter periods within an individual life span through nonassociative learning (e.g., sensitization, habituation), associative learning (e.g., classical conditioning, operant conditioning), and social learning (see table 4.2).

Classical conditioning enables the formation of associations between pairs of stimuli that occur sequentially over time, the first of which is initially perceived as inconsequential (e.g., the appearance of a honeybee) and the second of which evokes a response (e.g., the sting of a honeybee). Classical conditioning promotes adaptive behavior by increasing the predictability of the likely effects of stimuli in the world and adjusting behavior accordingly.

Operant conditioning is a form of learning in which associations are formed between an organism's behavior and the consequences that follow from that behavior. Operant conditioning promotes adaptive behavior by promoting the expression of a behavior that is the most likely to produce desired effects or to avoid undesired effects. For instance,

TABLE 4.2. Five types of learning

Type of learning	Cognitive process
Sensitization	Increase in the magnitude of responses to a kind of stimulus
Habituation	Decrease in the magnitude of responses to a kind of stimulus
Classical conditioning	Form a new connection between pairs of stimuli and a response
Operant conditioning	Form a new connection between a behavior and an outcome
Social learning	Learning by watching the actions and experience of another

if a predator enters an area rich in prey, the predator is likely to return to the area when searching for prey.

Social learning refers to a variety of behaviors and mechanisms in which one organism learns by observing or imitating others, or through direct instruction from others even in the absence of imitation or direct reinforcement. Thus, social learning includes but is not limited to vicarious classical conditioning (e.g., observing a contingency between an initially neutral stimulus, such as a bumblebee on another's neck, and the expressions of pain that follow soon afterward) and vicarious operant conditioning (e.g., observing someone catch copious numbers of fish at a specific location on a pond). Social learning also describes learning based on the unreinforced observation or imitation of a social model as well as verbal learning. Forms of social learning have been observed in an array of social species including chimpanzees, monkeys, rodents, octopuses, and birds,[72,73] but the breadth and depth of social learning is greater in humans, and verbal learning—a form of social learning that depends on language—may be specific to humans.[74]

Research on social learning has begun to identify brain regions that are involved. In humans, for instance, social learning is associated with activity in regions that are involved in TOM mentalizing (DMPFC, TPJ) or executive functioning (dorsolateral prefrontal cortex, anterior cingulate).[73,75] There are various mechanisms underlying social learning, such as social signals regarding the attentional value of a location or stimulus (i.e., stimulus enhancement), social signals regarding the valence of a stimulus (i.e., stimulus appraisal), response priming, and active instruction.[73] The specific mechanism or mechanisms for social learning that are operative influence the additional neural regions that are engaged.

4.5 Concluding Remarks

Social animals are subject to forces that promote the alignment and understanding of others at a distance. These include emotional contagion, affective perspective taking, and empathy, each of which engages related but separable neural regions and helps align the emotional states of the individuals in an interaction or group.

Emotional contagion, affective perspective taking, and empathy can each promote survival and reproductive success, but each does so by motivating different behavioral responses. For instance, if an observer feels fearful as a result of seeing another person expressing fear but fails to recognize that the fear originated from observing the other person, it is emotional contagion. The observer has not identified the source of the fear, and his or her response is likely to produce self-protective behaviors such as increased vigilance for threats in the situation (i.e., stimulus enhancement). If, instead, an observer sees another person expressing fear, notices a speeding truck headed in the direction of the person, and the observer surmises that the other person is fearful, the observer would be showing affective perspective taking. The observer would have the advantage of understanding the needs of the other person, and how the observer chose to act on this advantaged perspective would depend on his or her own interests, concerns, and goals (e.g., cost-benefit analysis). Finally, if an observer sees another person expressing fear, feels the other person's fear, and realizes the fear is not a personal emotion but an other-oriented emotion, the observer would be showing empathy. The observer in this instance is more likely to engage in other-directed behavior, such as altruism.

Imitation refers to an individual copying the appearance, behavior, or goal pursuit of another person. Imitation is especially likely when an individual identifies with the other person, in which case the person assimilates aspects or characteristics of the other person to form a persona that overlaps with that of the other person. Imitation and identification not only promote the connection and alignment of individuals, they promote social learning even in the absence of formal instruction.

Specifically, social learning is promoted by an implicit recognition that what happens to another could happen to oneself—that is, by identification and by self–other overlap. There is ample evidence of this self–other overlap in humans, including studies showing (1) similar expressive movements occur when an individual feels an emotion and when the person observes another person feeling that emotion, (2) similar brain areas are engaged when imagining one's own pain and when imagining another person's pain, and (3) similar brain areas are activated when imagining one's own actions and another person's action (e.g., premotor and posterior parietal cortices). In addition to social learning, these overlapping responses are thought to promote social processes including intersubjectivity (e.g., the same feelings, focus of attention), mutual understanding, behavioral predictability, and effective coaction.

Understanding the thoughts, emotions, intentions, and goals of others can be advantageous whether the intentions of the other are benevolent or malevolent, and we discussed three perspectives on neurocognitive mechanisms that function to approximate the mental states of others: (1) the egocentrism perspective, which is based on using one's own mental state is a proxy for the mental state of others; (2) the simulation perspective, which covers a family of theories (e.g., mirror neuron system) that share the notion that understanding the mental states of others results directly from neural simulations of the observed or imagined actions of others; and (3) the theory of mind perspective, which is based on the notion that we think about the mind like naïve scientists, developing a theory of another's mind based on observations and relationships between observations to explain our prior beliefs and observations about the person and to predict the person's future behavior.

Although mentalizing may be so ubiquitous in humans that it may seem to be a defining feature, it is *not*. Mirror neurons, which appear to underlie certain forms of simulation, were first identified in studies of macaque monkeys, presumably to support intention understanding and joint action.[48] TOM mentalizing, however, was thought to be unique to humans. Research with macaque monkeys has identified a dedicated brain network for the processing of social interactions that may be a forerunner of the TOM mentalizing network,[76] and research with great apes suggests that they possess a form of mentalizing that falls under the TOM mentalizing perspective. For instance, using eye-tracking, chimpanzees, bonobos, and orangutans showed evidence that they, like humans, correctly anticipate where an individual will look for a hidden item even though the apes knew the item had been moved.[77,78]

5

SOCIAL PERCEPTION: READING THE FACE

Social perception refers to the impressions and expectations drawn about others. Social perception is to social interactions what the oxygen atom is to water—a fundamental element. While forming impressions of others can occur automatically, a complex set of neural mechanisms operates in the background efficiently generating social perceptions. In chapter 4, we discussed mechanisms such as mentalizing for forming social impressions and inferences. The social cues used in social perception include physical appearance as well as verbal and nonverbal communication. In the current chapter, we focus on the social cues from the face as a model system for the neurobehavioral mechanisms underlying social perception.

5.1 Face Perception

The face is special in several respects. First, the face is a distinctive part of the body, and it serves as an important cue for distinguishing among individuals and for social recognition. The prototypic face consists of a mouth and its encircling lips, a nose, and two eyes, eyelids and lashes, and overarching brows, all framed within a more or less oval surround that includes the forehead above, cheeks and ears on the sides, and chin below. Beneath the skin is a set of striated muscles called the muscles of mimicry, which in humans is innervated by the seventh cranial nerve (see figure 5.1). Nearly all striated muscles are connected to tendon and bone and, when contracted, move the limb to which each is connected. The muscles of mimicry, in contrast, are connected to the skin and other muscles, and the contraction of these muscles creates folds, lines, and wrinkles in the facial skin and the movement of landmarks such as the brows and corners of the mouth. Between the skin and the muscles of mimicry is a layer of fibrous connective tissue called the fascia, as well as adipose (fat) tissue, facial tissue, septa, and ligaments distributed across distinct fat compartments.[1] Variations in facial appearance include the size and symmetry of facial features (e.g., eyes, nose, brows, lips), the reflectance (e.g., pigmentation, texture) of the skin, and the convexities and concavities of the underlying facial bones (e.g., high cheekbones reflect the convexity and projection provided by the zygomatic bone).[2]

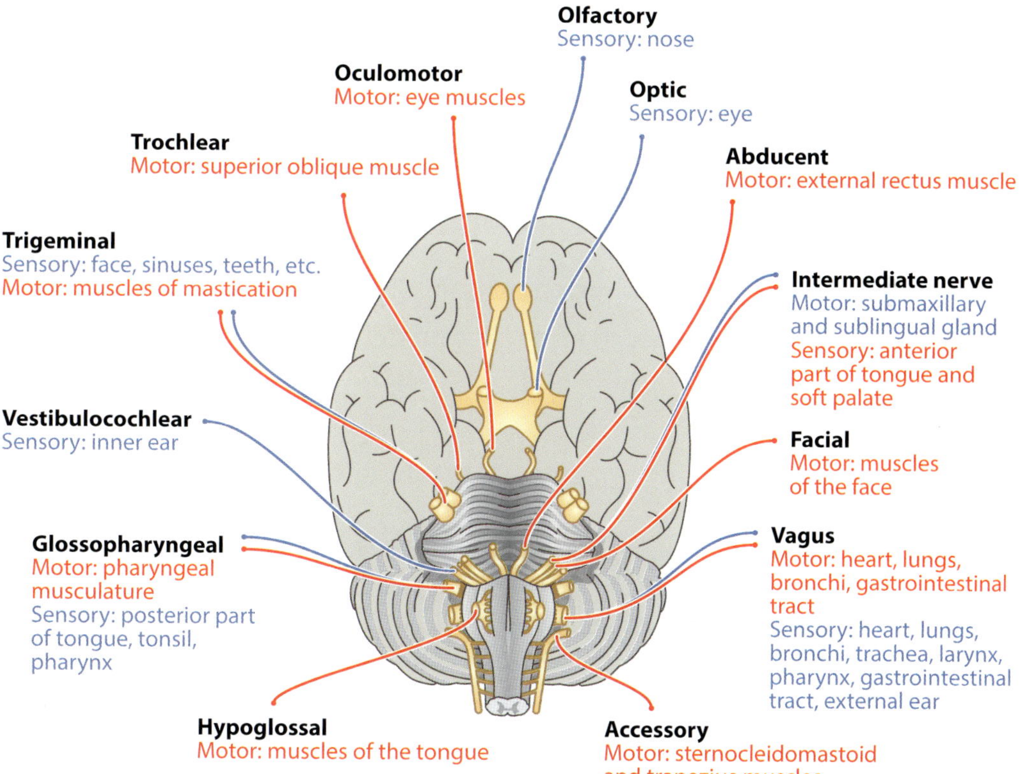

FIGURE 5.1. The cranial nerves of the human brain. Each has a specific role. For instance, the seventh cranial nerve innervates the muscles of mimicry. iStock.com /normaals.

A second way the face is special is the processing of the face as a unitary percept rather than as the sum of a set of features. For instance, psychologist Martha Farah has shown that, despite the importance of individual variation in facial components for facial recognition, faces are recognized "holistically."[3] In an illustrative study, participants were exposed briefly to pairs of faces, one beside the other, followed immediately by the name of a facial part (e.g., nose). The task was to specify whether or not the facial part was the same between the pair of faces they had just seen. To the extent that the face is represented in visual memory in terms of its component parts (e.g., nose, mouth, eyes), judgments of the similarity between the pair of faces in any single facial part should be unaffected by the similarity or differences between the faces in other facial parts. However, to the extent that faces are represented as a unitary configuration of parts, judgments of the similarity between faces in a component part should be influenced by the similarity between the faces in other component parts. Farah and colleagues found that judgments of similarity of a component part were, in fact, influenced by the similarity of other facial components. Farah and colleagues also used inverted faces in their study because the hypothesis that facial perception is holistic applies only to upright faces (see figure 5.2). The similarity in other facial components did not affect the judgments of similarity of a component part between the pairs of inverted faces. Together, these results indicate that faces are recognized holistically,[4] a finding observed in chimpanzees as well.[5]

A third way the face is special, though not unique, is the selective activation of a specific region of the brain involved in visual processing in response to a face. Figure 5.3 illustrates regions of the brain involved in face and body processing (areas in green), mirror processing (areas in yellow), and theory of mind (TOM) mentalizing (areas in red).[6] Neuroscientist Nancy Kanwisher, who had been performing functional magnetic resonance imaging (fMRI) studies of shape perception without much success, was aware of studies of patients with brain damage and positron emission tomography (PET) and fMRI studies indicating that there appeared to be an area in the back of the right hemisphere that was involved in facial recognition.[7] The neuroimaging studies, however, had compared the activation evoked by faces in contrast to stimuli such as letter strings or visual blobs, so it was unclear how special faces were in evoking a localized brain response. She and her colleagues began to contrast the response to intact faces in contrast to scrambled faces, houses, human hands, and so forth.

Two important results emerged from this early work: (1) a region of the brain, dubbed the fusiform face area (FFA) and typically larger in

FIGURE 5.2. Illustration of inverted faces. In their experiment using inverted faces like the faces shown here, Farah and colleagues found that judgments of similarity of a component part were, in fact, influenced by the similarity of other facial components.[3] Farah and colleagues also used inverted faces in their study because the hypothesis that facial perception is holistic applies only to upright faces. Hypothetical example of familiar faces. Courtesy of Getty Images Entertainment. Photo of Anne Hathaway/Ben Gabbe. Photo of Gwyneth Paltrow/Dan MacMedan.

the right hemisphere, was located in the fusiform gyrus (see figure 5.4); and (2) the specific location of this region within the fusiform gyrus was replicable within a participant but varied across participants. To identify the location of the FFA within an individual brain, Kanwisher used a functional localizer task—such as viewing the presentation of a series of faces and objects while fMRI was recorded to determine the specific location of the brain region activated by faces but not objects for that particular individual.[8–10] Additionally, patients who suffer from brain damage to the fusiform gyrus often suffer a neurological disorder

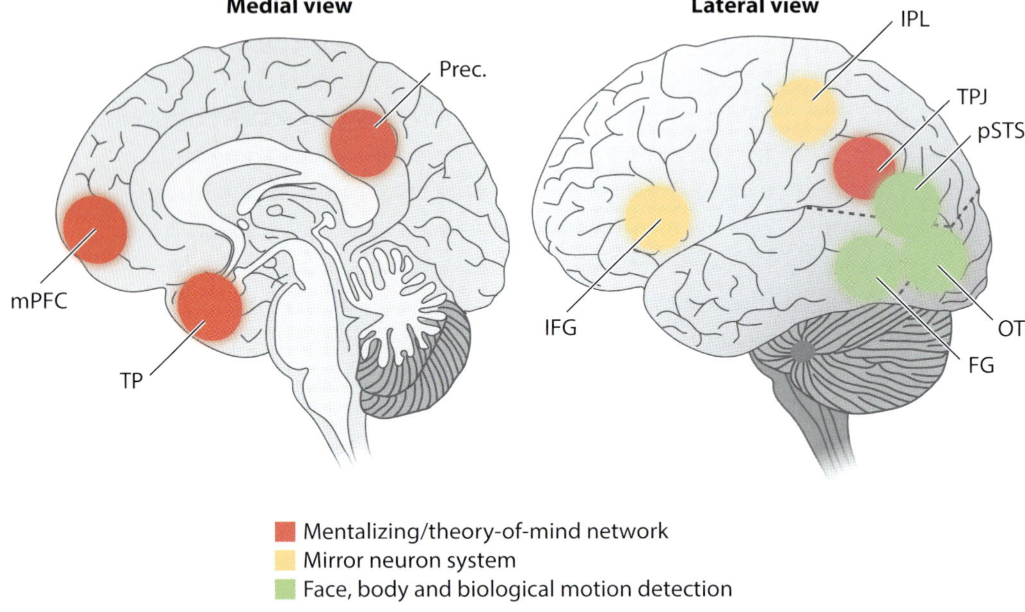

FIGURE 5.3. Three brain networks. Medial and lateral view of the human brain showing the mentalizing network (red), the mirror neuron system (yellow), and the biological motion detection network (green).

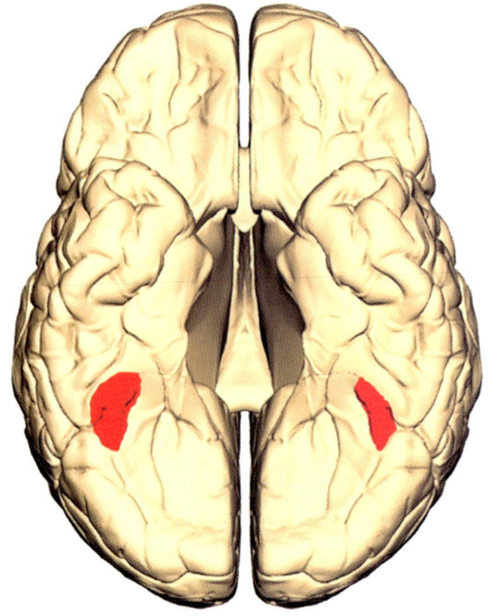

FIGURE 5.4. Fusiform gyrus. Fusiform face area shown in isolation (highlighted in red). From Grens (2014).[86]

known as prosopagnosia in which patients have difficulties in recognizing faces of familiar individuals (e.g., their spouse). These patients remain able to recognize other kinds of objects, but not faces.[11,12]

Since this early work on the FFA, a few other circumscribed brain regions have been identified that respond to a stimulus category, including: (1) the occipital face area (OFA), which along with the FFA is primarily involved in distinguishing between faces; (2) a face-specific region in the superior temporal sulcus (STS), which is activated in response to changes in gaze or expression; (3) an area in the parahippocampal region that is selectively activated in response to images of places (dubbed the parahippocampal place area, PPA), (4) a region in the extrastriate visual cortex that is selectively activated to images of the human body and body parts (dubbed the extrastriate body area, EBA), and (5) a small region near the FFA in the left hemisphere that is selectively activated to visually presented letter strings and word-like stimuli, but only in individuals who know how to read (dubbed the visual word form area, VWFA)[13–15] (see figure 5.5). Efforts to identify other selective regions have proven unsuccessful so far, and the identification of the VWFA in the ventral visual stream points to the potential importance of experience.

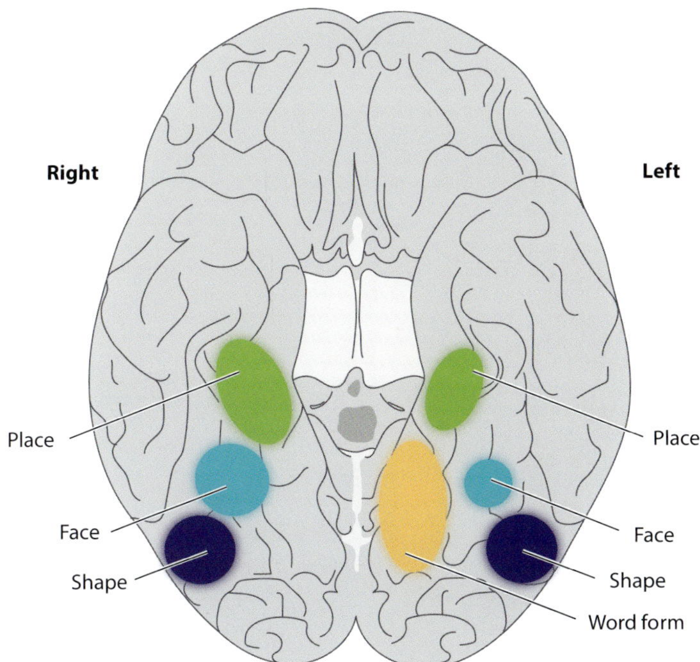

FIGURE 5.5. Fusiform face area (FFA). View of the human brain from below showing the FFA with other specialized regions within the fusiform gyrus.

BOX 5.1. Familiar Faces Area in Primates

Multiple face areas have been found in the temporal and prefrontal cortices of the macaque monkey.[88] Within this network, information about the face is first processed in the early picture-based area (middle face areas) into an identity-based representation in the anteromedial face area, that is, an area known to process familiar faces in humans.

In human primates, a model of the distributed set of areas that mediate familiar face recognition has been described by Gobbini and Haxby. The core system deals with the encoding of the view of a familiar face while the extended system extracts further information from a face.[50] In this functional model for face perception, the structures that participate in the retrieval of different aspects of knowledge about a person (such as biographical information and personality traits) and in the emotional response are highlighted.

The face plays an important role in social perception in animals as well as humans. Neuroimaging research on the macaque monkey has identified six patches of cortex in the temporal lobe involved in face processing.[16,17] Microstimulation of four specific face patches indicated that face processing engages a series of dedicated processing stages (e.g., facial detection, facial identification) rather than relying solely on the visual regions for object processing. These results show that functional specialization in the temporal lobe extends beyond regions of the cortex to include distributed networks of cortical regions.[16] Areas associated with facial recognition in the macaque brain have also been identified (see box 5.1).

5.2 Static Signals

The face is a stimulus that attracts attention early in life and remains one of the most salient visual stimuli we process across our life span.[3,18] The face, for instance, carries information about a person's gender, age, and identity as well as a person's attentional state, emotional state, and behavioral predisposition. *Static signals* represent relatively permanent features of facial appearance such as the size and symmetry of facial features, the pigmentation and texture of the skin, and the convexities and concavities of the underlying facial bones that give the face shape and contours. People draw rapidly and automatically upon static signals from the face to infer a great deal about a person, including their identity, personality, emotions, and intentions. The inferences that are drawn have been found to sometimes be alarmingly inaccurate, but people nevertheless persist. More than 70 years ago, a pioneer in impression formation, Solomon Asch, summarized the impact of the static signals of the face as follows:

> We look at a person and immediately a certain impression of his character forms itself in us. . . . Subsequent observations may enrich or upset our view, but we can no more prevent its rapid growth than we can avoid perceiving a given visual object or hearing a melody.[19] (p. 258)

To investigate the static signals associated with trait inferences, psychologist Alex Todorov and colleagues used computer modeling to create a large set of expressionless faces that varied in the static signals of shape (e.g., rounded face, small jaw) and reflectance (e.g., brightness, color, texture variations). Naïve participants then rated the faces along different trait dimensions.[20,21] They found that judgments of two general traits—valence/trustworthiness and power/dominance—were associated with variations in the shape and size of the face. For instance, when the shape of a neutral face is amplified by 3 standard deviations on the trustworthiness dimension, it is perceived as trustworthy. When the shape of the same face is reduced by 3 standard deviations on the same dimension, it is perceived as being untrustworthy (see figure 5.6).[22,23]

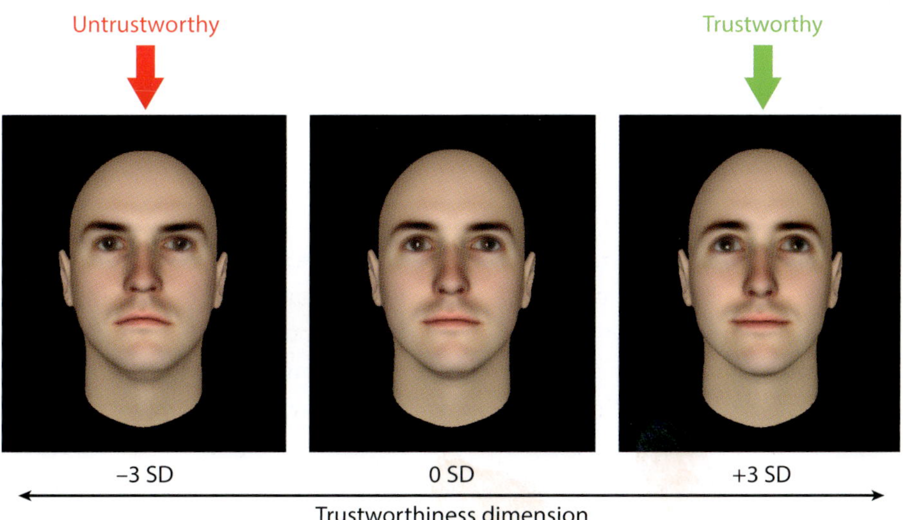

FIGURE 5.6. Facial expression of trust. Example of stimuli used in Todorov et al.'s study on trust.[22,23] Copyright © 2012 Rezlescu et al.

Related studies indicate that static facial signals associated with strength, maturity, or anger tend to elicit trait inferences of dominance and aggressiveness, whereas static signals associated with babies ("baby-faced") or happiness tend to elicit trust and affection.[20,24] Inferences about other traits, such as how threatening a person was perceived to be, were explicable in terms of a combination of the static signals in these two dimensions, for instance, with high dominant and low trustworthy faces perceived as threatening and low dominant and high trustworthy faces perceived as nonthreatening.

Todorov and his group showed that these social judgments based solely on static signals of the face occur rapidly, spontaneously, and effortlessly.[25,26] For example, a 33-millisecond exposure to a face (a small fraction of the time required to blink) is sufficient for people to decide whether the face looks trustworthy or not.[27] Even though the traits that are inferred from facial appearance are not necessarily accurate, they are thought to be adaptive. Psychologist Leslie Zebrowitz proposed that the effects of baby-faced shapes in adults on social perception reflect an overgeneralization to adults. As long as the cost of attributing childlike traits to baby-faced adults is less than the cost of not being responsive to the needs of an infant, this overgeneralization can be viewed as adaptive. Similarly, faces that are perceived to differ in terms of expressions of hostility or hospitability (e.g., a subtle frown or scowl versus a subtle smile; see figure 5.7) may reflect an overgeneralization that may be adaptive. Inaccurately inferring a stranger is hostile when their facial cues are reminiscent of strength or anger may be less costly overall than inaccurately inferring such a stranger is hospitable when they are, in fact, hostile. Consistent with the overgeneralization hypothesis, static facial features that resemble specific emotional expressions tend also to evoke trait inferences that are associated with that emotional state.[28–31] For instance, the facial appearance associated with the trait inference of extroversion is reminiscent of a welcoming smile, whereas the facial appearance associated with the trait inference of introversion looks more sullen (figure 5.7).

Neuroimaging studies of differences in trait inferences as a function of variations in static signals are also consistent with this reasoning. For instance, a meta-analysis of fMRI studies of faces that evoked negative impressions (e.g., untrustworthy, unattractive) identified activity in the amygdala to be elevated.[32] The amygdala is a brain region that plays a crucial role in fear conditioning and in orchestrating the neuroendocrine response to stressors, and the emotional expression of anger has been shown repeatedly to activate the amygdala. Positive impressions, in contrast, were associated with activity in dopamine-rich

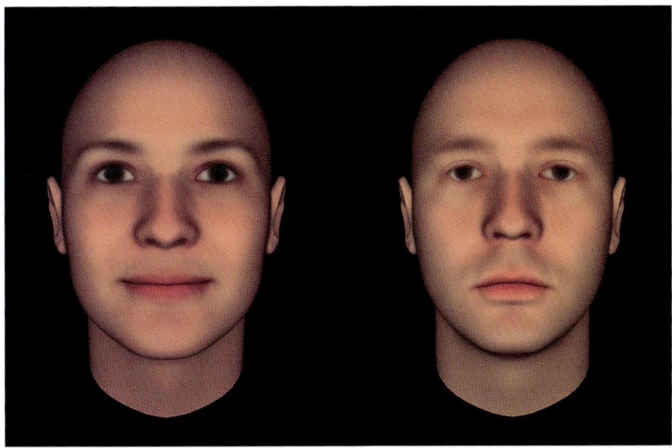

FIGURE 5.7. Examples of stimuli used in Todorov and Oosterhof.[87] Reprinted with permission from "Modeling Social Perception of Faces" by Alexander Todorov & Nikolaas N. Oosterhof, *IEEE Signal Processing Magazine* (March 2011). Copyright © 2011 by IEEE.

limbic regions (caudate extending to the nucleus accumbens/medial orbitofrontal cortex) as well as the ventromedial prefrontal cortex (VMPFC), pregenual anterior cingulate cortex, and right thalamus—a pattern of activation similar to that found in response to the emotional expression of happiness.[32]

5.3 Slow Signals

A second type of facial signal that affects social perception is slow signals. Slow signals represent changes in the appearance of the face that occur gradually over time, such as age-related changes in folds, wrinkles, and skin texture. As people age, the skin becomes thinner, less elastic, and drier, resulting in wrinkles, creases, and lines on the skin. Aging is not the only influence, though. For instance, repeated exposures to the sun (specifically, ultraviolet light) break down the connective tissue (collagen and elastin fibers) in the skin, causing the skin to become weaker and less flexible. As a result, the skin begins to droop and wrinkle. Repeatedly expressed displays of emotion (e.g., smiling, frowning) also contribute to the development of wrinkles due to the grooves under the skin that are created by the muscles pulling on the skin to create these expressions. As the skin loses its flexibility, these grooves become more apparent, leaving a record of the most frequent and intense emotions that were previously expressed.

The record of expressive behaviors captured in slow signals is typically faint, but not always. In *The Expression of Emotions in Man and*

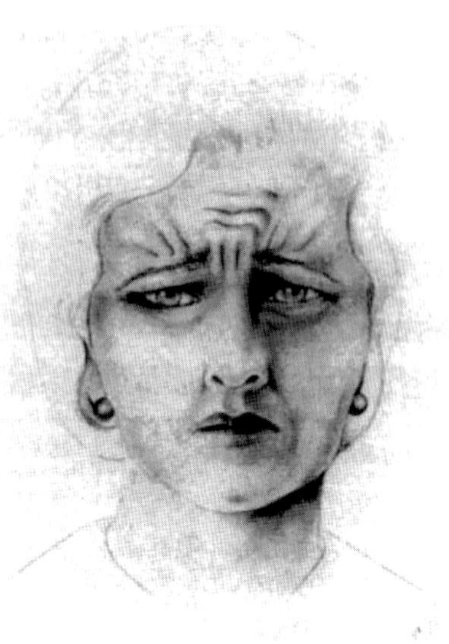

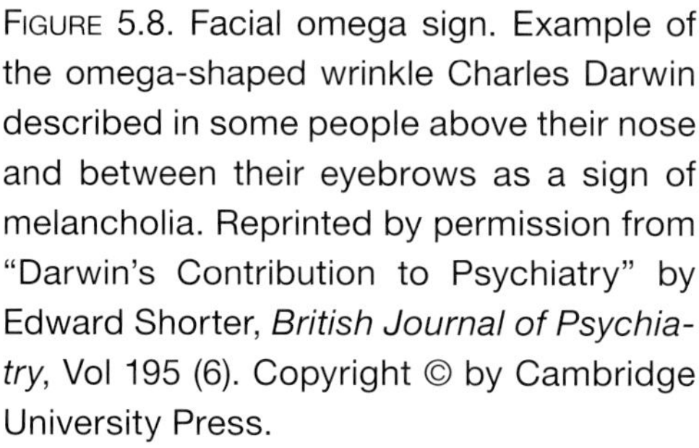

FIGURE 5.8. Facial omega sign. Example of the omega-shaped wrinkle Charles Darwin described in some people above their nose and between their eyebrows as a sign of melancholia. Reprinted by permission from "Darwin's Contribution to Psychiatry" by Edward Shorter, *British Journal of Psychiatry*, Vol 195 (6). Copyright © by Cambridge University Press.

Animals, Charles Darwin refers to the omega-shaped wrinkle some people have above their nose and between their eyebrows as a sign of melancholia (see figure 5.8):

> After the mind has suffered from an acute paroxysm of grief, and the cause still continues, we fall into a state of low spirits; or we may be utterly cast down and dejected. . . . The eyebrows not rarely are rendered oblique, which is due to their inner ends being raised. This produces peculiarly-formed wrinkles on the forehead, which are very different from those of a simple frown.[33] (p. 178–79)

Slow signals, like static signals, can rapidly and effortlessly affect the impressions people form about an individual. These effects can also be explained in terms of the overgeneralization hypothesis. If a person's past behavior is characterized by long, intense feelings of grief, leaving a visible omega-shaped wrinkle between their brows, then it may be adaptive to infer future mental states and behavior to be similarly predisposed.

Acetylcholine is a neurotransmitter responsible for muscular contraction. The injection of botulinum toxin A (also known as Botox) into the muscle underlying a facial wrinkle blocks the release of the neurotransmitter acetylcholine, blocking activation (i.e., paralysis) of the muscle. This generally results in a smoothing of the overlying skin. Injection of botulinum toxin into the right and left corrugator supercilii muscles (the muscles that pulls the eyebrows together) has become a common cosmetic treatment for wrinkles associated with concentration, sadness, worry, and other negative affective displays. Moreover, the injection of botulinum toxin into the corrugator supercilii muscles reduces a person's depressive feelings as compared to placebo in randomized controlled studies.[34–36]

The notion that afferent (sensory) facial feedback can, at least under certain circumstances, influence affective states is further supported by work showing that the imitation of facial expressions activates brain areas involved in emotional processing, such as the amygdala within the limbic system,[37–39] which have reciprocal connections to the hypothalamus and brain stem regions involved in autonomic control.[40] In an illustrative study, Andreas Hennenlotter and his colleagues asked participants to perform a facial expression imitation task while their brain activity was recorded with fMRI both before and two weeks after they received Botox injections into the corrugator supercilii.[39] Results showed that after two weeks of Botox injection, participants showed reductions in the level of activity in their amygdala and brain stem during the intentional imitation of angry faces. These findings show that blocking the activation of facial muscles not only reduces the ability to express emotions but also reduces the brain's response to congruent emotional stimuli.

5.4 Artificial Signals

Artificial signals represent exogenous features that change the appearance of the face. These include exogenous stimuli that can be donned or removed such as make-up, eyeglasses, or colored contact lenses. Research has shown that the facial appearance of a target person, including an appearance that is influenced by artificial signals, can affect not only the inferences a person draws about the target person but also their behavior toward the person.

In a classic study on this effect,[41] men ("perceivers") were shown a photograph of a woman before or after cosmetics had been applied and her hair had been prepared to accentuate her attractiveness. Each participant was told he would be speaking by phone with the woman in the photograph. In fact, each perceiver spoke to an unfamiliar female, and the conversations were recorded. When participants believed their conversation partner was attractive, they and their conversational partners were more likely to speak in a friendly, poised, warm, and socially skilled manner. On the other hand, for the participants who believed their conversational partner was unattractive, both they and their conversational partners were more likely to speak in a serious, uninteresting, awkward, or other socially inept manner. This is another example of *behavioral confirmation* because, based on the differences in artificial facial signals they saw, the participants acted in a way that elicited *confirmatory* behavior from their conversational partners.

5.5 Rapid Signals

Rapid signals represent visually detectible movements in facial features or skin and fascia by the contraction of muscles of mimicry that we describe in the beginning of this chapter. Although there are a relatively small number of muscles in the face (see figure 5.1), their connections to the surface of the skull bone, skin, and other muscles make it possible to produce some 6,000 to 7,000 appearance changes, including possibly more than 20 distinct emotional states.[42,43]

Neurons through which the brain innervates muscles are called motoneurons and are distinguished from those called sensory neurons, which bring information to the brain from sense receptors. Motor neuron circuits have two parts: upper motoneurons (UMNs) carry motor impulses from motor centers in the brain to the brain stem or spinal cord, and lower motoneurons (LMNs) carry the impulses from the brain stem or cord to the muscle itself. As illustrated in figure 5.1, the LMN tract that innervates the muscles of facial expression is called the seventh cranial nerve, or simply, the facial nerve. In humans, the primary function of the facial nerve is to control the motor expression of facial muscles in meaningful configuration. Similarities and differences exist across phylogeny. In humans, the motor nucleus of the facial nerve corresponds to a four-millimeter-long column of cell bodies located in the brain stem.[44] It contains some 7,000 to 10,000 nerve cells plus numerous glial cells and is considered to be the largest of the cranial nerve motor nuclei.[44]

The basic mapping of facial muscles onto the nucleus shares much in common with that found in other mammals. However, the facial nerve nucleus differs from that of lower mammals in several important respects here.[45] First, the volume of the facial nucleus is larger in the great apes and humans than predicted based on phylogenetic regression, suggesting greater control of the muscles of mimicry in great apes and humans than in other primate species.[46] Second, the clusters of cell bodies in the lateral portions of the facial nerve nucleus are much larger than in other animals, and the differentiation of the muscles, and therefore the specificity of the control of the facial muscles they innervate, is much finer. The higher level of differentiation and neuronal control of the facial muscles is especially evident in the muscles controlling movements of the mouth and lower face, making it possible to produce the fine motor movements needed for speech and subtle expressions of emotion. Third, the clusters of cell bodies in the medial and dorsal medial regions of the facial nerve nucleus, which innervate the upper face and auricular muscles, are much smaller in humans than in lower mammals. Motor neurons emanating from these regions are responsible for movements of the external ear in lower mammals, a function that is far less important in humans.[45]

5.5.1 Expressions of Emotion

There are two separable UMN pathways that project to the facial nerve nucleus and control the expression of emotions on the face. The evolutionarily newer of the two pathways is the pyramidal tract, with the axons from cell bodies (which look like inverted pyramids) in the cortical motor strip projecting without interruption to the facial nerve nucleus in the pons. The evolutionarily older pathway is called the extrapyramidal tract and includes the neurons originating in subcortical areas that project directly or indirectly to the facial nerve nucleus. Generally, the evolutionarily newer pyramidal UMN pathway sustains voluntary facial actions and expressions, whereas the older extrapyramidal UMN pathway controls involuntary actions and expressions.

The pyramidal and extrapyramidal upper UMN pathways project in parallel to the facial nerve nucleus, and a single LMN pathway projects from the facial nerve nucleus to the motor end plates of the facial muscles of mimicry. There is a facial nerve nucleus on the left and right side of the pons in the brain, and the UMN and LMN pathways are similarly organized bilaterally. If the LMN pathways are severed, facial paralysis results because there is no longer a neural connection between the brain and the facial muscles. If the pyramidal UMN pathway is severed, the person loses the ability to make voluntary movements such as intentionally posing an expression of a smile. Nevertheless, the person retains the ability to express a spontaneous smile when exposed to something humorous. The opposite tends to be seen in patients with damage to the extrapyramidal UMN pathway. They generally retain the ability to voluntarily retract the edges of the mouth to form a smile, but their ability to spontaneously smile when exposed to something humorous is diminished.[45]

The display of emotional expressions plays an important role in social perception and in nonverbal communication in social interactions. Psychologist Paul Ekman dedicated most of his career to the study of facial expressions and emotions. He has argued that at least seven emotions are basic and universal, each of which is characterized by a specific combination of rapid facial signals (facial actions) resulting from the contraction of underlying facial muscles: happiness, sadness, fear, anger, surprise, contempt, and disgust.[47,48] These emotional expressions are illustrated in figure 5.9, and illustrations of the rapid signals (facial actions) underlying a few of these emotions are depicted in figure 5.10. In the expression of disgust, for instance, the upper lip is curled up, the nose is wrinkled up, and the eyebrows are pulled down to the nose. In the expression of anger, the eyebrows are pulled down and together. In fear, the eyebrows are raised and the eyes are pulled wide open while the mouth is pulled back a bit and may open slightly. In sadness, the corners of the mouth are pulled down while the eyebrows are pulled together and up in the middle. And in surprise, the eyebrows are raised and the eyes are pulled wide open. Fear and surprise may appear similar, but the lip pulls back and the mouth falls open a bit in the expression of fear (see figure 5.11), whereas the lips do not pull back and the mouth opens wide in the expression of surprise.

The computer science community has made significant progress in automating facial actions to produce recognizable expressions of emotion in social robots and computer-generated images.[49] For instance, in collaboration with Hanson Robotics of Dallas, a group at Calit2 (University of California Institute for Telecommunications and Information Technology) implemented a facial expression system in a robot named "Einstein." The face of this robot has about 30 linear actuators serving in the place of facial muscles, and the robot has been programmed to recognize and display human facial expressions. The development of algorithms and programs for the reading and producing of facial expressions may shed light on the computational processes that the human brain performs to accomplish these functions, and social robots may prove valuable in a number of applications.

The majority of fMRI studies show that the human neural network underlying the perception of neutral faces, in general, includes superior temporal sulcus, the anterior fusiform gyrus (including the fusiform face area), the inferior occipital gyrus, and the amygdala[50–55]—areas involved in the perception of visual stimuli, the face, biological motion in particular, and perceptual binding (i.e., the coupling of characteristics between perceived items). The neural network involved in the processing of facial expressions of basic emotions also includes the amygdala and anterior fusiform gyrus, but additionally includes the posterior fusiform gyrus, parahippocampal gyrus, middle temporal gyrus, medial prefrontal cortex, inferior frontal gyrus, and middle and superior frontal gyri—areas associated with functions such as visual processing, memory, and self-awareness in coordination with sensory processing.[56]

FIGURE 5.9. Examples of basic facial expressions. iStock.com/ozgurdonmaz

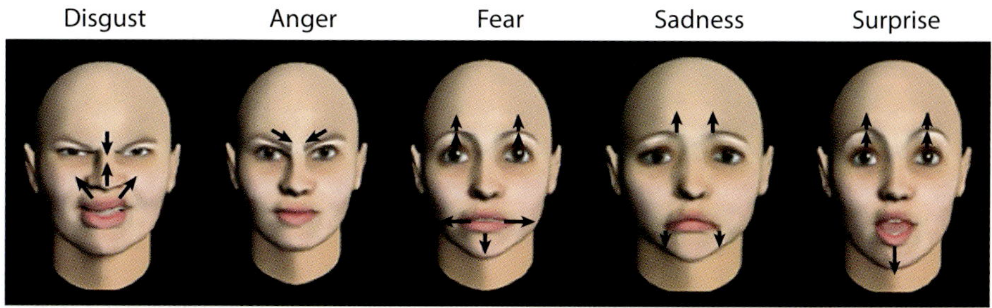

FIGURE 5.10. Examples of basic emotions and their visual facial characteristics. Excerpted from: https://www.youtube.com/watch?v=TrgNKGjSyxA.

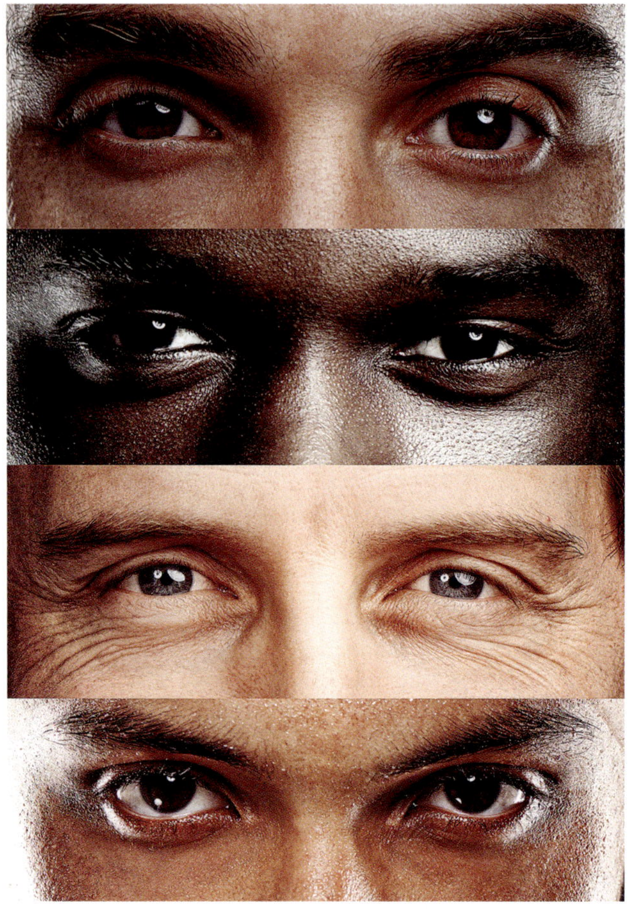

FIGURE 5.11. Examples of emotions perceived in the eyes. iStock.com/Yuri_Arcurs.

5.5.2 Patient and Animal Research

Social neuroscience is characterized by the study of humans and animals, and of neurological patients as well as neurologically healthy individuals. Investigations of the neural mechanisms of facial processing and social perception are no different. For instance, a patient with a rare disease was found to have bilateral amygdala damage due to severe atrophy of the right and left amygdalae with little damage to the surrounding white matter (see figure 5.12).

First described in 1995 by Ralph Adolphs and his colleagues, Patient SM not only had difficulties in recognizing fearful faces, she had also lost the ability to experience fear in general.[57–60] When she was asked to draw pictures depicting different emotions, she was able to provide reasonable cartoons of a range of states, except when asked to depict fear. This astonishing finding led Adolphs and colleagues to consider

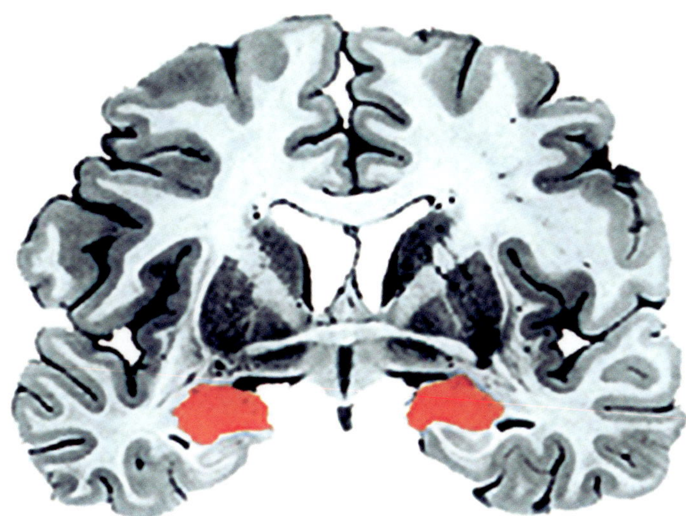

FIGURE 5.12. Amygdalae (shown in red).

the amygdala as being sufficient and necessary to experience and recognize fear. However, Patient SM had no difficulty in labeling fearful tones in voices. Further study of Patient SM indicated that her inability to recognize fearful faces was mainly driven by a strong tendency to avoid looking at the eyes. The recognition of the expression of fear is driven in large part by the eyes (e.g., enlarged sclera, i.e., white portion of the eyes). When asked to view faces, Patient SM rarely fixated on the eyes. When Patient SM was directed to look specifically at the eyes, however, her recognition of fearful faces became normal. This finding delineates the specific role of the amygdala in the perception of expressions of fear and, perhaps, in social perception more generally.[57–60]

There is evidence in nonhuman primates that, as in humans, behavior is strongly influenced by the facial gestures (actions and expressions) of others. Studies of these animals have provided additional information about the biological basis of expressions of emotion. Psychologist Lisa Parr and colleagues, for instance, adapted the facial action coding system used to study the rapid signals underlying emotional expressions in humans to investigate emotional facial expressions in nonhuman primates.[61]

These investigations have shown that chimpanzees accurately discriminate among the expressions of emotions observed in chimps. Moreover, as in humans, both individual facial actions and the holistic percept of the configuration of facial actions influence social perception in nonhuman primates.[61–64] Some facial expressions, such as the

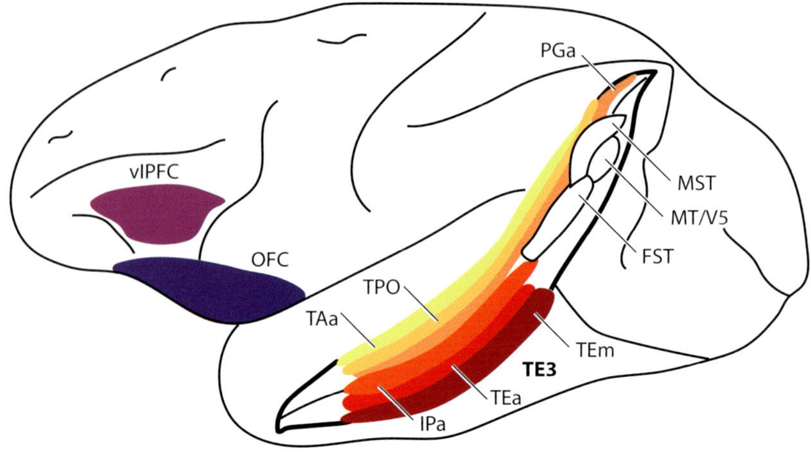

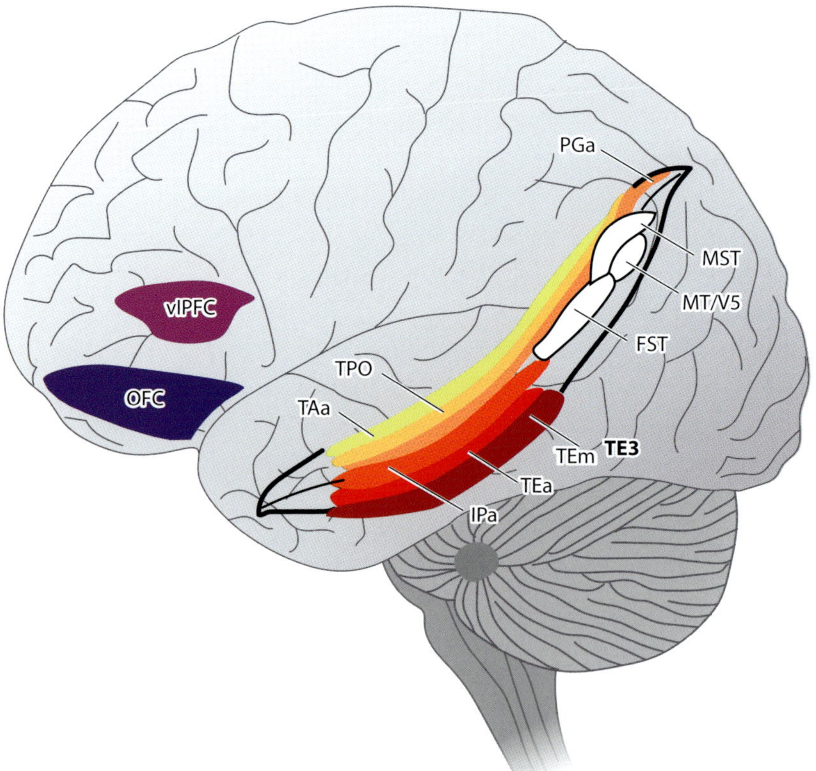

FIGURE 5.13. Illustration of some of the monkey brain areas with face-selective cells (in color). Areas in white background, such as the floor of the superior temporal sulcus (FST) are hypothesized to contain no face-sensitive cells. Rather, FST is considered to be part of the motion-detector system including MT/V5 and the medial superior temporal sulcus (MST).

"bared-teeth" display, also called the "fear grin," are well represented across diverse taxonomic groups, while other facial expressions, such as the "play face" are more species specific.[65] Importantly, the same facial expression may have different meaning and communicative functions under certain conditions in nonhuman primates as well as humans.[65,66] For instance, the function of the bared teeth, which has been compared to the human smile, has different communicative functions depending on the species and their type of social organization. In species that have an egalitarian social system, such as chimpanzees or Gelada baboons, the bared-teeth display corresponds to a signal of peaceful intentions and functions to increase social attraction and affiliation (e.g., grooming) and to facilitate social cohesion. In other species with strict dominance hierarchies, such as the rhesus macaque, the bared-teeth display is a signal of submission, often displayed in response to the approach of a dominant monkey.[65,67–70]

Animal studies have also provided information about face-sensitive neurons (see figure 5.13 and box 5.1). For instance, neurons in the superior temporal sulcus fire in the rhesus macaque in response to the opening of the mouth during the observation of a fearful face but not during the opening of the mouth in other conditions (e.g., chewing).[71–74] These findings indicate that some neurons code for the content rather than merely the visual features of a facial expression. Neurons distinguishing between similar categories of expression have also been found within the amygdala.[75] Whether or not the neurons within specific regions of the amygdala in monkeys and humans play the same role in social perception has yet to be determined.

5.5.3 Negative Moral Emotions

Humans appear capable of producing and distinguishing among a larger number of expressions of emotions than other animals. In addition to the basic emotions of happiness, sadness, fear, anger, disgust, contempt, and surprise, people are also able to express and distinguish compound and complex emotions. For instance, at least 21 different emotional expressions were perceptually discriminable when computer modeling was used to combine the facial signals of basic emotions to create compound expressions.[43]

Although less is known about negative *moral emotions* such as embarrassment, guilt, and shame, these emotions play a role in the development and maintenance of interpersonal relationships and in adaptive functioning in groups and societies. Psychologist Dacher Keltner asked participants to describe the situations that led them to feel embarrassment, guilt, and shame. Coding of these descriptions indicated that embarrassment is typically associated with transgressions of conventions that govern public interactions, guilt is associated with actions that violate duties or harm others, and shame is associated with a failure to meet important personal standards. In a follow-up study, Keltner found that participants could identify expressions of embarrassment and shame, but they could not reliably label any expression as guilt.[76] The difficulty in identifying these expressions may be due in part to their modulation in intensity and/or duration to minimize detection. Spontaneous expressions of basic emotions tend to last from 0.5 to 4 seconds and manifest as macroexpressions across the entire face.[77] However, the display of facial expressions of emotion can be inhibited or terminated to avoid detection, a process that is especially likely to apply to the expression of guilt (and, to some extent, shame and embarrassment), which not only can diminish an individual's standing within a community but can lead to significant, potentially life-threatening sanctions.[78,79]

Research on the neural correlates of moral emotions have implicated areas involved not only in emotions generally but also in (1) self-referential processing—consistent with the importance of self-consciousness in these emotions, and (2) the theory of mind mentalizing network—consistent with the importance of thinking about what other people are thinking about oneself.[80] Lesion studies indicate that damage to the VMPFC—an area involved in self-referential processing and decision making—is associated with inappropriate social behavior largely in the absence of embarrassment, guilt, or shame.[81] Differences in the pattern of neural activation exist across these self-conscious emotions, as well. Neuroimaging studies suggest that embarrassment is correlated with activity in the ventrolateral prefrontal cortex and amygdala; guilt is correlated with activity in the ventral anterior cingulate cortex, posterior temporal regions, and precuneus; and shame is correlated with activity in the dorsolateral prefrontal cortex, posterior cingulate cortex, and sensorimotor cortex.[82] The number of studies on moral emotions is still relatively small, however, and additional neuroimaging research, focal lesion studies, and animal studies are likely to improve our understanding of the neurobiology subserving each of these emotions.

5.6 Concluding Remarks

Social interactions can range from safe to threatening, trustworthy to exploitive. Alliances among people are not static, either, but can change abruptly at any given moment. An important aspect of social functioning, therefore, is the capacity to assess a social situation, including the goals and intentions of others. We need to be able to discriminate friend from foe, to appreciate the potentially duplicitous and volatile nature of social relationships, and to anticipate the actions of others. The human face is one of the first clues of the character and emotions of another individual, and it automatically attracts visual attention and impacts social perception.

The neurobiology of the human motor system has evolved to make it possible to modulate emotional expressions. There are two separable UMN pathways that project to the facial nerve nucleus and control the expression of emotions on the face: the pyramidal tract and the extrapyramidal tract. The existence and independence of these two different tracts, then, explain why some patients who have a pyramidal disorder have difficulties in performing voluntary facial expressions (such as a smile upon command), but can smile spontaneously when they hear a funny joke.

Perhaps you have started to laugh in a situation where it was socially inappropriate to do so, and you found yourself struggling to stifle your laughter. When healthy individuals experience an intense emotional situation, but need to control their facial expression, both UMN tracts can activate, which may lead to a tug-of-war over the neural control of the face. When this occurs, rapid or low-level leakage of rapid signals, or microexpressions, can occur. The first objective evidence for the existence of microexpressions was found in a study in which films of psychotherapy sessions were viewed in slow motion and replicated in a study in which the rapid appearance and disappearance of microexpressions were identified in a frame-by-frame analysis of films of interviews with depressed inpatients. Since that time, studies have verified the existence of microexpressions[83,84] as well as subthreshold expressions of affect resulting from low-level activation of only one or a small subset of the facial muscles underlying rapid signals and facial expressions of emotion.[78,85]

6

SOCIAL DECEPTION: READING THE EYES

Marcus Cicero (ca. 106–43 BCE), one of ancient Rome's greatest orators, is quoted as saying, "Ut imago est animi voltus sic indices oculi"—"The face is a picture of the mind as the eyes are its interpreter." We focused in the last chapter on facial signals that can operate spontaneously and rapidly to influence social perception. In this chapter, we continue our focus on the face as a model system for social perception, with an emphasis on the eyes and the implications for social cognition and the detection of deceptive nonverbal signals.

As described in chapter 5, facial expressions represent a complex nonverbal means of communicating and of making inferences regarding a person's traits, intentions, and behaviors, such as a person's benevolence or malevolence, and are thought to benefit the observer by improving behavioral prediction and interpersonal interactions. The inferred traits and behavioral dispositions represent one form of phenotype, the observable characteristics of an individual that result from genetic constitution (genotype) and environmental influences.

The genotype represents the molecular machinery out of which phenotypes and behavior are expressed, but the environment—including the social environment—operates through natural selection on phenotypes to shape the genotype. For instance, mutual aid and protection among social primates is fostered by social signals regarding environmental events, such as resources or threats, and internal events, such as fear, pain, or anger.

The adaptive value of these phenotypes led to the evolution of a neuroarchitecture that increased the capacity for communicative actions. In turn, the emergence of expressive communications made it easier for others to predict an individual's behavior, which promotes social coordination, cooperation, and caretaking. However, expressive communications can also make an individual's behavior more predictable to a foe, thereby threatening survival and reproductive success. The evolution of deception minimized these costs and provided the advantage of misdirection and exploitation. The evolution of deception represented yet another social challenge, leading to an adaptive value in the capacity to detect deceptive displays and counteract any associated malicious or exploitive behavior by others.

We begin this chapter with a discussion of deceptive expressive signals, the detection of deceptive nonverbal signals, and the underlying neural mechanisms. These evolutionary developments—and others like them over millions of years—have influenced both brain structure and brain function to enable more flexible, adaptive, and contextually determined behavior in response to an increasingly complex social environment.

6.1 Deceptive Expressions

As Cicero opined more than 2,000 years ago, the eyes (and the region around the eyes) are an important source of information about the state of mind of another person. Consider the images in figure 6.1. When participants are asked what the person in each image is feeling, the impression typically reported is that the person feels happy. This may seem obvious because the same person is depicted smiling in both figures. However, the expression in the top panel of the photograph in figure 6.1.A is called a fake smile, whereas the expression in the bottom panel in figure 6.1.B is called a genuine smile.[1,2]

The smile in both photographs is identical. We created the faces in figure 6.1 by using the lower half of the same photograph in both images. These images differ in the top half, however. The top half of the image in the top panel comes from a photograph of the person when he felt neutral, whereas the top half of the image in the bottom panel comes from a photograph of the person when he felt happy.

The facial signals from the area lateral to each eye differ between these two images. This is more evident when you separate the eyes from the smile, as we have done in figure 6.2. The upper part of the

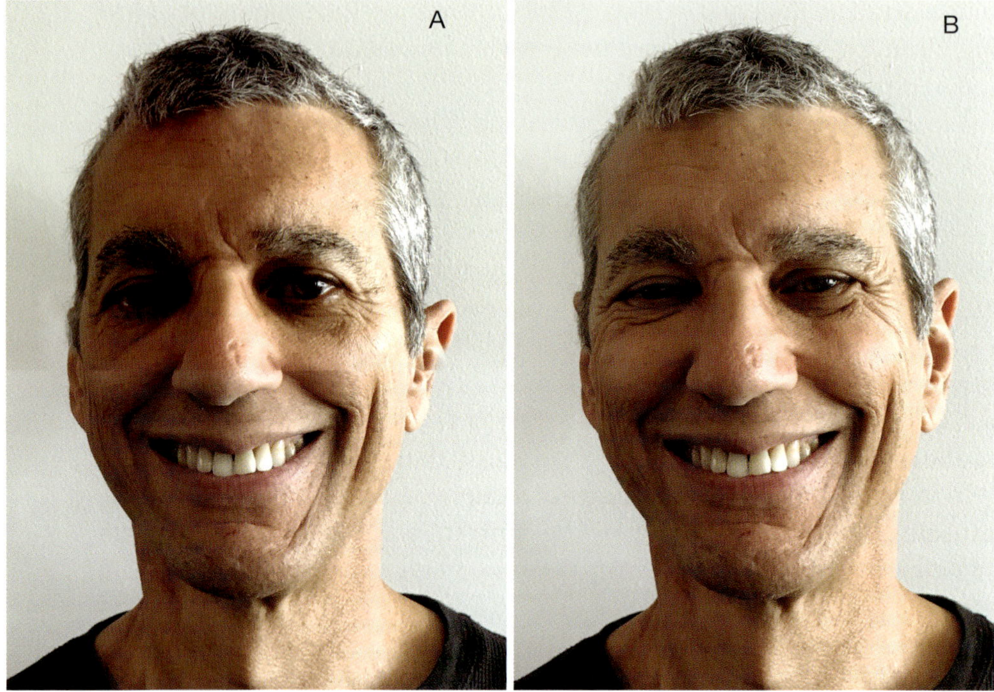

FIGURE 6.1. Real or fake smile. A. Fake smile. B. Genuine smile.

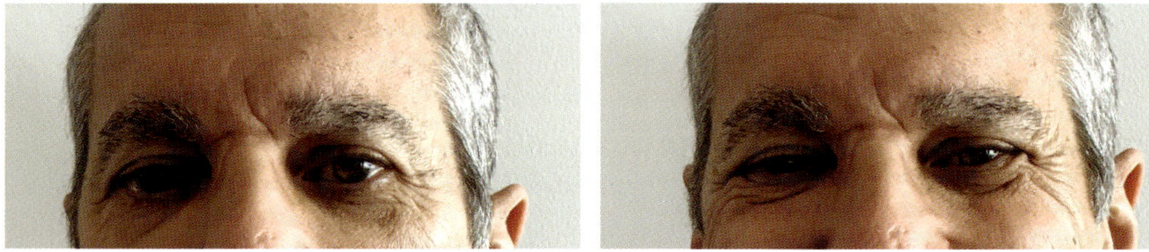

FIGURE 6.2. Smile with your eyes. Eyes movements observed during a fake smile (left) or genuine smile (right).

face in the left panel of figure 6.2 is used in the top panel of figure 6.1 (i.e., when he felt neutral), and the upper part of the face in the right side of figure 6.2 is used in the bottom panel of figure 6.1 (i.e., when he felt happy). The wrinkles visible in the region just lateral to each eye represent a feature called crow's feet.[3]

The fake smile per se does not mark malicious intent but rather it marks an explicit or implicit intention to deceive. People express fake smiles for various reasons including to be socially agreeable, to hide a selfish goal, or to be pleasant when one feels miserable.[1] A parent may express a fake smile to a child to provide encouragement whether or not the parent is happy at that moment. The same parent may express a fake smile to the child's coach in an attempt to secure additional playing time for the child, or to a business competitor to lure them into a false sense of security.

Genuine expressions of emotions do not reflect the sum of many individual efferent commands acting on individual muscles, but instead they reflect a configuration of facial actions (i.e., expressions) that, at least for basic emotions, may be innate.[4,5] The deliberate expression of a fake smile, unless extraordinarily well practiced, typically does not operate through the same prepared motor pattern but rather is constructed by activating one or more pieces of the configuration. The expression of a genuine smile typically involves facial contraction of the zygomaticus major muscle, which stretches from the side of the mouth to the zygomatic arch anterior to each ear, and contraction of the orbicularis oculi, which encircles each eye. The contraction of the zygomaticus major muscle creates a smile, whereas the contraction of the orbicularis oculi creates crow's feet at the outer edge of each eye.

As we described in chapter 5, there are two upper motoneuron (UMN) tracts that project to the facial nerve nucleus, the pyramidal UMN tract and the evolutionarily older extrapyramidal UMN tract. When a person deliberately seeks to express a fake smile to deceive another person, the efferent (motor) commands travel from the cortex to the facial nerve nucleus via the pyramidal tract, and from the facial nerve nucleus to the facial muscles of mimicry down a common lower motoneuron (LMN) tract.[6] The neuroarchitecture of this LMN tract innervates the lower part of the face more densely than the upper part, providing greater control over the muscles in the lower face. This difference in innervation density makes it easier to display a smile than crow's feet when seeking to control the expressive display shown to another person. Consequently, the rapid signals resulting from the activation of the orbicularis oculi muscle around the eyes are less involved in fake than genuine smiles. This difference makes it possible to detect deception, simply by looking at the person's eyes.

The neural machinery underlying the rapid processes of social deception can place undue weight on a broad smile and, consequently, perceivers can be deceived by the fake smile.[3,7] This is especially the case at a distance where the smile is still clearly visible to an observer whereas the crow's feet (or their absence) is not. In studies in which participants are shown photographs of smiles, fake and genuine smiles are more likely to be distinguished when people attend to the region lateral to each eye.[3] Furthermore, participants who are exposed to a person smiling, in contrast to a sad or neutral expression, make more fixations of longer duration to the crow's feet area,[8] suggesting that the human brain has evolved some capacities for detecting deceptive emotional displays.

In sum, the information from the area around the eyes is a source of emotional leakage in part because the underlying neurophysiology of the facial muscles of mimicry gives us less sensory feedback from and less voluntary control over the actions of the muscles in the upper part of the face than the lower part of the face. Because we have greater voluntary motor control over the expressive muscles in the lower part of the face, we are also more likely to voluntarily control the facial signals that are shown in the lower as opposed to the upper part of the face when we want to deceive someone regarding our actual emotional state. The result is that the expressive movements in the area of the face around the eyes tend to be a more accurate source of information about what a person is feeling than the expressive movements elsewhere in the face.

6.2 Reading the Eyes

Individuals differ in how effectively they use the physical signals from the eyes and surrounding region to recognize the mental or emotional state and predict the behavior of another person. Psychologist Simon Baron-Cohen developed the "Reading the Mind in the Eyes" (RME) test to determine how accurately an observer could recognize the emotional state of a person from a photograph of the eyes and surrounding facial regions.[9,10] For instance, if you look at figure 6.3, you can guess quickly that it depicts an expression of panic, just by looking at the eyes of the subjects. As in the RME test, the region around the eyes is a prototypical facial expression for the basic emotion of fear.[11] This information is not part of the RME test but is presented here to illustrate an evidentiary foundation for the RME.

Baron-Cohen has proposed a mindreading system that consists of four building blocks.[12] The first is an intentionality detector, which interprets self-propelled motion in terms of desires and goals. The second is an eye-direction detector, which has three functions: (1) detect the presence of eyes, (2) compute the direction of gaze, and (3) attribute the mental state of "seeing" to an individual whose eyes are directed toward itself or toward another person or object. The third component is the shared-attention mechanism, which identifies when the self and another are attending to the same thing—that is, the individuals have joint attention. The final component is what he called the theory-of-mind mechanism, which serves two functions: (1) infer the mental state of another from observable behavior, and (2) integrate

FIGURE 6.3. Facial expressions of fear. iStock.com/bowie15.

mental state information into a theory of the person (e.g., a person's traits) to predict the person's behavior. Baron-Cohen's system represents one of the specific models that fall under the theory of mind (TOM) mentalizing perspective described in chapter 4.

What differentiates Baron-Cohen's system from others is that the physical signals from the eyes play an especially important role in the mental states that are inferred. For instance, Baron-Cohen proposed that the eye-direction detector is a functionally specialized module instantiated in a dedicated neural mechanism for the detection of eyes and on what in the environment the eyes are focused. In the next section, we review brain regions involved in the detection of eyes and gaze direction, but their operation falls short of constituting the posited specialized module.

6.3 Gaze Direction

Goal-directed intentions and actions are functionally coupled with selective visual processing before the action.[13] For instance, predation of primates by snakes and felids (e.g., lions) has represented a major source of natural selection for millions of years.[14–16] Survival was fostered by the rapid orientation (overtly or covertly) to, detection of, and response to potential danger.[17]

The primate visual system, including eye gaze, has been shaped to rapidly orient gaze to detect significant stimuli (e.g., potential predators). In an eye-tracking study in humans, participants were exposed to a matrix of photographs of animals. The animal depicted in a photograph was either dangerous (e.g., lion or snake) or nondangerous

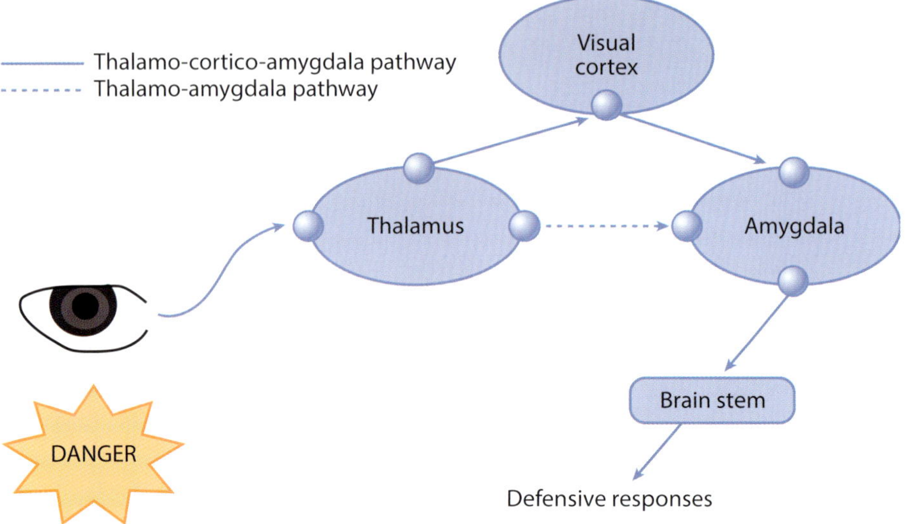

FIGURE 6.4. Schematic representation of the classical thalamo-cortical visual pathway, where afferent information goes to the cortex via the thalamus. Alternate thalamo-amygdala route mediating fast detection of potential threat is also represented here. From LeDoux[20] and Norman et al.[21]

(e.g., impala, lizard). Results showed that dangerous animals were visually located more quickly and retained eye gaze longer than nondangerous animals, even when physical features of the photographs, such as luminance, were controlled.[16] Animal studies suggest the amygdala plays an important role in the rapid visual detection of potential threats and the notification of the visual system that a potentially significant stimulus has been detected[18–21] (see figure 6.4).

Eye gaze (what people look at) is also an important social signal in collaborative activities that require joint attention.[22–26] In an analysis of the exposed sclera size in the eye outline, the width-height ratio of the eye outline, and the sclera coloration in 88 primate species, humans were found to have more visible eyes (e.g., white sclera) than other primates[27,28] (figure 6.5). The visibility and position of the eyes provide observers with a salient stimulus regarding the direction of another person's eye gaze. Neuroimaging and lesion studies have found the amygdala is engaged in an observer exposed to fearful facial expressions[29] or dangerous animals,[30] and is also activated in an observer who is exposed to fearful eye-whites even when masked[31] (see figure 6.6).

Attentional shift by the gaze of another person occurs rapidly, effortlessly, and even when a person is unaware they were exposed to the

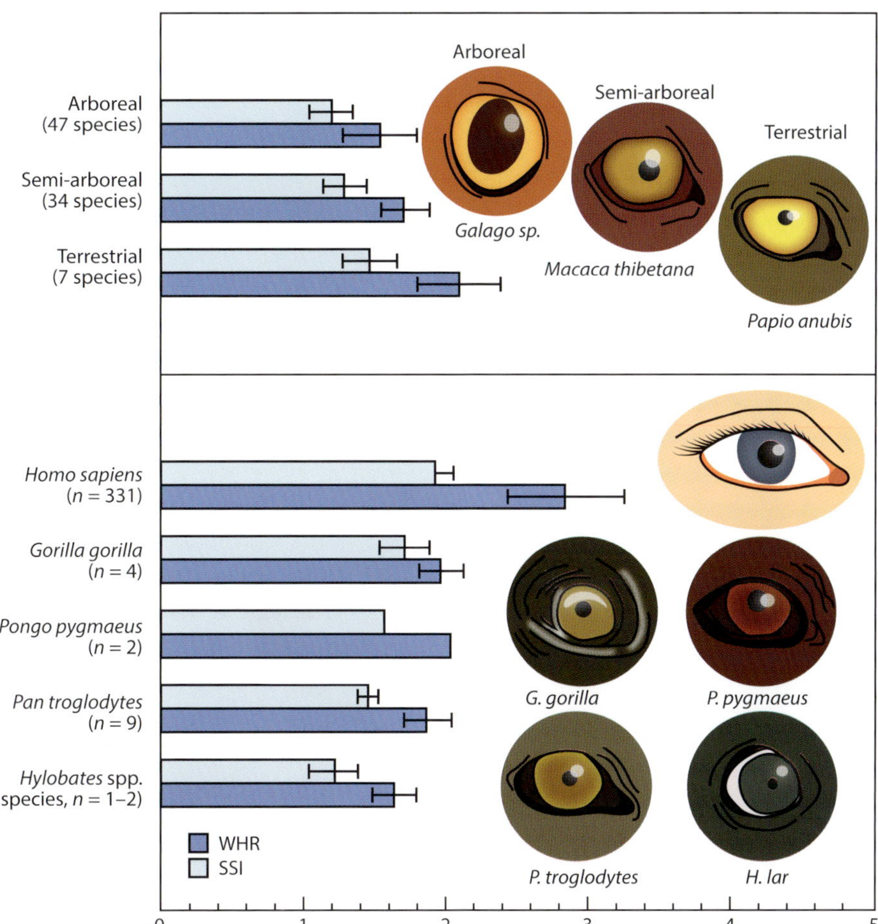

FIGURE 6.5. Variation of WHR and SSI across different species. Significant difference between habitat types. WHR = distance between the corners of the eye/longest perpendicular line between the upper and lower eyelid; SSI = width of exposed eyeball/diameter of iris. From Kobayashi & Kohshima.[28] Fig. 1 from "Unique Morphology of the Human Eye" by H. Kobayashi & S. Kohshima, *Nature*, 387 (19 June 1997). Copyright © 1997. Reprinted by permission of Springer Nature.

gaze direction of another person.[32] Chimpanzees are also influenced by the eyes of another. An experimenter offering food to chimpanzees who observed the experimenter displayed one of four behaviors: eyes closed, eyes open, hands over eyes, hands over mouth. The chimpanzees showed more visible behavior in the conditions in which the eyes were open, and more vocalizations in the conditions in which the eyes were closed than open.[33] Human infants, however, show much greater sensitivity to eye gaze than great apes, presumably to support mutually beneficial social interactions.[25]

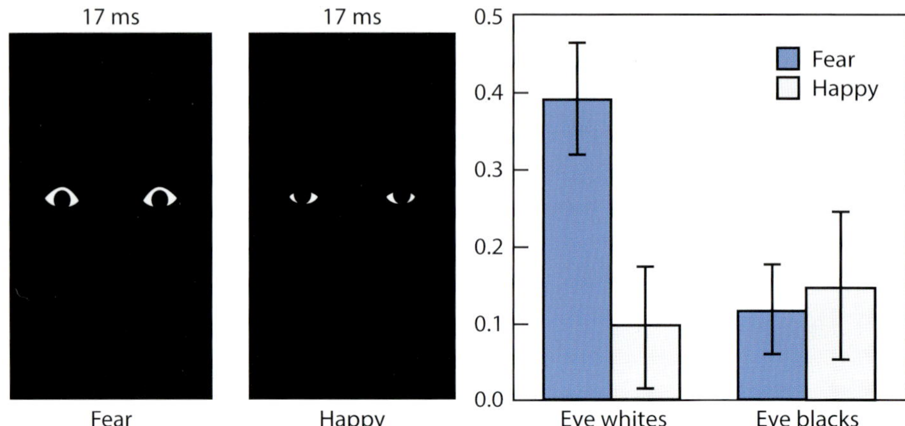

FIGURE 6.6. Fear in the eyes. Examples of the eye-white stimuli (left) and amygdala recruitment (right) in response to fearful eye-whites vs happy eye-whites. From "Human Amygdala Responsivity to Masked Fearful Eye Whites" by P. J. Whalen, et al., *Science*, Vol. 306, No. 5704 (2004).[31] Copyright © 2004 by AAAS. Reprinted by permission of AAAS.

When two individuals are speaking, the listener may follow the gaze direction of the speaker, especially when the listener is unclear about how to construe what the speaker is saying.[34,35] When this occurs, the visual information the listener collects from socially directed gaze influences the listener's social perceptions and construal of the speaker's thoughts, feelings, or intentions.[36,37] The primate brain includes neural mechanisms that are involved in gaze processing,[24,38,39] including localized neurons that are specialized for gaze.[39]

A similarly automatic gaze-directed process has been found for other forms of biologically significant stimuli, including potential mates[40] and newborns.[41] For instance, rapid changes in eye gaze in response to the appearance of a potential partner occur quickly and without a deliberate decision to do so. In an illustrative study, eye-tracking measurements were recorded from college students. Participants were shown photographs of individuals judged by the participants to have the potential to elicit feelings of romantic love (a sentimental and tender state of longing for union with another that is not necessarily associated with sexual feelings) or desire (the presence of feelings of sexual interest and of sexual thoughts or fantasies related to the image depicted in the photograph). Figure 6.7 illustrates a sample stimulus presentation.

Results showed that when participants indicated the person depicted in a photograph could elicit feelings of romantic love, the participant's

FIGURE 6.7. Example of desirable stimulus. In one experimental block, the participants were asked to look at each photograph and decide as rapidly and as precisely as possible whether they perceived the photograph as eliciting feelings of lust (sexual desire). In the other block, the participants were asked to look at each photograph and decide as rapidly and as precisely as possible whether they perceived the photograph as eliciting feelings of romantic love. The same photographs of attractive individuals were presented in each block. Responses were made by pressing one of two response keys ("K" for "yes" and "L" for "no") on a keyboard with fingers of the right hand (response "yes" with the index and response "no" with the middle finger). The order of these experimental instructions was counterbalanced across participants (See Bolmont et al.[40] for details).

eye gaze immediately following the presentation of the photograph was generally directed toward the face. However, when participants indicated the person depicted in the photograph could elicit feelings of desire, the eye gaze was directed more toward the torso and, to a lesser extent, toward the face.[40] Male and female participants showed the same pattern of eye gaze. Humans may not be unique in this regard, either. Similar results were reported in a study of freely moving peahens (*Pavo cristatus*) during courtship[42] (see box 6.1).

Literature, films, and divorce court documents are rife with interpretations of a partner's intentions based on eye gaze toward a competing potential partner. Eye gaze is a rich source of information, but eye gaze per se is not sufficient to ensure accuracy. Additional cues from head orientation, eye closure, pointing gestures, and contextual stimuli (e.g., gaze directed toward or independently of a specific

BOX 6.1. Visual Attention and Courtship in Animals

Using a miniaturized eye-tracker, Yorzinski and colleagues tested visual attention of peahens (*Pavo cristatus*) before and during courtship.[42] Results showed that peahens look at the body of the peacock before/during courtship. They look at the lower train and mostly ignore the head, crest, and upper train—a gaze pattern that is similar to that observed in humans.

person) are also computed to increase the coupling between eye gaze and visual attention. Even when gaze direction and visual attention coincide, the impressions or inferences an observer may glean may not be accurate, especially as in this study, when the eye gazes are so rapid that they do not reflect a deliberate allocation of visual attention. After discussing the neural mechanisms underlying gaze processing, we turn to the integration of dispositional factors such as signals from the person with situational (contextual) and behavioral factors.

Neurophysiological studies of macaque monkeys have identified specialized neurons in the lateral and ventral surfaces of the inferior temporal cortex for the detection of faces[43] and in the superior temporal sulcus (STS) that respond to gaze direction,[43–45] and neuroimaging studies of humans have identified a cortical network in the occipito-temporal cortex (i.e., fusiform gyrus, inferior temporal gyrus, bilateral middle temporal gyri, and parietal lobule) and regions of the STS that are engaged in gaze processing.[46,47] The fusiform gyrus and inferior occipito-temporal regions are more strongly activated during judgments of identity than gaze detection, whereas the STS is more active during judgments of gaze direction than judgments of identity.

The neural regions associated with the perception of gaze share more in common with those underlying visually triggered orienting—that is, involuntary eye movements—than those underlying voluntary visual orienting. These neuroimaging results are backed by behavioral evidence, which has demonstrated that reaction time to a gaze cue is more similar to the reaction time to an exogenous cue than to an endogenous cue.[48]

When conditions elicit not only the detection of a person's gaze direction but also the mental states of the person—as in gaze aversion—the regions of the medial prefrontal cortex involved in TOM mentalizing and joint attention are also activated.[24] Neuroimaging and studies

BOX 6.2. Visual Attention and Joint Attention

Anthropologist Michael Tomasello and colleagues proposed that the crucial difference between human cognition and that of other species is the ability to participate with others in collaborative activities with shared goals and intentions: shared intentionality.[25] Participation in such activities requires not only especially powerful forms of intention reading and cultural learning, but also a unique motivation to share attention and psychological states with others and unique forms of cognitive representation for doing so. Tomasello and colleagues also argued and presented evidence that great apes understand the basics of intentional action, but they still do not participate in activities involving joint intentions and attention.

of patients with localized brain damage suggest the amygdala is involved also in the aversion of eye gaze[49] in addition to detecting threats and the eye-whites in expressions of fear and surprise,[19,29,31,50,51] suggesting that the amygdala reflects the emotional significance of the eyes or gaze direction to the perceiver.[52,53] Thus, gaze processing involves a large network of brain regions involved in social cognition, and the various degrees of activation of each region depend on the specific task. It is noteworthy that the more the situation or task in which a person is observed demands response flexibility or contextual control, the more likely medial prefrontal and orbitofrontal regions are to be involved.

6.4 Production and Detection of Deception

As social capacities (e.g., deception) evolved to exploit others, capacities to detect and counteract exploitive interpersonal interactions became advantageous for survival and reproductive success. The iterative process of natural selection in brain evolution reflects a general principle in the evolution of the human brain, resulting in layers of neural mechanisms that evolved over millions of years. The differences between the pyramidal and extrapyramidal UMN tracts described above (section 6.1) hinted at the general differences in neural function as one moves from lower, evolutionarily older neural architecture (e.g., spinal reflex arc) to higher, relatively recent evolutionary developments (e.g., neocortex). In the next section, we briefly outline some of the principles of this functional organization.

6.4.1 Functional Organization of the Central Nervous System

The central nervous system (CNS), which includes the spinal cord and brain, is characterized by a bottom-to-top organizational pattern composed of simple reflex-like circuits at the lowest levels (brain stem and spinal cord) and distributed neural networks for more integrative processing at higher levels[54–57] (figure 6.8, top panel). Through parallel processing at various levels of the CNS, these interacting hierarchical structures allow neural systems to rapidly respond to threatening stimuli through low-level processing (e.g., startle reflex) while higher neural substrates permit elaborated processing of potential outcomes and future responses. Based on hierarchical interconnections, higher-level systems may depend heavily on lower-level systems for the transmission and preliminary processing of stimuli.

Neuroscientist Gary Berntson and colleagues noted that descending projections from the highest levels of the CNS are able to bypass intermediate levels and directly synapse onto lower level structures, such as the brain stem[21,55,57] (figure 6.8, bottom panel). This organization of the CNS contains the components of hierarchical systems, as higher levels are in continuous communication with lower systems via intermediate levels, but has the additional capacity to interact over widely separated levels via direct connections with the lowest levels. The capacity for direct interactions across widely separated organizational levels appears to be, at least in part, a recent evolutionary development. For example, within the primary motor cortex, thought to represent the highest level of motor responses, differentiated regions are able to entirely bypass intermediate levels—as in the case of the pyramidal UMN tract, which travels from the motor cortex to the facial nerve nucleus. The highest and most recently developed levels of the cortical motor system are able to directly interact with the lowest, evolutionarily older levels, allowing for novel patterns of motor output that are essential for flexible, highly skilled movements.[21,58] Thus, there continues to be an evolutionary re-presentation and elaboration occurring even within the highest levels of the CNS.

At the highest levels, beyond the primary motor cortex, cerebral systems must process a tremendous amount of sensory information and integrate this information with associative networks, emotional and motivational substrates, and expectancies, in the contexts of strategic goals and tactical plans. As Berntson and colleagues note, this

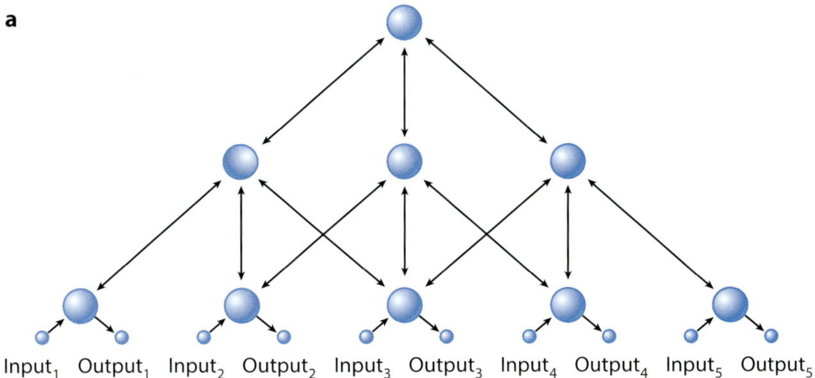

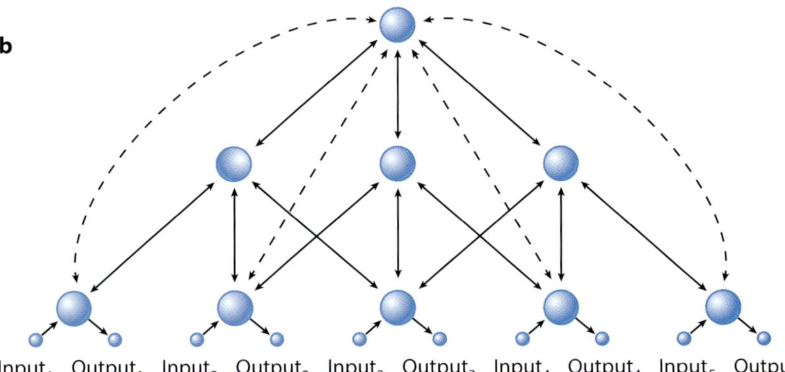

FIGURE 6.8. Hierarchical organization of the CNS that contains the components of systems. From Norman et al.[21]

organization enables enhanced information processing but can impose a processing bottleneck that necessitates a slower, more serial mode of processing and selective attentional mechanisms.

It is the neural systems at the highest levels of the brain that confer the greatest cognitive and behavioral capacity, temporal persistence, and contextual control, but they also generally operate more slowly and more serially than lower neural mechanisms, partly because these higher neural systems do not operate in isolation but depend upon and interact with lower levels in the hierarchy.

Social psychologists have emphasized the importance of the situation (context) as a determinant and as a moderator of human behavior.[59–61] It is the neural systems at higher levels of the neuraxis that permit behavior to be modulated by the context.[54,56]

6.4.2 Deceptive Signals and Expressions Revisited

The production and detection of facial expressions of basic emotions are thought to be older capacities than the intentional expression of an emotion for the purpose of deception. For instance, the evolutionarily older extrapyramidal UMN tract is sufficient for the expression of basic emotions, whereas the newer, higher pyramidal UMN tract is necessary, at least initially, for the expression of fake emotions.

Neuroimaging research on the neural correlates of a person who is responding untruthfully, in contrast to truthfully, has identified a set of brain regions including the anterior cingulate and STS (regions involved in social attention), as well as the medial prefrontal cortex, ventrolateral prefrontal cortex, and dorsolateral prefrontal cortex (more recently evolved brain regions in the frontal cortex involved in mentalizing and executive functioning).[62] The role of the medial prefrontal cortex in deception fits nicely with its role in an important aspect of TOM mentalizing—understanding that the knowledge of others *can* deviate from what one knows. Deception involves the additional demand of continually gauging the extent to which the knowledge of a person with whom one is interacting not only deviates from what one knows but also from what is being displayed. This additional requirement involves executive functions (e.g., inhibition of emotional responses, planning and decision making) that are subserved by the ventrolateral prefrontal and dorsolateral prefrontal cortices.

The perception that a person is untruthful is reflected in judgments of a person's trustworthiness. Specifically, a judgment that a person is untrustworthy reflects the belief that the person's expressions, statements, or behaviors are not to be accepted as true—that is, the person may engage in deception. Research on judgments of trustworthiness based on images of faces indicate such impressions can be formed rapidly and effortlessly and are associated with the activation of the STS, amygdala, fusiform face area, and ventromedial prefrontal cortex (VMPFC).[63,64] Unlike the amygdala and STS, however, the activation of the VMPFC is modulated by the context. The VMPFC represents a higher level of neural organization than the other regions in the model. As would be expected from our description of the functional organization of the CNS, higher neural regions proved to be less responsive to the specific physical features of a stimulus and more responsive to contextual factors such as task demands.[65] Neuroimaging studies of trustworthiness that also require participants to evaluate a person's intention—such as whether an expressed emotion is intended to deceive—activate the regions of the medial prefrontal cortex involved in TOM mentalizing.

6.5 Integration of Physical, Contextual, and Behavioral Influences

We have emphasized how the brain processes the physical signals from others. There are far more types of physical signals (e.g., verbal, gestural, postural, and kinematic cues; height, weight, vocal pitch) than we can cover here, so we have used the face as a model system. The perceptual cues discernible from another person represent an important starting point for understanding social cognition and the underlying neural mechanisms. For instance, the less trustworthy a face is judged to be, the greater the activation of the amygdala. The activation of the amygdala, however, is predicted better by consensual than individual judgments of trustworthiness, indicating that the amygdala automatically codes faces in terms of facial properties (e.g., eye whiteness, facial structure, and texture) rather than an individual's evaluation of the stimulus.[64]

Social perception and cognition are influenced not only by bottom-up features such as facial properties and expressions, but also by top-down features such as an observer's motivation, goals, expectations, and cognitive load.[64,66–69] The dimensions along which inferences are made vary across cultures, but two dimensions have emerged that appear to be universal. These dimensions are warmth and competence,[70,71] dimensions that are critical for social behavior to yield mutual benefits.

As we discussed in section 6.4.1, a general principle of the functional organization of the central nervous system is that lower levels of neural organization (e.g., spinal cord, in contrast to brain stem) reflect older evolutionary developments through which higher levels of neural organization now can operate. The neurobehavioral processes at lower levels of organization (e.g., the spinal cord reflex arc) are characterized by their automaticity, speed, and low energetic demands, but are also characterized by less behavioral flexibility or contextual control. Some of these evolutionarily older heuristic processes are evident in social perception, including perceptual fluency, overgeneralization bias (e.g., halo effect), availability heuristic, and egocentric bias.[72–74; cf. 75]

Perceptual fluency refers to the ease with which a perceptual cue is processed. The visibility of the whites of the eyes in the human face, for instance, makes this feature easier to process.[27,28] Any such feature biases impressions of the inferred characteristics and behavioral predispositions of the observed person.[73,76–78] Overgeneralization biases in social cognition refer to impressions and inferences that are influenced by the similarity between a feature of an observed person and the category of individuals for whom that feature is associated. For instance, the facial cues that represent babies (baby-faced) tend to increase inferences about an adult's warmth and decrease inferences about the individual's competence.[72]

Heuristic processes are thought to dominate in social cognition when a person has low motivation or ability to engage in effortful elaboration and analysis of relevant information.[79,80] The amygdala, basal ganglia, and lateral temporal cortex are regions involved in conditioning and associative learning and have been posited to underlie heuristic processes, whereas the higher brain regions of the prefrontal cortex, anterior cingulate, and medial-temporal lobe—areas involved in memory, problem solving and executive functioning—have been posited to underlie slower, more deliberate, effortful, flexible, and contextually controlled information processing.[81] Alternative models have been proposed, but these models are also in accord with the notion that automatic heuristic biases engage lower levels of neurobehavioral organization than slower, more deliberative, effortful, conscious processes do.[82]

6.6 Concluding Remarks

Social living provides opportunities and challenges, and the responses of an individual to opportunities and challenges can determine the benefits or costs that are realized. This, in turn, places a premium on social perception and cognition. Some of the neuroarchitecture underlying these social capacities evolved over millions of years and represents exaptation, whereas others have evolved more recently to serve social perception and cognition. Consequently, social perception and cognition are influenced not only by bottom-up features of social stimuli, but also by top-down features such as an observer's motivation, prior knowledge, and expectations.

The emergence of expressive communications made it easier for others to predict an individual's behavior and optimize mutual aid and protection, but expressive communications can also make an individual's behavior more predictable to antagonists. The evolution of deception minimized these costs and provided an individual the advantage of misdirection and exploitation, but deception also represented a potential threat, which created an adaptive value for the capacity to detect deceptive displays and counteract any associated malicious or exploitive behavior by others.

New social challenges placed increasing value on social connection and communication, and on computational capacity for multimodal stimulus processing, associative processing (memory load), and integration; temporal persistence (e.g., inhibition of impulsive responses); planning; and contextual control. These capacities, which characterize the evolutionarily newer, higher levels of neural organization, generally operate more slowly and more serially than lower neural mechanisms, partly because these higher neural systems do not operate in isolation but depend upon and interact with lower levels in the brain and spinal cord.[54,83]

7

GROUP PROCESSES

A group refers to two or more people, animals, or things that are categorized together. The term used to describe a group of animals differs across species, and groups across social species have common and unique adaptations that generally increase the fitness of their members.[1,2] An important biological explanatory concept for the evolution of social species is inclusive fitness, which refers to the ability of an individual organism to pass on its genes to the next generation, taking into account the shared genes passed on by the organism's close relatives (see chapter 1). For instance, *Pseudomonas aeruginosa* is a common bacterium that can cause disease in plants, animals, and humans (e.g., pneumonia in humans). It is also a social species that engages in altruistic cooperative behavior. Experimental evolutionary research has shown that the evolution of cooperative behaviors in this pathogenic bacterium varies as a direct function of kinship relatedness in the group—the stronger the kinship relatedness, the greater the selection for higher levels of cooperative behavior. However, kinship has its limits. When there is local competition among relatives for resources in this bacterium, there is a selection for lower levels of cooperative behavior.[3]

In this chapter, we focus on the origins of group structures and processes in humans. A general principle in this chapter is that culture exerts a stronger influence on group processes in humans than in other species. Culture, in turn, rests on and interacts with biological predispositions and capacities that have evolved over millions of years. Although group processes are more complex in humans than in bacteria, phylogenetically ancient processes such as inclusive fitness form a foundation on which more recently evolved processes are built.[4]

7.1 Interspecific and Intraspecific Competition

Anthropological investigations of *Ardipithecus ramidus*, who lived approximately 4.4 million years ago, indicate that certain adaptations were exhibited by this early hominin species. These included the diminution of male canine size, upright walking, the absence of ovulatory signaling, reduced intrasexual conflict among males, and fostered pair-bonding and male parental investment.[5–7] In the human pair bond and nuclear family, members are part of a larger collective within which the nuclear family is fully integrated. Pair-bonding in humans raises the certitude of paternity, thereby increasing the parental investment in dependent offspring. In addition, it lessens rivalry for sexual partners among males, with the consequent reduction in sexual rivalry lessening the competition among males, promoting cooperation in hunting and warfare, and altering the organization and operation of the group to make it more resilient.[cf. 8,9]

Social hierarchies in nonhuman primates are rigid and, in species such as the macaque, they are determined by birth. In humans, social hierarchies are more likely to reflect flexible alliances and coalitions based on cooperative agreements. Across human history, coalitions tend to form when males compete.[5,10] Moreover, early warfare among modern humans altered the evolutionary costs and benefits and, consequently, contributed to the selection for larger brains and social behaviors, especially altruistic behaviors in humans.[11,12]

The effect of competition between groups on cooperation within groups (e.g., figure 7.1) was demonstrated in a classic study of boys at summer camps by psychologist Muzafir Sherif and colleagues in the mid 1950s[10] and in more recent research by primatologist Frans de Waal and colleagues[5] in studies of chimpanzees:

> Chimpanzee males conduct warfare against neighboring groups and do so in a highly consolidated fashion; if they did not, then foreign groups would invade their territories. We see how competition and cooperation are not mutually exclusive aspects of society, be it human or ape. (p. 40)

Most animal communities are geographically localized and are relatively closed to contact with members of other communities. Modern humans, by contrast, have historically been mobile and relatively open to contact with outside communities and are now found globally. The contact between groups increases access to resources and reproductive opportunities, but it also can increase the potential for conflict. Genetic evidence suggests limited interbreeding between modern humans

FIGURE 7.1. Women in tug-of-war in India.

and Neanderthals dating back about 100,000 years ago, and anthropological evidence suggests that interspecific competition between a smaller population of invading modern humans and a larger, resident Neanderthal population contributed to the extinction of the Neanderthals. Modern humans emerged victorious from this interspecific competition in part due to their ability to learn, their capacity to expand and share knowledge through culture, and their level of cultural development.[13–15] Thus, the evolution of the social brain in humans transformed what group structures and processes were possible and increased the importance of culture in human survival and reproductive success.[15,16]

7.2 Reciprocity

From an evolutionary perspective, indiscriminate altruistic or cooperative behaviors are not an effective individual strategy. Cooperative contributions can be an effective individual strategy only when others can be trusted generally to reciprocate in some fashion. This reciprocity need not be direct or immediate, however (chapter 1).

Reciprocity is an exchange: it manifests as the tendency to respond in kind to actions expressed toward or observed by us. With reciprocity, a small favor creates an unspoken sense of obligation to return the favor. It also can stimulate a sense of gratitude and mutual respect, promote cooperation, and strengthen the bonds between people.

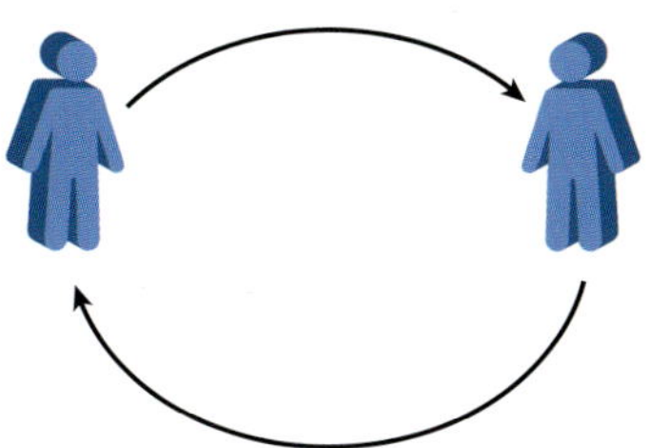

FIGURE 7.2. Schematic representation of direct reciprocity.[17]

There are different types of reciprocity.[17] *Direct reciprocity* is an exchange between two people (figure 7.2). One of the reasons salespersons at grocery stores hand out free samples is to have you try the product, but also because it produces a feeling of obligation in the shopper to purchase the product. Direct reciprocity leads to cooperation when both individuals can provide help that is less costly for the donor than it is beneficial for the recipient. Although direct reciprocity increases the likelihood that one individual is likely to cooperate as long as the other person continues to cooperate, if either fails to cooperate, direct reciprocity may lead to reciprocal refusals to cooperate thereafter.

A variation on direct reciprocity specifies that there is a tendency to respond in kind to actions toward one person by another person and that thereafter they follow a win-stay, lose-shift strategy.[18,19] For instance, if another person acts in a cooperative way toward you, you are inclined to act cooperatively toward that person. If the person continues to act cooperatively, you tend to act similarly (win-stay). However, if the other person changes to act in an uncooperative fashion toward you, you also change to reciprocate the uncooperative act (lose-shift). If the person continues to act in an uncooperative fashion, however, you may be able to reestablish a more positive exchange by showing forgiveness and responding cooperatively (lose-shift).

Direct reciprocity requires repeated interactions between the same two people, but many human interactions are fleeting and unequal, as when one person helps a stranger with no possibility of direct reciprocation or donates to a charity that does not provide direct benefits in return. Altruistic behaviors of this form can be supported by indirect reciprocity. In *indirect reciprocity*, whether you choose to help (or hurt) someone establishes a reputation, which increases the odds of others acting toward you in a rewarding (or punishing) fashion[17] (see figure 7.3). For instance, you might choose to help or cooperate with someone so others see that you're a good person. A more colorful example comes from baseball player Yogi Berra's book, *When You Come*

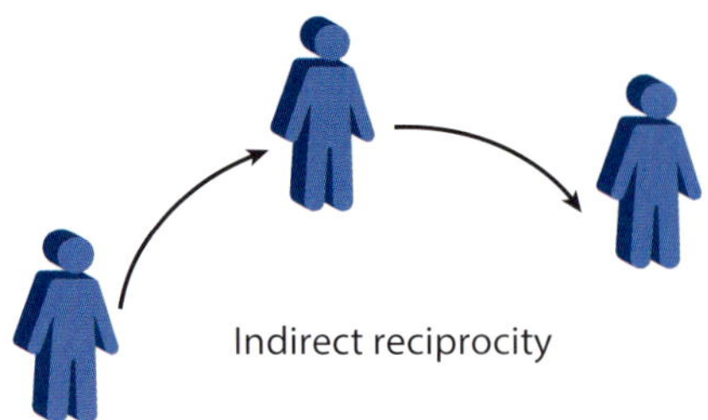

FIGURE 7.3. Schematic representation of indirect reciprocity.[17]

to a Fork in the Road, Take It!: "Always go to other people's funerals, otherwise they won't go to yours." These examples illustrate a form of indirect reciprocity that produces downstream effects. One individual helps a second individual, which in turn makes it more likely that a third person will help the first individual.

Indirect reciprocity has also been shown to produce upstream effects. One individual helps a second individual, making it more likely that this second person will help a third individual. Scientific studies have shown that people who are more helpful are also more likely to receive help. Indirect reciprocity may also function like an insurance policy against a catastrophic loss. If you provided a neighbor with food when their crops failed or shelter when their home burned down, you are more likely to receive care and support should you or your family fall victim to such a calamity.

A third type of reciprocity is *network reciprocity*.[17] People are not equally likely to interact with everyone, but rather people are much more likely to interact with some individuals over time than others. Now assume that some of the people in a population interact in a way that benefits others around them whereas some people do not; would you not prefer to be around and to interact with the former set of individuals? Network reciprocity refers to the formation of clusters of people within the population who engage in prosocial behavior in which cooperation and mutual aid and protection are the norm among members (see the blue images in figure 7.4).

In sum, individual reciprocity promotes cooperative arrangements between individuals, whereas indirect and network reciprocity promote cooperation and prosocial behavior within a group. When prosocial behavior and cooperation can be assumed based on group membership rather than personal friendship, the transaction costs of interactions within the group can be reduced significantly. The strength of the connection to the group, in turn, benefits the individual members and increases the ability of the group to adapt to new challenges and to marshal an effective defense.

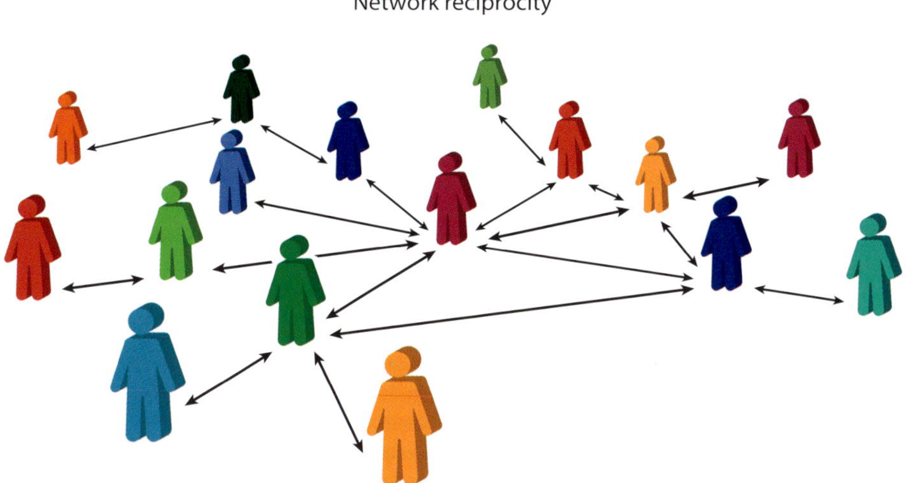

FIGURE 7.4. Schematic representation of network reciprocity.[17]

7.3 Ingroups and Outgroups

Group processes, norms, and culture play an important role in the realization of the benefits of cooperative enterprises while minimizing the risks of incurring excessive costs. The development of network reciprocity, for instance, assumes that people have the capacity to recognize a distinct group whose members can be trusted to reciprocate altruistic acts in some fashion. Kinship may have contributed to the formation of such groups until around 4000 BCE, when environmental opportunities and challenges led to the development of ingroups that were too large to rely on kinship for cohesion. The establishment of clear group boundaries and group norms, and allegiance to a group that shares one's interests, concerns, and goals—all of which are strongly influenced by culture—contributed to cohesion and the realization of the benefits of cooperative enterprises while minimizing the risks of incurring excessive costs.[20] These groups vary greatly across human history and in modern times. A group that consists of individuals who share interests, goals, and social identity is termed an ingroup, and everyone who is not a member of this ingroup is referred to as the outgroup.

Identification with and investments in an ingroup increase the likelihood of the continuity of the ingroup as a whole and the survival of its members.[21,22] A mortal threat may arise from a member of an outgroup or from a member of one's ingroup. Across human history such threats disproportionately came from interactions with members of outgroups, especially when the groups were competing over limited resources or political power.[20] Ingroup-outgroup distinctions do not always involve competition or conflict over limited resources, but people tend to favor those who have favored them in the past or who are likely to favor them in the future.[23] As psychologist Marilynn Brewer notes, ingroup members are generally perceived as more likely to be hospitable than outgroup members:

> At its most basic level, the apparent universal preference for ingroups and ingroup ways over those of the outgroup stems from the simple observation that one can expect to be treated more nicely by ingroup members than by outgroups.[20] (p. 435)

Consequently, individual learning, cultural processes, and evolutionary processes have contributed to individuals preferentially cooperating with and providing aid and protection to ingroup members as a means of minimizing the risks and costs of betrayal and exploitation.[20,22,24]

Ingroups in contemporary society are larger and more transient than in earlier periods, but ingroup and outgroup categorizations still introduce powerful biases. To study the effects of ingroup and outgroup categorization per se and to rule out the influence of any objective differences between an ingroup and an outgroup, psychologists Henri Tajfel developed the minimal group paradigm.[25] Briefly, participants in this paradigm are randomly divided into one of two groups, seemingly on the basis of trivial criteria (e.g., a coin flip, the color of the name tag they are wearing). Participants then take part in a task in which they distribute a valuable resource (e.g., money) between other participants who are only identified by code number and group membership (e.g., a member of the group to which they are assigned or a member of the other group). Participants are told that, after the task is finished, they will receive the total amount of the resource that has been allocated to them by the other participants. In this way, what a given participant receives is unrelated to what that participant decides in terms of allocating resources to others in the two groups.

Results revealed a significant tendency to allocate more resources (e.g., money) to ingroup than outgroup members.[26] This ingroup favoritism extends beyond the allocation of resources to include the liking of and prejudicial feelings toward a putative ingroup versus outgroup member. Subsequent research has identified additional biases in how people perceive and treat ingroup versus outgroup members even when there is no factual basis for differential treatment.[20,27,28]

1. People are generally more cooperative toward ingroup than outgroup members, and they work harder to accomplish ingroup goals.
2. Outgroup members are generally perceived to be more homogeneous ("all the same") and are individuated less than are ingroup members.
3. Outgroup members tend to be perceived to be more threatening to ingroup members than are other ingroup members.
4. Outgroup members tend to elicit less empathy than ingroup members, and outgroup members tend to be less influential than ingroup members.

These biases create strong ingroup cohesion, and the cooperation and collective investments this promotes can be beneficial. However, these biases can also lead to poorer decisions in some circumstances and can create feelings of intergroup tension, hostility, and competition.[20,27,28]

Social identity theory posits that people show ingroup biases to maintain a positive social identity with the ingroup.[29] An alternative formulation, the bounded group reciprocity perspective, posits that these ingroup biases reflect a human adaptation to cooperate with ingroup members to maintain a positive reputation within the ingroup and to avoid exclusion from the ingroup. Both positions have support in the literature. A meta-analysis of research on ingroup favoritism provided evidence that ingroup biases are produced by categorization and identification (as posited by social identity theory), and by expectations of reciprocation (i.e., interdependence, as posited by the bounded group reciprocity perspective).[30,31]

The human adaptation for individuals to preferentially cooperate with and provide aid and protection to ingroup members has helped manage the risks and costs of betrayal and exploitation when groups were relatively small and stable.[20,22,24] Ingroup favoritism in modern global and diverse societies, however, may contribute to intergroup resentment, hostility, conflict, and violence.[32]

7.4 The Neuroscience of Prejudice and Stereotyping

The human brain evolved to sort people, animals, objects, and events into categories, typically automatically, to provide maximum information with the least cognitive effort.[33,34] Across human history, social knowledge has played an important role in survival and reproductive success. The mutual aid and protection on which people within groups depend to survive and prosper, for instance, places a greater value on individuated cognitive representations of ingroup members across human history—with whom interactions were generally more frequent and varied—than outgroup members.[20] For instance, the faces of ingroup members undergo greater visual encoding than the faces of outgroup members.[35] Neuroimaging research has found that the perception of the face of an ingroup member, compared to an outgroup member, leads to a larger early attentional component (the N170) in the event-related brain potential, greater activation of the fusiform gyrus, and greater facial recognition.[36]

Less individuated cognitive representations of outgroup than ingroup members, when shared and aggregated over individuals in a society, result in cognitive representations that oversimplify and overgeneralize the traits and characteristics of outgroup members relative to ingroup members. These cultural stereotypes, in turn, can have unintended and unrecognized influences on judgments and behaviors toward members of an outgroup.

A stereotype refers to an oversimplified cognitive representation of a group or its members as defined by culture or society, prejudice refers to a preconceived affective or evaluative response toward a social group or its members, and discrimination refers to differences in behavior or actions toward an outgroup or toward a member of an outgroup. We begin by addressing the neuroscience of stereotyping.

7.4.1 Stereotyping

Stereotyping has been the subject of numerous neuroimaging studies with the goal of providing insights into stereotyping and providing new ways to reduce its deleterious societal effects. In a review of this literature, social neuroscientist David Amodio proposed a stereotyping network that consists of the lateral temporal lobe, anterior temporal lobe, dorsomedial prefrontal cortex (DMPFC), and inferior frontal gyrus[36] (see figure 7.5). According to this model, the encoding and storage of a stereotype involves cortical structures underlying semantic memory generally, such as the lateral temporal lobe, and semantic memory for people and groups in particular, such as the anterior temporal lobe. The formation or modification of stereotypes in response to trait diagnostic information involves cortical structures that support impression formation such as the DMPFC. Finally, the activation and use of stereotypes to guide goal-directed action is associated with activity in the left inferior frontal gyrus.

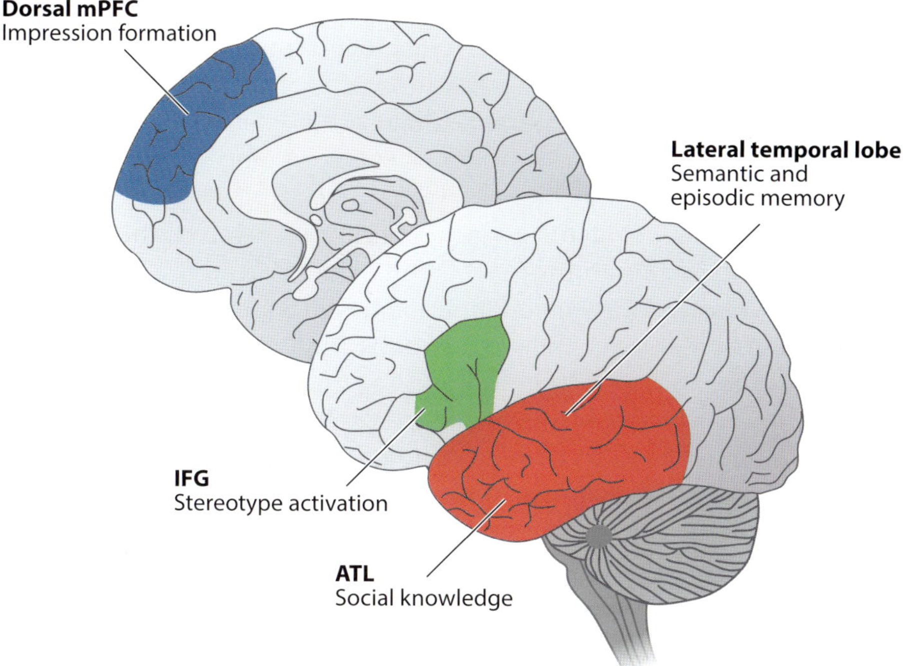

FIGURE 7.5. Stereotyping network. Brain structures underlying components of intergroup stereotyping. Semantic and episodic information stored in the lateral temporal lobe—stereotype-related knowledge about people and social groups processed by the anterior temporal lobe (ATL)—impression formation in dorsal mPFC.

In an illustrative study, participants made judgments about whether or not an element in a category (e.g., gender or string instrument) would be characterized by a particular feature (e.g., a stereotypical feature of the social category). Activity in the anterior temporal lobe was associated with judgments about features of social, in contrast to nonsocial, categories.[37] Behavioral indices of implicit stereotyping were also correlated with activity in the anterior temporal lobe, and the disruption of the anterior temporal lobe by transcranial magnetic stimulation diminished the behavioral expression of implicit stereotype associations.[38,39]

Research on social stereotypes has differentiated stereotype activation, which refers to the extent to which a stereotype is activated in semantic memory and is capable of influencing a person's thoughts about a member of the stereotyped group, from stereotype application, which refers to the extent to which the stereotype is used to judge the member of the stereotyped group. According to the model of stereotyping depicted in figure 7.5, the activation of stereotype-related information associated with the anterior temporal lobe is typically automatic and provides input to the DMPFC, where it influences the impressions that are formed. The extent to which these stereotypes are *applied* to goal-directed behavior is guided, in turn, by activity in the inferior frontal gyrus.

7.4.2 Prejudice and Discrimination

Amodio also proposes a prejudice network that consists of the amygdala, orbitofrontal cortex, insula, striatum, and ventromedial prefrontal cortex (VMPFC)[36] (see figure 7.6). According to this model, the amygdala receives early input from sensory organs and responds rapidly to a stimulus that, based on early perceptual processes, may represent a motivationally self-relevant stimulus (e.g., an imminent threat or potential reward).[40] Specific cells and nuclei within the amygdala respond quickly to any such stimuli based on these rudimentary perceptual analyses to guide adaptive behaviors, and the amygdala updates these responses based on input from subsequent more elaborative processing of the stimulus. The outputs from the amygdala guide adaptive neural and hormonal responses to aversive stimuli via projections to the hypothalamus (paraventricular nucleus) and brain stem (e.g., periaqueductal gray) and instrumental responses via projections to the ventral striatum.

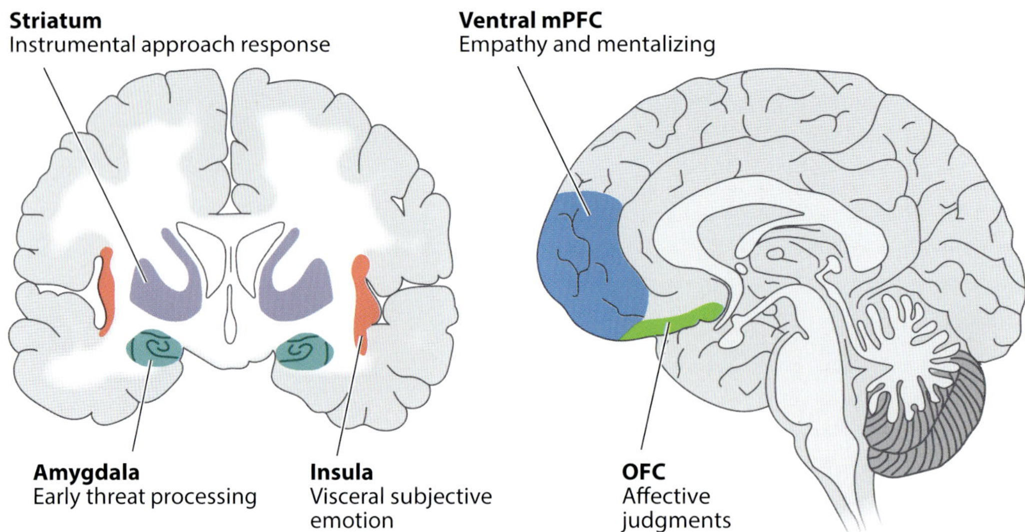

FIGURE 7.6. Prejudice network. Several interactive brain structures underlie components of a prejudiced response.

Studies of the response of the amygdala to faces of racial ingroup versus outgroup members suggest that prejudice reflects rudimentary threat processing based, at least in part, on aversive conditioning.[36,41] This threat processing can be affected by context, in some studies reflecting a perceived threat from an outgroup member and in others a threat of appearing prejudiced in the presence of others who may disapprove.[42] The response of the amygdala to people from different racial groups also depends on an individual's goals. For instance, when the goal was to identify whether individuals from different racial groups were members of one's own team or another team, activity in the amygdala, fusiform gyrus, orbitofrontal cortex, and dorsal striatum was greater when viewing members of one's own team, regardless of racial category.[43] Together, these data suggest that the role of the amygdala is not limited to threat processing but rather responds to motivationally relevant cues to organize adaptive responses.[36]

The orbitofrontal cortex, including the VMPFC, supports more complex and idiosyncratic evaluative representations, and more flexible, contextually determined behavioral responses. For instance, activity in the orbitofrontal cortex is associated with (1) slower, more deliberative judgments about befriending a black, relative to white, individual; and (2) the identification of members of one's own sports team rather than members of an opposing team, regardless of the individual's racial category.[38,43]

Visceral and somatosensory states are represented in the insula, including bodily responses associated with both negative and positive emotions. Activity in the insula has been found to be associated with implicit prejudice toward blacks, rewards given to a member of a disliked outgroup, and painful stimuli given to a member of a liked outgroup (but not to a member of a disliked outgroup).[cf. 36] What these studies appear to have in common is an uncomfortable affective response (e.g., dissonance) aroused by the perception of an ingroup or outgroup member within a specific social context. As Amodio notes:

> It could be speculated that the representation of this affective response in the anterior insula may—through its connections with the ACC [anterior cingulate cortex] and PFC [prefrontal cortex]—facilitate the ability to detect and regulate one's behavior on the basis of a prejudicial affective response.[36] (p. 673)

The striatum is involved in instrumental learning and reward processes, including goal-directed and habitual responses. For instance, the value placed on a potential action or anticipated outcome is associated with activation in the striatum.[44,45] The striatum has been proposed to play a role in guiding positive social interactions through instrumental and approach-related responses. Consistent with this proposition, neuroimaging studies have found activity in the striatum is associated with viewing the implicit preference toward a member of a racial ingroup, and the trust shown toward a member of a racial outgroup.[36]

Finally, the DMPFC and VMPFC have been shown to be associated with self-reference processing and the formation of impressions of other people (see chapter 4). In a study of gender bias, men who reported sexist attitudes showed lower activation of the medial prefrontal cortex when viewing sexualized images of female bodies than men who reported less sexist attitudes. These results were interpreted in terms of the medial prefrontal cortex being less active in response to outgroup than ingroup members because the former is dehumanized.[46]

In sum, prejudice involves a set of cognitive and affective processes subserved by a network of neural structures. Regions that reflect relatively old evolutionary developments, such as the amygdala, are thought to underlie rapid responses to potential threats—including but not limited to those that reflect implicit prejudices—and to other motivationally relevant stimuli. Regions that reflect more recent evolutionary developments, such as the prefrontal cortex, reflect the operation of slower, more deliberative processes, are subject to greater contextual control, and afford more flexible behavioral responses to others.

7.4.3 Mitigating Bias

Research on the activation versus the application of stereotypes and prejudices suggests that their activation may be more difficult to control than their application, just as it is more difficult to control the activation of the impulse to consume ice cream after an arduous day than it is to inhibit the behavioral expression of this impulse (e.g., discrimination).

According to Amodio's network model of the regulation of discrimination (see figure 7.7), the detection of a conflict between the activation of an unwanted social bias and either external cues such as social norms or internal goals such as being a fair and unbiased individual involves regions in the anterior cingulate cortex. Consistent with this notion, studies measuring an event-related brain potential indicative of error detection suggest elevated activity in the anterior cingulate in tasks that require the inhibition of a stereotype-based response.

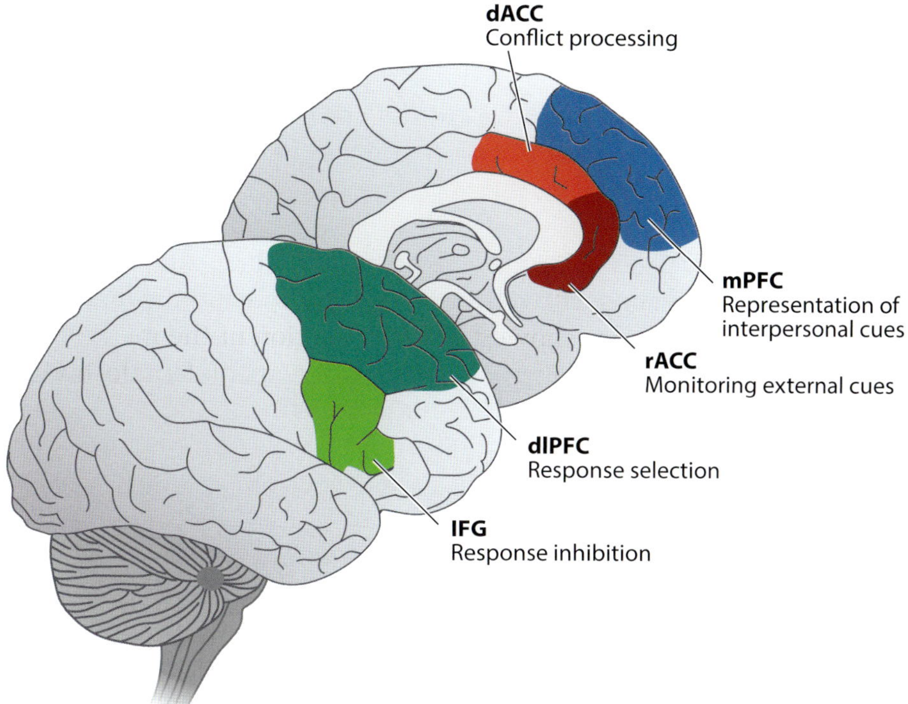

FIGURE 7.7. Regulation network. Brain structures supporting the regulation of intergroup responses. Conflicts between biased tendency social norms tend to be processed in the dorsal anterior cingulate cortex (dACC) and rostral ACC (rACC), respectively. Intergroup response selection is processed in the lateral PFC and implemented in association with the striatum and motor cortex. dlPFC, dorsolateral PFC; IFG, inferior frontal gyrus.

Moreover, the strength of effect is related to the participants' motivation to avoid prejudice.[47]

The dorsolateral prefrontal cortex and inferior frontal gyrus in this model are involved in the selection and control (including the inhibition) of the application of biased responses. Studies using event-related brain potentials, for instance, indicate that activity in the dorsolateral prefrontal cortex was associated with the control of stereotyping.[48,49]

The putative functions performed by the dorsal anterior cingulate cortex, inferior frontal gyrus, and dorsolateral prefrontal cortex reflect processes involved in the detection of conflict and cognitive control generally. The posited role for the medial prefrontal cortex and rostral anterior cingulate cortex, in contrast, are more specific to social information such as cultural norms or the social context, and the connections between the medial prefrontal cortex and the amygdala and orbitofrontal cortex support the top-down modulation of motivational and affective responses.

In sum, neuroimaging research on the regulation of stereotypes and prejudice has provided insights into why it is more difficult to control the activation of stereotypes and prejudice than it is to control their expression. Given that much of the storage and activation of stereotypes and prejudice is automatic, people are typically not fully aware of the nature or extent of the stereotypes and prejudices they hold toward an outgroup nor are they particularly aware of the influence of these stereotypes and prejudices on their behavior toward members of others. As a result, the self-regulatory processes exerted in an attempt to mitigate the potential biasing effects of stereotypes and prejudices on judgments and behaviors may not sufficiently correct for these biases, or they may overcorrect for these biases. Culture plays a major role in defining ingroups and outgroups. As more is learned about the development and activation of stereotypes and prejudice, new roles may be identified for culture to play in modulating the activation of stereotypes and prejudice.

7.5 Punitive Altruism and Cooperation

Although people tend to respond more positively and more cooperatively to ingroup than outgroup members, individuals within an ingroup are not uniformly positive or cooperative, especially when a person's self-interests and the interests of an ingroup are not aligned (see box 7.1).[50] Social dilemmas have been used to investigate what people do in these circumstances. In a social dilemma, an ingroup member benefits personally by acting selfishly unless everyone chooses the selfish alternative, in which case the whole group loses. Economists Özgür Gürek, Bettina Rockenbach, and Bernd Irlenbusch conducted an experiment that illustrates the importance of sanctioning selfish behavior for the development and maintenance of cooperation within a group.[51]

The experiment consisted of thirty repetitions of a three-phase process: (1) choose to participate in one of two groups (institution A or B), (2) make a contribution, and (3) decide whether or not to sanction others in your group for their contributions. In institution A, there were no sanctions. In institution B, each player had the options of penalizing people who acted selfishly and rewarding the generous. Each player could assign a penalty token worth three money units to selfish members, but at a cost to themselves of one money unit. Conversely, they could reward especially generous contributors with a token worth one money unit, at a total cost to themselves of that one money unit.

BOX 7.1. Hyperscanning Connecting Brains in (and out of) the Laboratory

The current challenge for brain imaging is to bring everyday human interaction, occurring in a complex natural environment between two or more subjects, into the laboratory.
—Fabio Babiloni and Laura Astolfi, 2014

Physiologists and bioengineers Fabio Babiloni, Laura Astolfi, and colleagues have developed several hyperscanning setups that are now considered standard for EEG hyperscanning. For instance, they measured simultaneous electrical brain activity from couples of pilots during flight simulation, from couples playing the Prisoner's Dilemma game, and in a group of four subjects playing a card game. The pattern of functional connectivity when the participants adopt a pure-cooperative behavior is different from the pattern of functional connectivity when the participants adopt a pure-defect behavior. During the cooperative behavior, the pattern of functional connectivity links is much denser than in the defect situation. Hyperscanning from multiple subjects opens a new avenue toward a better understanding of the neurobiology of social behaviors.

Each participant in the experiment started out with twenty units of money. After choosing which institution to trust with their investment, these players next chose what to contribute to the collective fund, a sum that could range from nothing to all twenty units. Whatever a participant chose not to contribute went into her own private account. Whatever went into the collective pot would increase in value, but at the end of play it would be divided equally among all the members of that institution, regardless of the level of their individual contributions. Those who contributed the least, and thus were guilty of a form of selfishness known as free riding, would receive just as much as someone who had made generous contributions. At the end of the contribution phase of each round all the players were told how much everyone else had contributed. They were also kept informed of everyone's earnings to date.

Although only about a third of the participants initially chose to join the institution that permitted sanctions with financial penalties, almost 90% of the participants had joined this institution by the tenth round of the game. Moreover, the participants in this institution were cooperating fully, whereas cooperation collapsed in the sanction-free institution. By the thirtieth (final) round, contributions to the sanction-free institution had dwindled to zero.

Altruism and cooperation are supported by the expectation that others are likely to reciprocate altruistic acts in some fashion. As this experiment showed, interactions with or observations of others can reinforce or extinguish these expectations, and the likelihood of cooperative behaviors is adjusted accordingly. The experiment also showed the importance of sanctions for selfish behaviors. In the absence of sanctions, selfish behaviors are reinforced. Free riders in the sanction-free institution earned the highest payoff in the first round. But after that their payoff sharply declined, leading eventually to the institution no longer serving anyone's interests.

In contrast, it did not take long for everyone to realize that high contributors in the sanctioning institution were earning more. As more people saw the benefits of this institution, more people abandoned the sanction-free institution to join the institution that permitted sanctions. With more members joining and contributing freely to this institution, the benefits of this positive social behavior became greater still.

The experiment also showed that much of the credit for high earnings in the sanctioning institution belongs to the participants who engaged in punitive altruism—those who incurred a significant cost to themselves in order to punish those who engaged in selfish behavior. In the first rounds there was no obvious financial benefit to slapping other participants with fines. But after several iterations, the establishment and enforcement of a social norm led to a continuing increase in efficiency, with more and more people becoming high contributors, so much so that the need for negative sanctions became minimal.

Neuroimaging research suggests that the aversiveness of unfair treatment and the anticipation of a pleasant emotional response motivate such policing and social control. Being treated unfairly is associated with increased activity in the anterior insula, an area that is involved in strong emotional responses.[52] In addition, activity in the right dorsolateral prefrontal cortex appears to modulate the behavioral response to unfairness based on context.[53]

When people cooperate, brain regions associated with pleasure and reward such as the ventral striatum and caudate nucleus are activated.[54,55] Because reciprocity evolved as the tendency to pay back in kind, the tendency is for cooperation to carry social benefits (e.g., cooperative responses) and noncooperation to carry social costs. Altruism and cooperation among nonrelatives are also supported by punitive altruism. Nonreciprocators and low reciprocators evoke strong negative emotions and increased activity in the insula of cooperators.[56] Cooperators are not only willing to incur a personal cost to punish a cheater, doing so increases activity in the punisher's dorsal striatum—a reward area in the brain similar to that activated when a person cooperates:[57,58]

> The ability to develop social norms that apply to large groups of genetically unrelated individuals and to enforce these norms through altruistic sanctions is one of the distinguishing characteristics of the human species.[57] (p. 1258)

In addition, natural selection favored moral indignation and retribution in response to being treated unfairly. Emerging from the same impulse that makes a monkey pelt a caretaker with the proffered cucumber slice after the caretaker rewarded another monkey with a grape, the foundation of our legal system is embedded in our genetic endowment and in our culture. So is the concept of honor, which, especially in traditional societies, can lead to duels, revenge killings, and many other forms of mayhem.[59,60] These responses emerge from the fact that it is usually maladaptive in terms of one's genetic legacy to allow oneself to be deceived, exploited, or betrayed. Stable, cooperative societies need what biologist Robert Trivers[61] called a strong show of aggression when the cheating tendency is discovered.[cf.17,62]

7.6 Concluding Remarks

Cooperative behavior has been observed in simple bacteria under conditions in which altruistic cooperative behavior increases the ability of an organism to pass on its genes to the next generation. Varied and complex forms of cooperative behavior have evolved in different social species, but the most varied and complex forms found in any species are observed in humans. Models have been proposed illustrating how kin selection, direct reciprocity, indirect reciprocity, network reciprocity, and altruistic punishment can explain cooperative behaviors in nonhuman and human populations.

However, humans cooperate with nonkin with whom they are unlikely to interact again and in situations in which the benefits of a good reputation are minimal.[63] Social learning and culture play an especially important role in human group processes, including altruistic and cooperative behavior. Economists Samuel Bowles and Herbert Gintis hypothesized

> that where members of a group benefit from mutual adherence to a norm, individuals may obey the norm and punish its violators, even when this behavior incurs fitness costs by comparison to other group members who either do not obey the norm or do not punish norm violators, or both.[64] (p. 18)

Strong reciprocity serves as a form of spite, discussed in chapter 1, that serves in groups to promote fairness and cooperation.[63,65,66]

Intraspecific competition also presents challenges for survival and reproduction, and across human history mortal threats disproportionately came from members of outgroups rather than members of ingroups. Membership in an ingroup, versus an outgroup, results in more positive expectations, more individuated social perceptions (e.g., less stereotyping), less negative affect and greater empathy (e.g., less prejudice), and preferential treatment (e.g., less discrimination) by other ingroup members. These effects can be observed even when the difference in membership between groups is meaningless. The human adaptation for individuals to preferentially cooperate with and provide aid and protection to ingroup members may have contributed to the management of the risks and costs of betrayal and exploitation when groups were relatively small and stable, but ingroup favoritism in global and diverse societies may contribute to intergroup resentment, hostility, and conflict within and across societies. To address these deleterious effects, investigations of the neural mechanisms underlying ingroup biases, stereotyping, prejudice, and discrimination—and underlying the mitigation of each—represent an active and important area of social neuroscience.

8

SOCIAL INFLUENCE

We noted in the introductory chapter that social animals are characterized by sufficiently stable interactions that structural features (e.g., relationships, hierarchies) can be identified. In the present chapter, we focus on another characteristic of social animals: social influence, which refers to the change in preferences or behavior that one individual or group causes in another, intentionally or unintentionally.

Genetic inheritance provides a basis for the development of certain capacities (e.g., language) and predisposes various behaviors (e.g., taste aversions). Individual learning builds on this inheritance system to improve adaptability within a specific ecological niche. Social influence contributes to social learning and cultural transmission, thereby complementing genetic inheritance and individual learning by using the knowledge of others to promote adaptive behavioral repertoires within a specific social group or circumstance.[1–3]

In addition, the social calculus of benefits to costs for individuals in a group can be improved though social influence. One simple form of social influence is retreating behind others for protection. To do so requires that an individual identify the position of conspecifics as well as potential threats, and adjust their own position accordingly. An example of this repositioning is seen in the behavior of shoaling fish when attacked by a predator.[4] This is not a decision that fish make after mentalizing but is instead an automatic response to predation that has evolved through natural selection. The likelihood of a fish surviving a predatory attack is enhanced when it can position itself in the middle of the group, where others are between itself and the predator. Given each member of the group has evolved to respond similarly, the middle of the group represents a moving target, and the path of each fish is influenced by the changing position of the others in the group. Social influences on foraging behavior have also been identified across species.[5,6] We turn next to a specific form of social influence—conformity.

8.1 Conformity

Conformity refers to a change in behavior or beliefs to be more similar to others around you, including what they are doing. Conformity refers to a form of social influence that is common in social species including honeybees, fish, birds, rodents, nonhuman primates, and humans.[7–12]

Two forms of conformity have been identified in humans[13,14] and animals.[15,16] Pan-species distinctions between normative and informational conformity are summarized in table 8.1. Briefly, normative conformity refers to conformity that occurs to fit in with a group whether or not one agrees with the actions or beliefs of the group. Normative conformity serves to maintain or increase social interactions, affiliation, and/or protection, and to maintain a positive evaluation of themselves. Informational conformity, in contrast, refers to conformity that occurs for accuracy, such as following the crowd when one is uncertain how to exit a venue. Normative conformity reflects either the abandonment of personal preferences or behaviors to match alternatives exhibited by a majority of others, or copying the behavior of the majority with a higher probability than the proportional bias that was observed. Informational influence, on the other hand, reflects an updating of preferences or behavior based on information highlighted by the behavior of others in the group.

Evolutionary analyses and comparative studies have tended to emphasize normative conformity, although evidence for informational conformity in animals first appeared more than four decades ago,[17] and psychologists Nicolas Claidière and Andrew Whiten have provided an integrative review of research on normative and informational conformity in animals.[15] In an illustrative study of conformity to local norms in vervet monkeys, groups of monkeys in the wild were provided with a tray of corn dyed blue and an adjacent tray of corn dyed pink. The corn in one of the two trays was made distasteful for several months so that the monkeys in the group developed a strong preference for the other tray of corn (see figure 8.1).

Afterward, the two trays of corn were provided without the distasteful additive. Observations confirmed that naïve infants conformed to their mothers' food preference despite both trays of food now being

TABLE 8.1. Summary of theoretical and empirical differences between information and normative conformity

Condition	Informational conformity	Normative conformity
	Theoretical differences	
Function	Function to adapt one's behavior to the nonsocial environment	Function to adapt to one's social environment
Content	Individuals motivated to find the best possible solution to a particular problem (e.g., foraging strategy)	Individuals motivated to build and maintain social interactions and to maintain positive evaluation of themselves (e.g., show group membership)
Evolutionary origin	Help find most appropriate behavior in an uncertain situation	Help manage social interactions as an honest signal of group membership
	Experimental differences	
Psychological uncertainty	Individuals face an unknown situation with unknown individuals	Individuals are in a known situation with familiar individuals
The other's awareness of one's behavior	The behavior continues in the absence of the group	The behavior stops in the absence of the group
Conflict between the individual and the group	The individual relies on social information ***only when*** her/his/its personal information is not reliable	The individual relies on social information ***even if*** it is not reliable.
Effect of varying the size of the influence group	Marginal effect **de**creases with group size for small group sizes	Marginal effect **in**creases with group size for small group sizes
Diversity of behavioral repertoire	Individuals learn and perform only one option	Individuals learn and perform several options and settle on one afterward
Resistance to the introduction of new behavior	Weak	Strong
Migration	Migrants preserve their own behavior, even when migrating in a new group	Migrants adopt the behavior of the group

Source: From Claidière and Whiten[15]

comparably palatable. Social influence was not limited to infants, either. Fifteen male vervet monkeys immigrated into the group during the study. Even when these monkeys had developed a preference for the other colored corn in their prior group, they too conformed by eating the colored corn eaten by others in their new group.[1–3] This result suggests their behavior represented a case of normative rather than

FIGURE 8.1. Example of foraging in the wild. From Claidière and Whiten.[15] iStock.com/Trevorplatt.

informational conformity. As this study of monkeys in the wild shows, conformity serves multiple functions: it aligns behavior within a group without centralized control, it contributes to social learning, and it promotes the transmission of cultural norms.

The original experiments of conformity date back to the middle of the twentieth century when psychologist Solomon Asch tested groups of eight students whose task it was to identify which of three lines they had seen previously. Specifically, participants viewed a card with a line on it, followed by a card with three lines (see figure 8.2, left and right panels, respectively). Each participant was then asked to announce which of the three lines matched the line shown on the first of the two cards.

Of the eight "participants" in the testing session, only one was actually the subject of the study. The other seven were confederates who worked for the experimenter and responded in a predetermined fashion. In some trials, all seven confederates selected the obviously correct answer before the participant responded. On other trials, all seven confederates would select an answer that was obviously incorrect.[18,19] Based on seating position, which unbeknownst to the true participant, had also been prearranged, the true participant always responded last.

There was also a separate, control group in which participants answered the same questions but without others expressing their answers first. Participants in the control condition chose the correct answer over 99% of the time. However, when preceded by seven others who selected an incorrect response, participants chose the correct answer only about 60% of the time. Interviews following the study indicated that these participants gave incorrect answers to go along with the others in the group.

Neuroimaging studies suggest that conformity includes decreased activity in the ventral striatum and increased activity in the posterior medial frontal cortex and insula.[20,21] Activity in the posterior medial

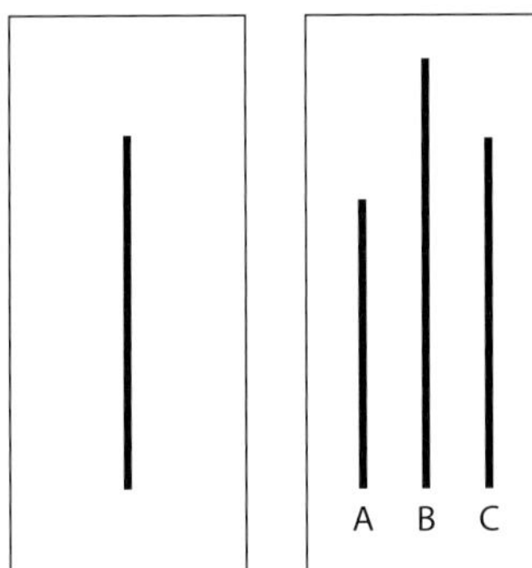

FIGURE 8.2. Illustration of the conformity task from Solomon Asch in which students' task was to identify which of three lines they had seen previously. From Asch (1955).[19] Adapted from Milgram (1963).[29]

frontal cortex also correlates with conformity with a reference group or experts.[22,23] The observed pattern of activity in the posterior medial frontal cortex and ventral striatum has been associated in prior research with components of reinforcement learning, such as prediction-error signaling and behavioral adjustment to minimize subsequent prediction error. Further evidence that conformity involves error signaling and behavioral adjustment comes from studies of the event-related brain potential, which showed that recognizing a difference between one's own opinion and those of others is associated with a feedback-related negativity.[24,25]

To investigate the causal role of the posterior medial frontal cortex in conformity, psychologist Vasily Klucharev and colleagues used transcranial magnetic stimulation to experimentally manipulate activity in this brain region during a conformity task.[26] Female participants viewed photos of female faces as part of a study on "Seeing Beauty." Their task was to rate the attractiveness of each face. Prior to this task, transcranial magnetic stimulation was applied to the posterior medial frontal cortex (experimental group), medial parietal cortex (control group), or subthreshold (sham group).[26,27]

Participants rated 222 faces during the session. Following each such rating, participants received (bogus) feedback regarding the average rating of the same face given by 200 students from the same university. In fact, these group ratings were adjusted such that in 33% of the trials they agreed with the participant's rating, and in 67% of the trials they were in moderate or strong disagreement with the participant.

In a subsequent session, participants again were instructed to rate the same faces in terms of their attractiveness. Conformity was computed as the change in attractiveness rating from the first to the second session. The results show that there was an overall conformity effect such that (1) faces rated more negatively by the group than by the participant in the first session were rated more negatively by the participant in the second session, and (2) faces rated more positively by the group than by the participant in the first session were rated more positively by the participant in the second session.

Importantly, the down-regulation of the posterior medial frontal cortex by transcranial magnetic stimulation reduced this conformity. This finding indicates that activity in the posterior medial frontal cortex is not only correlated with conformity, it contributes to conformity. There are several possible explanations for the specific role of the posterior medial frontal cortex in conformity, including (1) performance monitoring (e.g., error signaling), (2) recognition of a significant cognitive inconsistency (e.g., not being in agreement with a liked group), (3) the motivation to make behavioral adjustment based on performance monitoring or cognitive inconsistency, or (4) the ability to make behavioral adjustments based on performance monitoring or cognitive inconsistency.[27,28] Additional work is needed to determine which provides the most parsimonious explanation.

In sum, the extant evidence suggests that social conformity involves the detection of conflict and adjustments of attention and action that maximize reward or minimize punishment. When individuals find that their preferences are contrary to the responses of a group, conforming to the behavior of others is an efficient means of maximizing rewards within the group. Social conformity may also minimize costs or risks. Under conditions of uncertainty, copying the behavior of others may be adaptive because doing so is most likely to adhere to the locally optimal solution.[16] In addition, if everyone in a group is responding in a particular fashion, adhering to this norm may be adaptive because it minimizes the likelihood of being disliked, punished, or shunned.[15]

8.2 Obedience to Authority

Obedience to authority refers to compliance with an instruction or request, or submission to another's authority in the absence of physical coercion. This form of social influence drew worldwide attention in 1961, when a Jerusalem court tried Nazi SS Lieutenant Colonel Adolf Eichmann for his role in moving Jewish people from their homes into ghettos and then to their deaths in German concentration camps. In defense of his role in the Holocaust, Eichmann claimed that he was just following orders. The Holocaust represents a particularly dark period of human history, and the notion that Eichmann was a normal person who was simply obeying the instructions of an authority seemed

TABLE 8.2. Distribution of breakoff points

Verbal designation and voltage indication	Number of subjects for whom this was the maximum shock
Slight shock	
15	0
30	0
45	0
60	0
Moderate shock	
75	0
90	0
105	0
120	0
Strong shock	
135	0
150	0
165	0
180	0
Very strong shock	
195	0
210	0
225	0
240	0
Intense shock	
255	0
270	0
285	0
300	5
Extreme intensity shock	
315	4
330	2
345	1
360	1
Danger: severe shock	
375	1
390	0
405	0
420	0
XXX	
435	0
450	26

Source: From Milgram

inconceivable. After all, he may have received orders from superiors, but he was never physically coerced to take the actions he did.

Could a person by virtue of their authority influence people to harm others? To address this question, psychologist Stanley Milgram performed a series of studies to investigate the extent to which normal people from the community were willing to obey verbal instructions from an authority to harm another individual.[29–31] These studies revealed that people were willing to defer to the instructions of an authority and inflict significant harm to others on command. In Milgram's initial study of obedience, only one in three participants stopped at some point after the victim protested and refused to provide further answers (see table 8.2).

The vast majority of the participants were obedient and proceeded to administer increasingly intense electric shocks to the victim, including a final level labeled "Danger: Severe Shock." These obedient individuals took no pleasure or pride in their actions, and several attempted to help the victim by yelling out the correct answer. From various surveys and interviews, the participants seemed quite normal, as well. The social influence of an authority simply proved overwhelming for the vast majority of the participants.

The results of the Milgram studies stunned the scientific community, but they may be more understandable in light of the evolutionary significance of obedience to authority. The costs of communal living include competition for access to reproductive opportunities and limited resources such as high-quality food, water, and space. To maintain social order in communal living requires some system of authority, such as a dominance hierarchy. Chimpanzees and bonobos punish rule violations by subordinates and appear to feel guilty when they themselves violate these rules.[32] How dominance hierarchies are maintained varies across species, but two components of dominance in primates are (1) a competitive component based on physical coercion or psychological intimidation (e.g., eye contact) by an individual, partnership, or small coalition; and (2) a cooperative component based on competence-based attraction to high-ranking individuals.[33]

Physical coercion and aggression can be costly to the winner as well as the loser, so characteristics (e.g., age, size or strength of the body, power) that predict which animal would likely emerge as the victor often serve as a substitute for fighting in the establishment and maintenance of social order. Submission to an individual of higher rank—that is, obedience to authority—contributes to the maintenance of such a system without the risks of physical confrontation. Such behavior is not so much a reflection of individual preference as it is an adaptive predisposition that maintains position and access to resources and reproductive partners within a community.

Even though humans are more likely to cooperate as entire bands when it comes to sanctioning of rule violations, the distribution of food, and the provision of safety, social dominance hierarchies represent a major feature of our evolutionary past.[32,34] Moreover, the fragile and dependent nature of infants and children makes it adaptive for them to be relatively obedient early in life, and obedience to authority contributes to social organizations operating in an orderly and efficient manner.[35]

Follow-up research to Milgram's studies found that when people obey an authority, they feel less responsible for their actions. They also show a smaller negative attentional component of the event-related brain potential when participants were ordered to harm another person than when participants freely chose to inflict the same harm.[36] Obeying an authority to harm another has also been associated with increased activity in a distributed set of brain regions (e.g., amygdala, rostral prefrontal cortex) associated with personal distress rather than empathy.[37] Together, these findings paint a picture of obedience to authority as a means of minimizing risks to oneself while maintaining the status quo within a social group. This by no means excuses the behavior of Adolf Eichmann, but it does help us understand the importance of social learning and culture to ensure such behavior is not repeated.

8.3 Persuasion

Persuasion refers to the active and intentional attempt by an individual, group, or social entity (e.g., government, political party, business) to change a person's beliefs, attitudes, or behaviors by conveying information, rational or irrational beliefs, values, and reasoning. Not included under the rubric of persuasion are changes in knowledge or skill (i.e., education) or changes in behavior that require surveillance or sanctions by others (i.e., compliance). Persuasion, therefore, represents a form of social influence that relies on communication rather than physical intimidation or coercion.

8.3.1 Evolutionary and Cultural Contributions to Persuasion

In nonhuman primates, dominance hierarchies are based in significant part on physical prowess, and physical intimidation and coercion represent major means of maintaining social control within a group. Surveillance, physical intimidation, and physical prowess play important roles in maintaining order between groups, as well. For instance, the area that animals from a given community traverse during food gathering, mating, and caring for young varies as a function of the intrinsic fighting power of the community and the relative fighting power of neighboring communities.[38]

The evolutionary path of hominins, however, diverged from the heavy reliance on dominance hierarchies. According to economist Herbert Gintis, anthropologist Carel van Schaik, and anthropologist Christopher Boehm,[39] the emergence in hominins of cooperative breeding and bipedalism, the recognition by hominins of a diet of meat from large animals as a reliable food source for groups, and the incorporation of fire and cooking in hominin cultures created a niche in which there was a high return for coordinated, cooperative scavenging and hunting of large mammals:

> This was accompanied by an increasing use of wooden spears and lithic points as lethal hunting weapons that transformed human sociopolitical life. The combination of social interdependence and the availability of such weapons in early hominin society undermined the standard social dominance hierarchy . . . group success depended on the ability of leaders to persuade and of followers to contribute to consensual decision processes.[39] (p. 327)

The likelihood of successfully persuading others, in turn, increases with the development of capacities such as linguistic facility, logic, analytical abilities, and social cognition. The adaptive value of persuasion is therefore thought to have contributed to the progressive encephalization of the human brain and to the continued evolution of these capacities across human history.[39]

Using persuasion to influence a general behavioral predisposition (e.g., approach or avoidance) toward a class of stimuli is far more efficient in large, complex social groups than using persuasion to influence a single behavior in a specific context. Persuasive appeals, therefore, have historically targeted attitudes as well as behavior. *Attitudes* refers to general and enduring evaluative (good/bad, harmful/beneficial, wise/foolish) predispositions toward a stimulus or category of stimuli (e.g., person, object, issue, position, group), while *behaviors* refers to the observable (though not necessarily observed) actions by a person.[40,41] An important functional characteristic of attitudes is the breadth of their behavioral influence even if the effect on any single behavior may be small.

A second important feature that contributes to the adaptive value of persuasion and makes attitudes so functional is their capacity to *change* in light of new information, goals, and challenges. People not only form and use attitudes spontaneously to help guide their behavior through a complex world,[42,43] they also engage in persuasion in an attempt to change the attitudes and influence the predispositions, decisions, and behaviors of others to ensure they have the resources and social backing to accomplish their goals.[40,44,45] Persuasion across human history has primarily involved face-to-face communication, which permitted counterargumentation and negotiation. The give-and-take from such a dialogue increases the likelihood that the adopted position effectively serves the goals of the group rather than those of a single individual.[46,47]

In addition to having a directive function (e.g., differentiation of hostile from hospitable stimuli, predisposing positive or negative response), attitudes also have the dynamic function of energizing people to act. Individuals who come to feel strongly about a person, object, or issue, for instance, are more likely to channel their thoughts and behavior toward the target accordingly when given the opportunity, and to create opportunities to act in a positive or negative fashion toward the target.[48]

Finally, attitudes (1) serve as convenient summaries for beliefs, emotions, and preferences regarding specific categories of issues, objects, and people; (2) help individuals to know what to expect when exposed to a specific stimulus; (3) promote social interactions by increasing the predictability of a person's behavior; and (4) reduce the stress of decision making.[49–52]

Cultural developments over the past century have resulted in persuasion becoming a major economic, social, political, public health, and diplomatic/military enterprise. Advertisers spent more than $590 billion worldwide in 2015, the 2016 political campaigns expended about $6.8 billion on ads, and international conflicts and ideological wars are now being waged on the Internet as well as on the battlefield. In addition to advertisers and politicians, occupations that rely on persuasion in contemporary commerce include lawyers, public relations specialists, salespersons, lobbyists, actors and directors, counselors and managers, and social, recreational, and religious workers.[53]

As is the case for most social functions, persuasion can be used for malevolent as well as benevolent purposes. The weaponization of persuasion in the form of propaganda, the exploitation of individuals through false advertisements or scams, and the manipulation of public opinion through biased or fraudulent proclamations have made it easy to employ persuasion as a tool for manipulating the masses for the benefit of a few. The increased reliance on persuasion in the mass media and social media, coupled with the decreased reliance on face-to-face communications, has increased the audience size for persuasive appeals, including malevolent appeals. For instance, phishing scams are emails in which an individual claims to be a trusted source in an attempt to persuade the recipient into providing confidential information that can be used for identity theft. Studies suggest that 23–30% of recipients open phishing messages and about 10% open attachments, in part due to our use of heuristics and perception of being invulnerable to duplicitous manipulations.[33] Not surprisingly given these statistics, the number of unique phishing campaigns has increased from 173,063 in 2005 to 1,220,423 in 2016.[54] It is the potential for malevolent uses of persuasion that makes it so important to understand

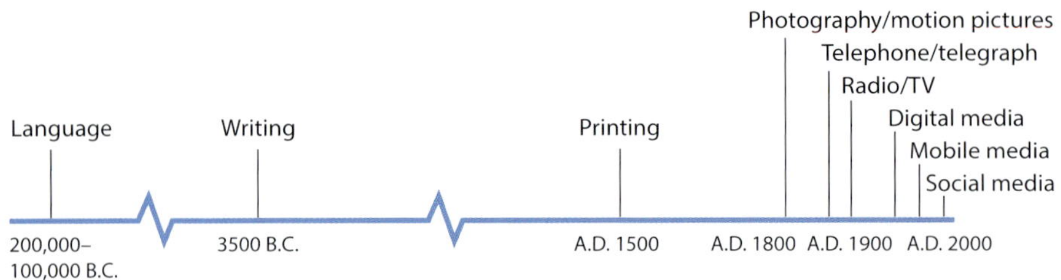

FIGURE 8.3. Time line from handwriting to social media.

the evolutionary and cultural origins of persuasion and the psychological and neural processes underlying persuasion.

Persuasion may have evolved as an adaptive form of social influence in hominins, and cultural developments over the past 3,500 years have made it easier for an individual to use persuasion to influence larger and larger numbers of others (see figure 8.3). The reliance on persuasion to resolve conflicts and maintain social order is also a relatively recent development. Persuasive appeals played a central role in resolving conflicts and maintaining social control in Athens in 427 to 338 BCE (during which time Plato and Aristotle considered the processes underlying persuasion), Rome from approximately 150 to 43 BCE (during which time Cicero wrote about oration and persuasion), in Europe from approximately 1470 to 1572 CE (during the Italian Renaissance), and the present period dating back to the middle of the eighteenth century (during which time the mass media formed and evolved). Social control through physical intimidation, coercion, and elimination, however, were far more common during the centuries surrounding these periods.

The cultural and sociopolitical changes over the past 150 years have contributed to a particularly rapid rise in the importance of persuasion in commerce, mass media entertainment, political campaigns, public health campaigns, and international conflicts.[51,52,55] These developments have emerged because persuasive appeals can be a powerful source of social influence. Although the influence of an appeal on any single individual's behavior in a specific context is small, the cumulative effect can be substantial. Meta-analyses have been performed to determine the extent to which experimental studies of persuasive appeals designed to change health-related attitudes actually led to changes in health-related intentions and behavior. The results showed that message-induced changes in attitudes produce significant changes in the advocated health-related intentions and behavior.[56]

8.3.2 Communication and Persuasion

Communication refers to the exchange of information, news, feelings, and ideas to create a shared meaning and mutual understanding. Primate communication is thought to have evolved from an evolutionarily older, relatively fixed, recurrent signaling (e.g., recognizing associations between cues) to flexible exchanges of information that rely heavily on inferential capacities.[57,58] Nonhuman primates and humans express impulsive (e.g., alarm calls) and habitual (e.g., affiliative displays) communications that involve minimal time, effort, and inferential load. In addition, humans, perhaps uniquely, express strategic communication that is slower, more cognitively effortful and controlled, and more demanding of cognitive representations and inferential processes.[55]

The processes underlying persuasion in humans have been characterized somewhat similarly in the Elaboration Likelihood Model (ELM) as varying along a continuum of cognitive processing (a dimension termed "cognitive elaboration").[41,59] According to the ELM, individuals can reach the same attitude position via different sets of information-processing operations ("routes"), each of which involves both automatic and controlled components (see figure 8.4).[60] The evolutionarily older, less effortful, and less inferentially demanding route to persuasion is termed the peripheral route to persuasion, and the slower, more effortful, and more cognitively elaborative route to persuasion is termed the central route.

Whether the peripheral or central route operates depends on an individual's motivation and ability to engage in an effortful inferential analysis of the evidence or arguments for a specific position. When motivation (e.g., the personal relevance of an appeal) and ability (e.g., knowledge on the topic of the appeal) are high, persuasion depends on an individual's cognitive elaboration of issue-relevant information (i.e., the central route). When motivation and/or ability are low, however, the cognitively simpler peripheral route to persuasion rules.[52] For instance, rather than persuasion depending on the cognitive resources required for idiosyncratically evaluating the merits and relevance of the information (e.g., message arguments) for or against an advocated position, the peripheral route involves simpler affective and associative processes such as cues that trigger simple heuristics (e.g., "experts understand the topic so their position must be correct"). Although the central route involves more issue-relevant thinking than the peripheral route, the idiosyncratic issue-relevant thinking can range from relatively objective to deeply biased (motivated reasoning).[61] Finally, according to the ELM, attitudes changed through the central, in contrast

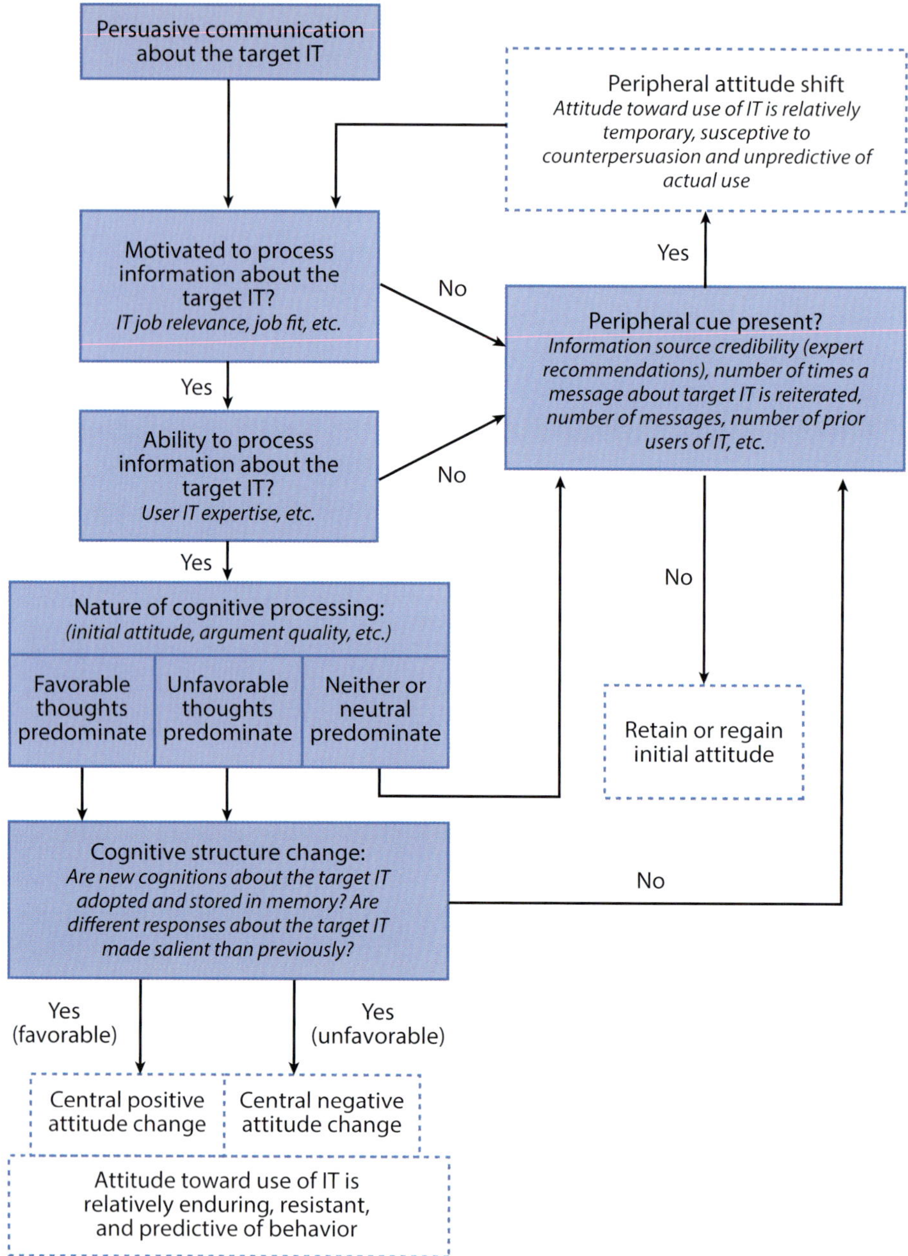

FIGURE 8.4. Elaboration Likelihood Model (ELM).[41,59] According to the ELM, individuals can reach the same attitude position via different sets of information processing operations ("routes"), each of which involves both automatic and controlled components. The evolutionarily older, less effortful, and less inferentially demanding route to persuasion is termed the peripheral route to persuasion, and the slower, more effortful, and more cognitively elaborative route to persuasion is termed the central route. Adapted from Petty and Cacioppo.[59]

to peripheral, route are more likely to persist, be resistant to counter-persuasion, and guide behavior.[62]

Attitudes and persuasion are products of the operation of the brain. The elaboration continuum is reminiscent of fundamental ontogenetic and phylogenetic trends in CNS development (see section 6.4.1). The highest levels of the CNS show the greatest expansion and elaboration through the development of both the individual (ontogeny) and the *Homo sapiens* species (phylogeny), and differentiate the adult human from the infant and from other animals. The central route to persuasion is characterized by more cognitively controlled, effortful, and demanding cognitive operations (e.g., analytical and inferential processes) than the cognitively simpler and perhaps evolutionarily older peripheral route.

In the past decade, persuasion has become the focus of a growing number of patient and neuroimaging studies. Attitude formation and change in response to a persuasive appeal were once thought to be closely associated with how much an individual learned from a message or appeal, but subsequent research has demonstrated that memory and attitude change are separable,[63,64] and that the cues people use and the idiosyncratic cognitive elaborations (e.g., issue-relevant thinking) people generate in response to a message are more important determinants of attitudes than is message learning per se.[52] For example, people can learn virtually nothing from a message and still be influenced if they simply count the number of arguments in the message without encoding their meaning, and they can show no persuasion even if message learning is high if all of the arguments are counterargued.[65] Consistent with the dissociation of learning from persuasion, the brain areas underlying declarative memory are not necessary for attitude change. Studies have found that patients with Korsakoff's syndrome show impaired memory for the content of a persuasive message but intact attitude change, and amnesic patients with medial temporal lobe damage show dissonance-induced attitude change in a free choice task in the absence of memory for the stimuli or their choices.[66,67]

Psychologist Hanna Faye Chua was the first to use functional magnetic resonance imaging (fMRI) to investigate the regional brain activity associated with persuasive messages that varied in personal relevance.[68] Smokers who had expressed a desire to quit smoking were exposed to message arguments that were tailored to each individual's smoking behavior (e.g., "smokes 20 cigarettes a day"; high personal relevance), message arguments that were not tailored to each individual's smoking behavior (e.g., "smokes a lot of cigarettes"; moderate personal relevance), or generic statements (e.g., "quitting is not easy").

Behavioral research on persuasion has shown that personally relevant persuasive appeals produce greater *motivation* to engage in idiosyncratic issue-relevant thinking (i.e., central route) than appeals that are not. When recipients are also *able* to do so (e.g., high prior knowledge about the issue, low-distraction conditions), they generate higher levels of idiosyncratic issue-relevant thinking,[69,70] processes that comprise the central route to persuasion. The neuroimaging results of Chua and colleagues identified differences in activation of the rostral medial prefrontal cortex and precuneus/posterior cingulate in response to the high, relative to low, personally relevant appeals.[71]

In a follow-up investigation, Chua and colleagues investigated the prediction of smoking cessation by the neural responses to persuasive messages in smokers who were interested in quitting within the next 30 days.[72] The messages were similar to those used in their initial study. Participants also completed a self-relevant adjective task (adjective does or does not describe you versus adjective is positive or negative) to identify regions involved in self-relevant processing. Following the scanning session, participants were given a 10-week supply of nicotine patches, completed a web-based tailored smoking-cessation program, and were instructed to quit smoking. Four months later, participants were called and responded to a seven-day point prevalence abstinence measure (cigarette free for the past seven days). Participants were categorized as quitters and nonquitters.

Analyses of the self-relevant adjective task and persuasion task identified brain activation in both tasks in the dorsomedial prefrontal cortex, precuneus, and angular gyrus. Activity in the dorsomedial prefrontal cortex during the personally relevant, in contrast to neural, messages was found to predict smoking cessation. Together, these studies are consistent with the notion that activity in the medial prefrontal cortex is associated with the central rather than peripheral route to persuasion. Subsequent replications have provided additional evidence for an association between activity within the medial prefrontal cortex in response to a persuasive communication and changes in behavior following the communication.[68,73–77]

A few neuroimaging studies have investigated the persuasive effects of an association between a product and a cue (e.g., an expert or celebrity endorser). In an illustrative study, a picture of the face and name of a female celebrity was paired with pictures of shoes on some trials, and a picture of the face and name of a female noncelebrity was paired with pictures of shoes on other trials. The personal relevance was generally low, so these conditions are more likely to engage the peripheral than central route to persuasion. Results revealed an increase in activity in subcortical as well as cortical regions associated with reward processing, conflict detection, and self-awareness (e.g., caudate, superior frontal gyrus, anterior cingulate cortex).[68,78]

Overall, these results show that persuasion does not depend on a single brain region or even a single network of regions, but instead involves different networks of regions depending on specific source, message, and channel factors as well as the motivation and ability of the recipient to evaluate the evidentiary merits for an advocated position.

8.4 Concluding Remarks

Individuals seek to achieve their goals in an effective and rewarding fashion. Social species evolved because there is some benefit to individual members that exceeds the cost of the individual being a member of the social group. Although there are controversies about how to evaluate this fitness benefit,[79–82] social influence serves generally to increase the benefits and/or lower the costs of membership in a social group. Among the ways in which it does this is by promoting the efficiency and stability of social groups, the predictability and coordination of individual actions, affiliation and protection among members of the group, and accurate perceptions of reality in a dynamic and potentially dangerous social environment. Social influence also contributes to the alignment of behavior within a group even in the absence of centralized control, promotes social learning, and fosters the establishment and transmission of cultural norms. Thus, social influence complements genetic inheritance and individual learning by using the knowledge of others to promote adaptive behavioral repertoires to challenges and stressors within a specific social group or circumstance.

In some cases, social influence operates through relatively fast, simple, heuristic processes that improve the likelihood of effective behavior within the group at minimal expense to one's cognitive resources. In others, social influence operates through slower, more controlled, effortful, and sophisticated cognitive processes. The former is fast and has little cognitive overhead, but it leaves individuals more susceptible to manipulation by an illegitimate authority posing as a legitimate authority (e.g., as in phishing), a nonexpert posing as an expert, or even an expert with selfish intentions representing themselves as working for the benefit of others. The scrutiny of the evidentiary and logical merits of a persuasive appeal is more costly in terms of the time and demands on limited cognitive resources, but it makes it less likely that an individual would fall victim to a specious appeal.[35,82] The costs and benefits of each are weighted differently depending on an individual's motivation and ability to effectively analyze the influence attempt. When time is severely limited, the rapid, simpler processes underlying social influence are likely to maximize the likelihood of achieving one's goals. On the other hand, when the personal consequences are high and the time and prior knowledge are sufficient to permit message scrutiny, the slower, more cognitively costly processes underlying social influence are likely to operate, thereby maximizing the likelihood of achieving one's goals. The more deliberative cognitive processes that are triggered by a persuasive message in these situations are associated with activation in the rostral regions of the medial prefrontal cortex, and the extent of this activation predicts the extent to which an individual's subsequent behavior aligns with the persuasive appeal.[cf. 68]

9

SALUTARY SOCIAL CONNECTIONS

According to the US Census, approximately 50% of the marriages that were celebrated between 2000 and 2015 ended in a divorce or an annulment, without any states being spared (see figure 9.1). Because unhappy marriages are associated with premature morbidity and mortality (see chapter 2), researchers are dedicating time and efforts to understanding what predicts the long-term maintenance of salutary close relationships. As discussed in prior chapters, there is growing evidence that we are motivated to connect with and influence one another across our life span through a myriad of invisible forces,[1] despite the fact that people often think of humans as being unique compared to other species and as being unique and independent relative to those around them. For mammalian species to survive, infants must, for instance, instantly engage their parents in protective behavior, and the parents must care enough about these offspring to protect and nurture them.[2–4]

9.1 From "Me" to "We"

Being interdependent corresponds to a form of social self (or social identity) in which an individual's core values, beliefs, emotions, expertise, and goals are strongly connected with those of another individual or with a group of people bigger than one's self. The social self is one of many selves one can have. Neisser described five types of constructs for the self: (1) the *private* self (based on awareness that our conscious experiences are our own), (2) the *conceptual* self (based on semantic knowledge about oneself), (3) the *extended* (or remembered/autobiographical)

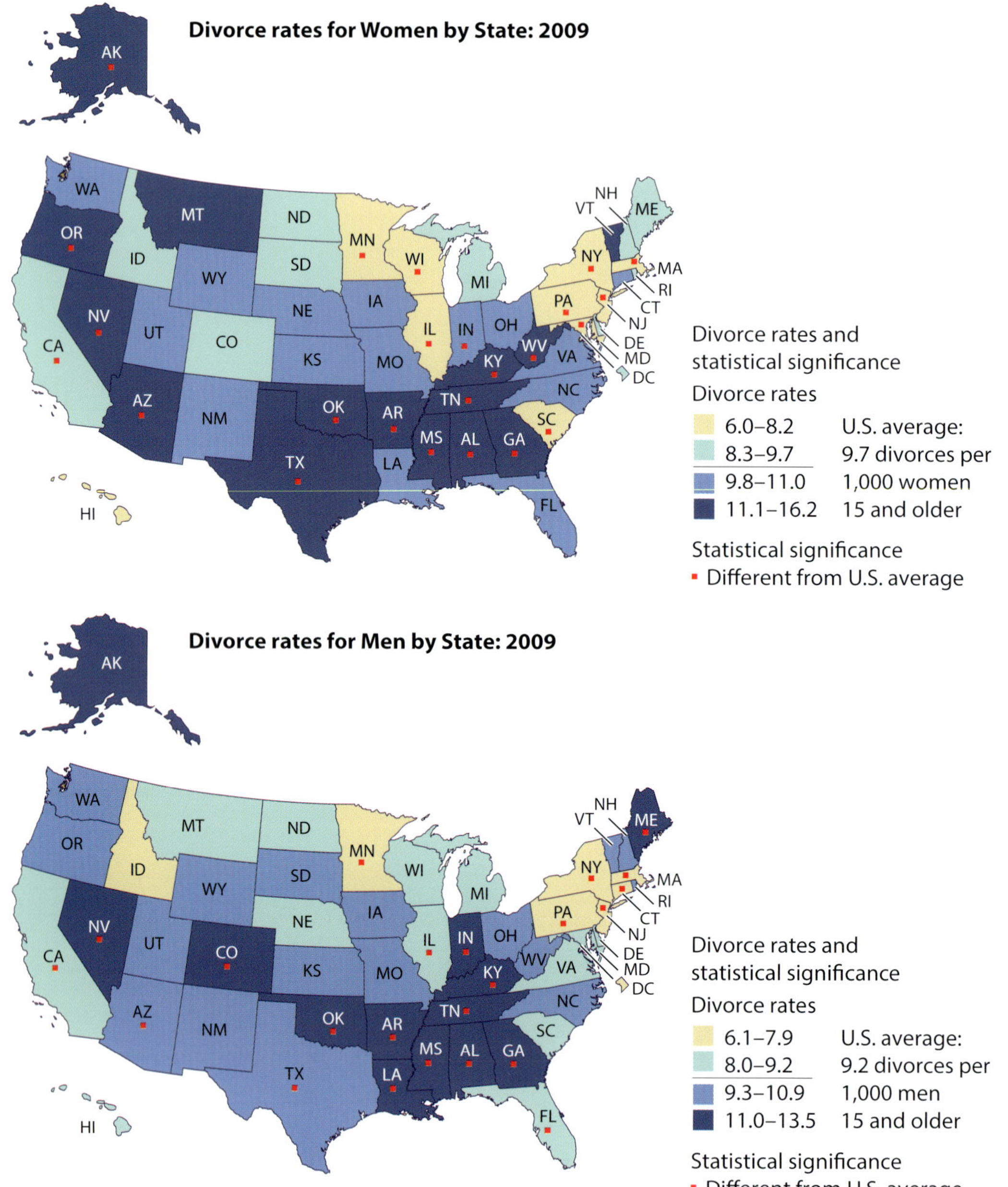

FIGURE 9.1. Divorce rates for men and women by state (From US Census Bureau, 2009). Adapted from Neisser.[5]

TABLE 9.1. Neisser's five constructs of the self

Construct of self	Description
Private self	Awareness of having own thoughts, feelings, intentions
Conceptual self	Awareness of having an abstract model of oneself
Extended self	Awareness of the self in time (past, present, future)
Ecological bodily self	Awareness of internal or external stimuli
Interpersonal self	Awareness of interactions with others

Source: From Neisser[5]

self (based on memories of existing over time), (4) the *ecological bodily* self (based on direct bodily-related perceptions and sensations), and (5) the *social/interpersonal* self (based on perceived communication signals during social interaction; see table 9.1).[5,6] Each construct of the self is sustained by neural mechanisms whose components are only partially overlapping with the neural mechanisms underlying the facets of the self. Many of these neural mechanisms also share neural pathways with other neural mechanisms, such as those underlying memory, familiarity, social identity, and visuospatial information processing. Although we are all interdependent to some degree, the notion of social self highlights the extent to which partners, friends, and family members who feel connected to one another share a common social self and tend to expand their mental representations about their own self into the mental representation of their significant others.

From a cognitive viewpoint, expanding one's self and including others into one's social self suggests that shared mental representations are created of one's self with significant others (e.g., romantic partners).[7,8] People in love share a common "transactive" mental representation of each other's self.[9–14] These shared mental representations include stored representations that are common across the individuals, leading to the construction of a social self, i.e., from "me" to "we." The expansion of one's self into a social self is an important part of salutary relationships in humans. Couples in love try to secure their unified social self and their strong self–other overlap. As an illustration, couples in passionate love are more likely than couples not in love to refer to each other in terms such as one's "better half" or "soul mate."[7,15–17] They also often use "we" instead of "I" or "you"; they stand close to one another, and consider themselves as "one" such that the usual give-and-take of relationships does not apply.[7,15,16,18–20] This quest for union among couples is consistent with evolutionary theory's claim that intense emotional experiences during a lifetime (e.g., passionate love) may be a central human motivation to expand one's self.[21,22]

9.2 Passionate Romantic Love

Not surprisingly to us, the expansion from "me" to "we" has a specific brain signature that moves from self-centered brain areas to areas that are involved in self-expansion, self–other overlap, and other social functions. Such a complex brain signature was, however, a surprise to most scientists who viewed love as a basic instinct, emotion, or addiction. Although philosophers, psychologists, artists, and poets have been interested in the nature and origin of passionate romantic love throughout the ages, only in the 1960s did social psychologists begin to systematically investigate love as a cognitive construct.[15,16,23–25] And only in the past two decades or so have social neuroscientists contributed to a better understanding of the neural bases of love and the social self.[26–30]

The first neuroscientists to use fMRI to identify the brain regions associated with passionate love were neuroscientists Andreas Bartels and Semir Zeki.[31] Participants were asked to write about their feelings of love and to complete the *Passionate Love Scale* ([16,18,23]; see appendix B). Participants were then placed in an MRI scanner and asked to passively view a color photograph of their beloved partner and pictures of a trio of casual friends.[31] This study sparked an interest in the fMRI investigation of love,[29] but because most such studies tested only a small number of participants, the findings across studies were inconsistent.

To address this problem, neuroscientist Stephanie Cacioppo and colleagues performed a meta-analysis of all the fMRI studies that had been done on the neuroimaging of love, and she compared the results with those obtained for different types of love. Figure 9.2 displays a schematic representation of the findings.[26,27] Results showed that love, independently of its type (e.g., passionate romantic love, maternal love, unconditional love), activates a specific brain network including a set of 12 main brain areas (i.e., caudate nucleus/putamen; thalamus; ventral tegmental area; anterior insula; anterior cingulate cortex; posterior hippocampus; occipital cortex; occipito-temporal/fusiform region, angular gyrus/temporoparietal junction, dorsolateral middle frontal gyrus, superior temporal gyrus, and precentral gyrus; figure 9.2).[29] These results indicated that the neural bases of love were more complex, and subject to more cognitive influence, than once thought, as they include brain areas beyond the basic emotional system. For instance, these 12 brain areas are not located only in subcortical dopaminergic-rich areas within

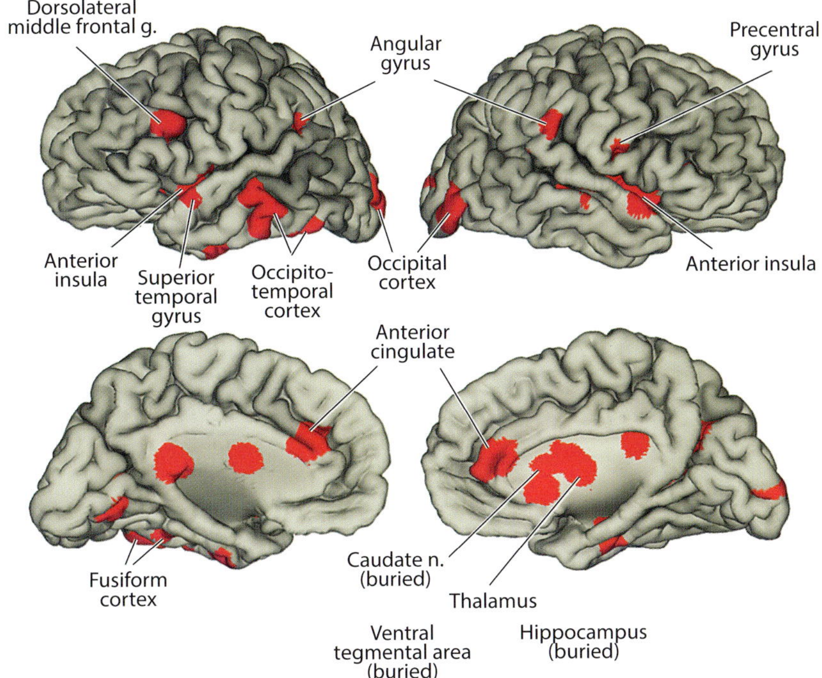

FIGURE 9.2. The love brain network. This figure depicts the 12 main brain areas that are systematically activated in response to a beloved. Adapted from Cacioppo et al.[27]

the limbic emotional system and the reward/motivation system; they also include cortical associative brain regions (e.g., angular gyrus/temporoparietal junction) that mediate more complex and cognitive functions, such as self-expansion, body image, self-representation, metaphors, attention, memory, and abstract representations.[32]

Based on these findings, Stephanie Cacioppo proposed a theoretical model in which the brain network underlying love includes three main functional systems (see figure 9.3):

1. Cognitive system, which includes an appraisal process that categorizes stimuli as loved and personally meaningful, and which is related to the extent to which a person in love perceives their and their partner's selves to be overlapping. Associative brain areas involved with self-image can be activated at this stage.
2. Reward/motivation system. This mesolimbic dopaminergic reward/motivational system is underpinned by specific areas, such as the ventral tegmental area, thalamus, and caudate nucleus. The caudate nucleus is an important area that is acti-

vated across species (including dogs, see box 9.1) in response to salient familiar stimuli that are mentally associated with a social bond or a past rewarding experience or incentive. This reward/motivational system detects the loved one as a salient stimulus with rewarding features.

3. Affect system, which in contrast, involves specific structures such as the precentral gyrus and the anterior cingulate cortex (i.e., brain region involved in various emotional functions, such as affect regulation). The anterior insula is at the intersection of the three systems, as it helps us become aware of our internal regulation responses, such as heart rate, hot flashes, respiration, and hunger, and also form more abstract cognitive representation of these feelings from the body across time.

A main difference between Cacioppo's and Adolphs and Stanley's systems resides in the fact that Cacioppo's cognitive system of love, for

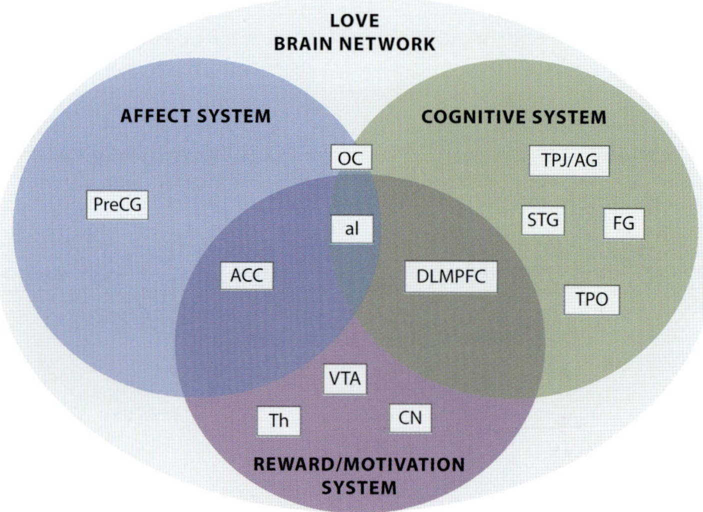

FIGURE 9.3. Three systems of the love brain network. The love brain network includes three functional systems: (1) the affect system, (2) the reward/motivation system, and (3) the cognitive system. These areas have been categorized here based on their assumed functional role in love. Each brain area may serve different functions in different conditions, contexts, or situations. Abbreviations: al: anterior Insula, VTA: ventral tegmental area, Th: thalamus, CN: caudate nucleus, ACC: anterior cingulate cortex, TPJ/AG: temporoparietal junction/angular gyrus, PreCG: precentral gyrus, DLMPFC: dorsolateral medial prefrontal cortex, STG: superior temporal gyrus, FG: fusiform gyrus, TPO: temporo-occipital area, and OC: occipital cortex.

BOX 9.1. Neuroimaging of Familiar Beloved Smells in Dogs

In 2015, Gregory Berns and colleagues tested the difference between self and familiar significant other in a cohort of 12 dogs that had been trained to remain motionless while unsedated and unrestrained in a scanner. Because olfaction is believed to be dogs' most powerful sense, the dogs were presented with five scents rather than pictures during the functional magnetic resonance imaging. These five scents were (1) self, (2) familiar human, (3) human stranger, (4) familiar dog, and (5) dog stranger. The authors focused their analysis on the dog's caudate nucleus because of its well-known association with reward and positive saliency, and because of its clearly defined anatomical location in dogs. Their results were clear-cut: "While the olfactory bulb/peduncle was activated to a similar degree by all the scents, the caudate was activated maximally to the familiar human." This caudate activation suggested that not only did the dogs discriminate that familiar scent from the others, they had a positive "social bond" with it.[130] Berns and colleagues reinforced their results a year later when they investigated individual differences in dogs' preferences for food vs. social interaction with their owner and found the dog's ventral caudate activation to serve as a stable neural predictor of individual differences in dynamic choice for food and praise by their owner.[131]

instance, bridges the gap between three of Adolphs and Stanley's systems (empathy, mirror neuron system [MNS], and mentalizing systems), rather than calling upon one system only. The fact that love recruits 12 integrated associative brain areas that work together across three different functional systems suggests that love is more than a basic instinct or emotion with dopaminergic-like rewarding experiences akin to those experienced during craving and addictive behaviors. Love is also cognition (e.g., self–other overlap).[33,34] Within the cognitive system of love, a few associative areas may play a role, as suggested by correlational analyses. For instance, the angular gyrus, which is located in the posterior part of the inferior parietal lobule (i.e., just behind your ear) and is involved in a broad variety of cognitive functions (see figure 9.4),[35] also plays a critical role in self–other expansion mechanisms experienced in the love relationship.

The more the participants report feeling passionately in love with their partner, the more the left angular gyrus (Brodmann area 39) is specifically activated (see figure 9.5). The activation of the angular gyrus/inferior parietal lobule is also related to the extent to which a person in love perceives their and their partner's selves to be overlapping, and activation of the angular gyrus is more intense for a passion for a person than a passion for a hobby (e.g., sport). The association

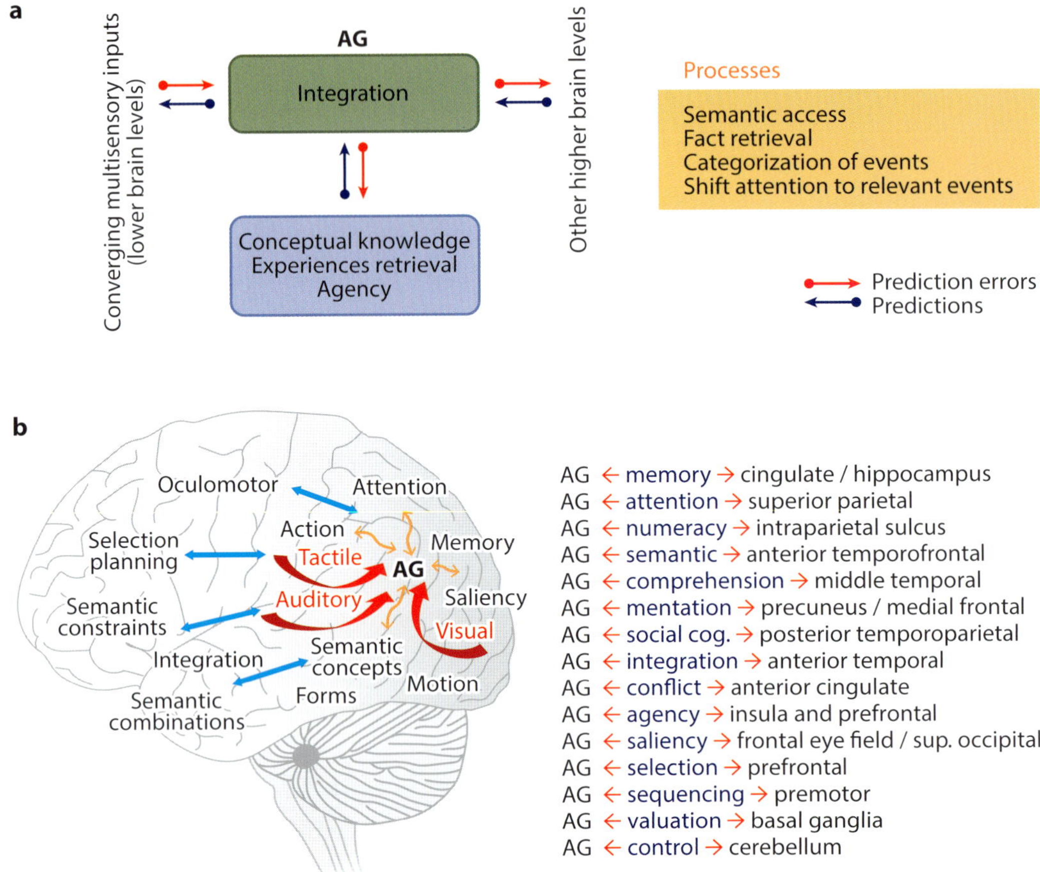

FIGURE 9.4. Angular gyrus and its multiple cognitive functions. A. Seghier's framework that could account for the multiple functions of the angular gyrus (AG).[35] AG is a crossmodal integrative area that is "converging multisensory inputs and integrate them (green box) in a context-dependent fashion." Both bottom-up (red arrows) and top-down (blue arrows) processes contribute in comprehending and reasoning about external events or internal mental representations and results in a set of core processes (orange box) that includes events categorization, semantic access, fact retrieval, and shifting attention to relevant information. B. Schematically illustrates the complex interplay between the AG and other distributed subsystems. Candidate regions that may strongly interact with the AG are listed with the most likely function/domain of interest. This is an oversimplified illustration because each domain/system contains several regions that can interact differently with the AG. Figure 3 from Seghier M. L. (2013), "The Angular Gyrus: Multiple Functions and Multiple Subdivisions," *The Neuroscientist*. 19(1), 43–61.

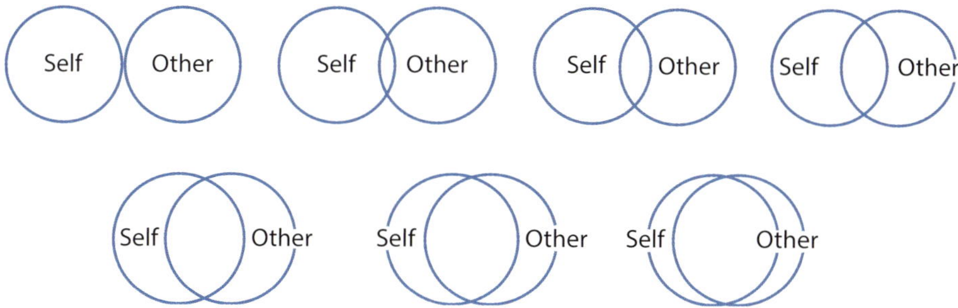

FIGURE 9.5. Inclusion of the other in the self scale. Adapted from Aron et al.[8]

between love and the angular gyrus is fascinating given the multifunctionality of this brain area and its role in self-expansion in space and time (see box 9.2). This finding is in line with the psychological model of self-expansion of love.[7]

According to this self-expansion model, people fall in love with another individual when they detect (consciously and/or automatically) a cognitive opportunity to expand their own self,[7,36] and include their significant other in their cognitive sphere (as indicated in the inclusion of the other in the self scale, figure 9.5).[8] The closer they feel, the more they represent themselves as sharing characteristics, values, traits, and qualities with their significant other.[7,8,37] In 1996, social psychologists Aron and Aron specified this model by suggesting that individuals who are in love seek to enhance their potential self-efficacy by *"increasing the physical and social and cognitive resources, perspectives, and identities that facilitate achievement of any goal that might arise"*[7,37] This self-expansion model of love is also in accord with a universal and key evolutionary purpose of love, which manifests itself in the maintenance and upholding of a species by ensuring the formation of firm bonds between individuals for mutual benefit.[38,39] By highlighting the spatial dimension of love in the human brain, neuroimaging studies of love have identified brain mechanisms that are involved in self-expansion.[7,37]

Interestingly, neuroimaging studies suggest that the duration of a loving relationship correlates with activity in the right insula, right cingulate cortex, and the right posterior cingulate/retrosplenial cortex, but not with activity in the angular gyrus.[29,40,41] Similar findings were obtained in a study interested in the functional connectivity among the areas of the love brain network and the length of time either in love or, for the lovelorn, since breakup of a romantic relationship. Neuroscientist Hongwen Song and colleagues measured the brain activity at rest from healthy college students who completed the Passionate Love

BOX 9.2. The Angular Gyrus and Its Involvement in Self-Expansion

Overlapping brain areas are activated when a person is passionately in love with a partner with whom they have a strong self–other overlap and when they allow their self to expand to another location in time and space. For instance, research shows that seeing oneself from above and feeling like an "outside observer with respect to one's thoughts, feelings, sensations, body, or action" and experiencing illusory own-body perceptions (or out-of-body experiences, OBE) involve the temporoparietal junction (TPJ) extending to the angular gyrus, that is, a brain area that is also related to the extent to which a person in love perceives their self and their partner's to be overlapping.

OBE is defined as "a feeling of spatial separation of the observing self and body" or "experiences in which the sense of self or the center of awareness is felt to be located outside of the body."[62,132,133] In OBE, localization of the psychological self to an extrapersonal space is completely dissociated from the perception of one's body, implementing the "dissociation of egocentric" from "body-centered-perspectives." OBE is characterized by three main components: (1) disembodiment (location of the self outside one's body); (2) the impression of seeing the world from an elevated visuospatial perspective (extracorporeal, but egocentric visuospatial perspective); and (3) the impression of seeing one's own body (autoscopy) from this perspective.[134,135] Two other types of dissociation from one's own body can occur when the individuals are standing or seated. These two forms are (1) autoscopy (from the Greek words *autos* [self], and *skopeo* [looking at]), during which an illusory visual experience consisting of the perception of a second body occurs without any changes in bodily self-consciousness as if one were looking into a mirror; and (2) heautoscopy, during which individuals report seeing a second body (one's double) with whom they feel a strong self-identification, often associated with the experience of existing at and perceiving the world from two places *at the same time*. Because in this case, the double is not a mere image, the person who experiences an heautoscopy may wonder whether it is the physical body or rather the reduplicated body that contains the real self.[136] Not only self-localization and first-person perspective but also self-identification may therefore be experienced as "split in two parts."[62,136]

Throughout the past 15 years, a growing body of work from neurology and neuroscience (using different techniques such as functional magnetic resonance, imaging electroencephalogram, electrical stimulation, and transcranial stimulation) has investigated the neural bases of dissociative states from one's own bodily self.[132–134,137] In brief, research on the neural bases of OBEs suggests a disruption of the following three main brain areas:

1. the temporoparietal junction (notably the right TPJ, including the angular gyrus),
2. the premotor cortex, a key area in the mirror neuron system,[138] and
3. the extrastriate cortex, an area involved in visual processing and attention.

Scale[42] (see appendix B). Three groups of participants were identified: (1) students who were passionately in love with a partner, (2) students who had recently had a loving relationship end and were no longer in love, and (3) students who were single and who had never been in love with a romantic partner. Between-group comparisons showed that the subjects in love had increased functional brain connectivity in the areas of the love brain network (e.g., insula, caudate nucleus, angular gyrus/inferior parietal lobule/temporoparietal junction [TPJ], dorsomedial prefrontal cortex [DMPFC], and anterior cingulate cortex).[42] Song and colleagues also performed correlation analyses between neural patterns and behavior. Their results revealed that the regional homogeneity (ReHo) in the left anterior cingulate cortex was positively correlated with length of time in love in the in-love group, and negatively correlated with the lovelorn duration since breakup. Other brain regions showed a similar pattern.

9.3 Other Types of Love and Biological Drives

Passionate love activates similar associative cortical areas as other types of love, including companionate love (feelings of calm, social comfort, emotional union, and the security felt in the presence of a long-term mate[16]), unconditional love, and parental (maternal and/or paternal) love (see figure 9.2).[26,29] For instance, when female participants view photographs of their own child (vs. a photograph of another child of the same age with whom they were acquainted), increased brain activity is observed in the anterior insula and other brain areas overlapping with activity observed with passionate love.[26] Like passionate love, maternal love recruits cortical brain regions mediating higher-order cognitive or emotional processing, such as the lateral fusiform gyrus, lateral orbitofrontal cortex, and the lateral prefrontal cortex, as well as subcortical dopaminergic-rich areas, such as the caudate nucleus. The main difference is at the subcortical level, such as the periaqueductal gray (PAG, an area involved in pain modulation), which is more strongly activated in maternal love than passionate love. Interestingly, both unconditional love and maternal love elicit activation of the PAG,[26,29] which is not surprising as this brain area is known to contain a high density of vasopressin receptors that are important in suppressing pain and increasing bonding.[43] The cognitive components involved in romantic and maternal love are more similar.

These brain activations are specific to loved ones rather than familiar faces. As discussed in chapter 5, neuroimaging studies on familiar faces (vs. neutral faces) show specific recruitment of brain areas specialized in faces, such as the fusiform face area,[44–46] and familiar faces, such as the precuneus,[47] rather than the love brain network. Moreover, neuroscientist Marie Arsalidou examined brain activity in adults in response to faces of their mothers and fathers compared to faces of celebrities and strangers.[48] Faces of mothers or fathers evoked more activity in core and extended brain regions associated with familiar or salient face processing than faces of celebrities or strangers.[48] Overall, these results complement the standard models of face processing that suggest that familiar faces are processed by a specific distributed network of brain areas, the constituent components of which may vary as a function of the emotional valence or salience of the person.[29]

In one of our studies examining the temporal brain dynamics of attachment mechanisms in a matrilineal society (a society where the children are raised by both their mother and their aunt, which then makes the face of the mother and the aunt equally familiar to the children[49]), we further showed that familiar faces of a mother and an aunt both elicit electrical neuroimaging signatures (i.e., left N170) that are typical of familiar faces. However, a mother's face elicited a set of three larger evoked components (i.e., N100, right N170, P300) compared to those observed in response to an aunt's face.[49] These findings suggest that the emotional attachment between mother and child has neural ramifications across three successive processing stages of face processing that are distinguished from the neural effects of facial familiarity.[49] (See box 9.3 regarding attachment styles.)

Compared to other biological drives, such as sexual desire, love has also a distinctive brain signature as demonstrated in fMRI with a diminished activity in the ventral striatum, hypothalamus, amygdala, somatosensory cortex, and inferior parietal lobule (figure 9.6). Those reductions are in keeping with sexual desire as a motivational state with a very specific, embodied goal, whereas passionate love could be thought of as a more abstract, flexible, and behaviorally complex goal that is less dependent on the physical presence of another person. On the other hand, love is associated with a more intense activation of subcortical areas, such as the ventral tegmental area (VTA) and the right dorsal striatum, two dopamine-rich regions involved in the affect system that generally correspond to motivation, reward expectancy, and habit formation. These findings are in line with hypotheses regarding the importance of specific goal-directed incentives when "head over heels in love."

Both sexual desire and passionate love spark increased activity in the subcortical brain areas that are associated with euphoria, reward, and motivation, as well as in the cortical brain areas that are involved

BOX 9.3. Neuroimaging of Attachment Styles

In 2012, neuroscientists Pascal Vrtička and colleagues reviewed the literature and proposed a functional neuroanatomical framework for the brain mechanisms involved in the perception and regulation of social emotional information, and their modulation by individual differences in terms of secure versus insecure attachment patterns described in the attachment theory, a theory first developed by the joint work of social psychologists John Bowlby and Mary Ainsworth.[142–147] According to the evolutionary attachment theory, attachment corresponds to a preprogrammed need to form deep and lasting emotional bonds with others to survive. While attachment does not need to be reciprocal to take place, it is characterized by specific behaviors between the two individuals who are connected with one another. Attachment style is thought to be developed in childhood and to influence subsequent attachment development in adulthood. In their review of the literature,[148] Vrtička et al. describe (1) the modulations within and beyond the limbic system that take place when an attachment style influences the encoding of approach vs. aversion behaviors during social interaction, and (2) how these neural modulations influence other brain networks sustaining more complex, cognitive processes. For instance, the authors describe how affective evaluation is decreased in anxiously attached individuals, and how mental state representations are enhanced with attachment insecurity and particularly anxiety.

in self-representation and social cognition.[27] The co-activation of subcortical emotion-related areas and higher-order cortical areas that mediate more complex cognitive functions (e.g., body image, mental associations, and self-representation) reinforces the top-down neurofunctional model of interpersonal relationships and the potential role of past experiences on future emotional feelings and behaviors.

Interestingly, neural differences also exist between desire and love. Notably at the subcortical level, the posterior-to-anterior insula pattern, from desire to love, suggests that love is a more abstract representation of the pleasant sensorimotor experiences than desire. The anterior part of the insula is activated significantly by feelings of love, whereas the posterior part of the insula is activated significantly by feelings of sexual desire. This posterior-to-anterior insular distinction between sexual desire and love again reinforces the neurofunctional characteristic of a posterior-to-anterior progression of integrative representations of affective bodily feelings to an ultimate representation of all feelings. This is in line with the view that love is an abstract construct, which is partly based on the mental representation of repeated past emotional

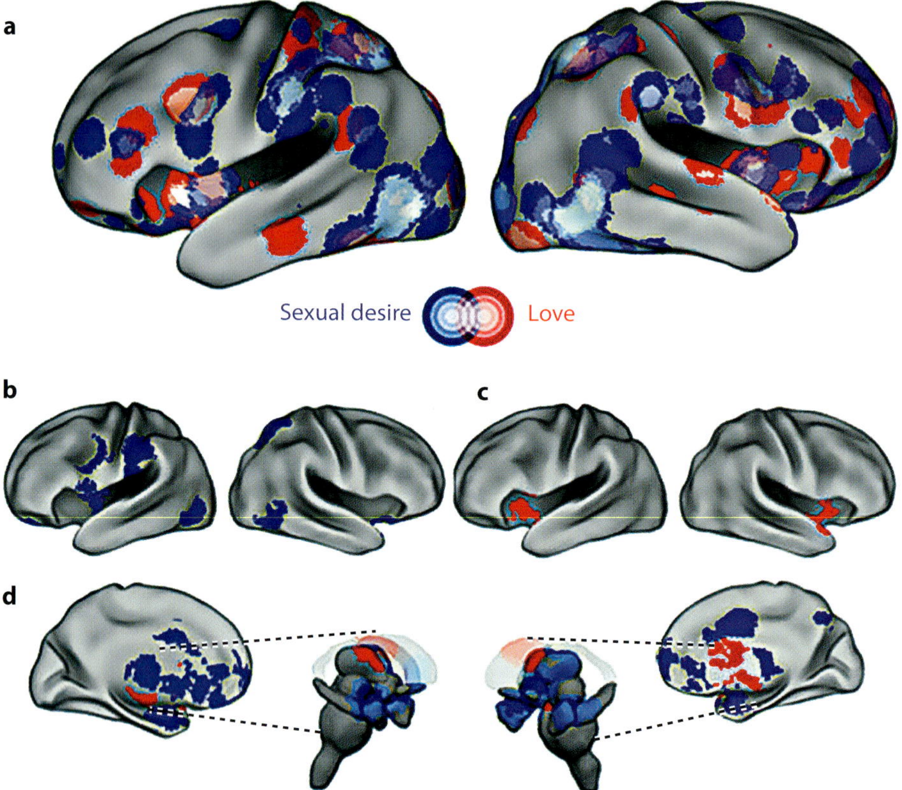

FIGURE 9.6. Neural comparison between love (red) and sexual desire (blue). A. Surface model illustrating results from all studies based on a qualitative analysis. The color hues in the "heat maps" depict an increasing number of paradigms that activated a given portion of cortex. B. Lateral view of regions uniquely activated by desire based on the quantitative multilevel kernel density analysis. C. Regions uniquely activated by love. D. Medial view and brain stem view of regions uniquely activated by desire vs. love. Adapted from Cacioppo et al.[27]

moments with another. In other words, this specific pattern of activation suggests that love builds upon a neural circuit for emotions and pleasure, adding regions associated with reward expectancy, habit formation, and feature detection. In particular, the shared activation within the insula, with a posterior-to-anterior pattern, from desire to love, suggests that passionate love grows out of and is a more abstract representation of the pleasant sensorimotor experiences that characterize desire. From these results, one may consider sexual desire and love on a spectrum that evolves from integrative representations of affective visceral sensations to an ultimate representation of feelings incorporating mechanisms of reward expectancy and habit learning. Although love is not a prerequisite for sexual desire, these recent neu-

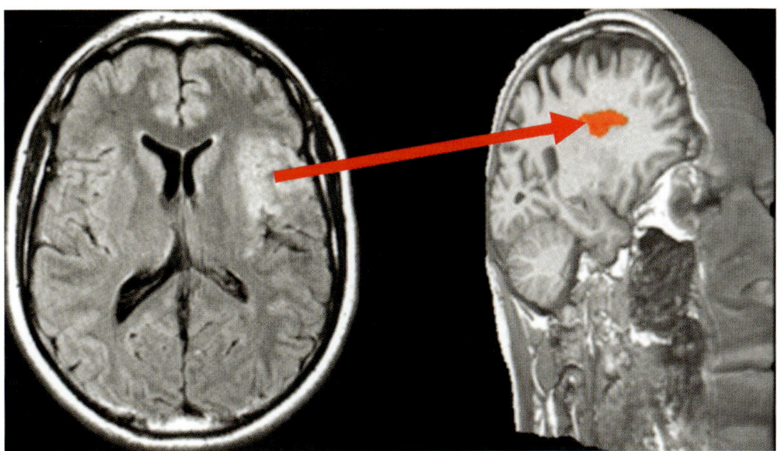

FIGURE 9.7. Patient with lesion in the anterior insula. Brain reconstruction of brain ischemic lesion located in the anterior insula (in red). Adapted from S. Cacioppo et al.[50]

roimaging meta-analyses[27–29] suggest that desire might be a prerequisite for love based on the researchers' interpretation of their results that desire is grounded in a relatively concrete representation of sensorimotor experiences, whereas love is a more abstract representation of those experiences in the context of a person's current and prior experiences. If this interpretation is correct, then a lesion in the anterior (or posterior) insula should be associated with a diminished capacity to ignite normal responses to love (or desire) in particular.

In 2013, we had the rare opportunity to test a 48-year-old heterosexual man from Argentina who suffered from a circumscribed ischemic lesion in the anterior insula (see figure 9.7).[50] Although the patient did not report any changes in feelings of love or desire, behavioral testing revealed a selective deficit for love (but not desire). The patient was not aware of any changes in his capacity for love, but sophisticated behavioral testing revealed he had a selective deficit—as compared to a control group—when making judgments about love. For instance, the patient was slower than a matched control group when making decisions about his feelings of love, but he was comparable to the control group when making decisions about lust. From a purely anecdotal (and nonscientific) perspective, the patient and his spouse divorced following his stroke.

These findings provide the first clinical (and causal, rather than correlational) evidence that the anterior insula contributes to love but not desire. The specific interference of the anterior insula damage on the speed of judgments of love (but not desire) further suggests that the anterior insula may play a role not only in concrete feelings but also in more abstract forms of human emotions. These data also support the notion of a posterior-to-anterior insular gradient, from sensorimotor to abstract representations, in the evaluation of anticipatory rewards in interpersonal relationships.[50]

9.4 The Speed of Love in the Human Brain

To understand brain function, it is important not only to specify what brain areas are recruited during a behavioral task, but also to specify when and in what specific combinations they are activated.[51–54] By providing detailed information about the relationship between neuronal activity (i.e., postsynaptic dendritic potentials of a considerable number of neurons that are activated in a pattern that yields a dipolar field) and the temporal resolution (millisecond by millisecond) of each component information-processing operation required for behavioral performance, high-density electroencephalogram (EEG) recordings and averaged EEG (event-related potentials, ERPs) have provided a useful additional tool in investigations of brain function. Whereas fMRI analyses are performed in source space, EEG/ERP analyses are usually performed in sensor space, with high-density sensor recordings producing more detailed information about changes in brain activity measured across time and sensor space. To date, only a few neuroscientists have investigated the spatiotemporal dynamics of love, however.[29]

The first modern-day neuroscientists to study passionate love were Niels Birbaumer and his Tübingen colleagues.[55] They performed a series of electrophysiological surface recordings from 15 different locations on the scalp of healthy participants. Participants' brain electrical activity was recorded during love-related imaging tasks (imagining a time in their past in which they had been joyously in love [without sexual imagery] and imagining the same scene [with sexual imagery]) compared to sensory control tasks.[55] Their results suggested the frontal and posterior parts of the electrode sites showed similar dimensions on the romantic imagery tasks, whereas smaller dimensions were found in the frontal as compared to the posterior electrode sites on the four sensory tasks. The authors then concluded that passionate imagery involves a significantly higher brain complexity than does sensory stimulation at all brain sites, but particularly at frontal regions.[55] Although these electrophysiological findings shed light on the temporal mechanisms of love, they don't provide enough high spatiotemporal resolution because of the limited method and number of electrodes available at that time.

Since this initial investigation by Birbaumer, there have been significant developments and refinements in spatial sampling (e.g., from 15 electrodes to current systems consisting of 128–256 electrodes)[56,57] and in quantitative techniques for investigating brain state dynamics (e.g., waveform analyses, Fourier analysis, independent component analysis, principal component analysis, *k*-means cluster analyses, high-performance electrical neuroimaging).[51,57]

In 2008, Başar et al.[58] investigated the oscillatory brain dynamics of love using facial stimuli of a "loved person" in women. Results showed that a specific frequency band (i.e., delta band) and a high amplitude generated by the brain were evoked by the photo of a "loved person" compared with an "unknown person," and with the picture of the "appreciated person."[58] Vico and colleagues[59] tested female undergraduate students while they were viewing black-and-white photographs of faces that belonged to one of five categories: loved ones, famous people (preselected by the participants), unknown people, babies from the International Affective Picture System, and neutral faces from the Ekman and Friesen system. Subcategories of loved faces included romantic partner, parents, siblings, second-degree relatives, and friends. Participants were informed that the purpose of the study was to examine physiological responses to familiar faces. One of the selection criteria was that participants were required to have a current romantic relationship and to reside in close proximity to five loved ones, including the partner, so as to be able to take their photograph.[59]

Heart rate, skin conductance, electromyography of the zygomatic muscle, and ERPs were recorded while participants passively viewed the pictures of their loved ones and control faces. Results indicated that both central and peripheral electrophysiological measures differentiated faces of loved ones from all other categories, and specifically showing faces of loved ones elicited higher heart rate, skin conductance, and zygomatic activity, as well as larger amplitudes of the late ERP components P3 and LPP.[59]

Over the years, some have argued that measuring peaks and troughs was sufficient for gauging the temporal processing in the brain, while others[60] argued that another approach (e.g., a statistical decomposition of the evoked brain response) was necessary. Using a microstate decomposition of the ERP, we analyzed high-density brain activity recorded from healthy participants while they were performing a cognitive priming paradigm known to activate the love brain network.[53] Our results showed that when a person is passionately in love, the subliminal presentation of their beloved's name evokes specific brain states that are mediated by generators located in the pleasure, reward, and cognitive brain pathways very quickly (within 200 milliseconds) following the onset of the stimulus. Estimation of the brain generators underlying this data set suggested that visual areas were activated first, followed by activation of higher-order associative brain areas, such as those involved in self-related processes (e.g., the angular gyrus/temporoparietal junction). Finally, a flow of backward activation occurred from these associative brain areas to the primary visual and emotional brain areas. These results reinforce the neurofunctional top-down model of love, suggesting that associative brain areas (such as the angular gyrus) may prime more basic brain areas at a preconscious stage and detect significant others and differentiate them from strangers.

9.5 Romantic Rejection

While the perceptions of social isolation, ostracism, and romantic rejection fall under the broad rubric of social disconnection or exclusion, *being* socially rejected by a significant other recruits different brain areas than those associated with *feeling* socially isolated or lonely (see chapter 3) or with being rejected by strangers. In other words, feeling dissociated from a romantic partner, in contrast to being rejected by a stranger, are associated with overlapping but not identical regions of brain activation.[61] Romantic rejection causes a profound distress and a sense of loss of self that may be accompanied by signs and symptoms associated with self dissociation,[62] including suicide and depression.[63] For instance, sociologist Jack Mearns showed that 40% of men and women who had been rejected by a partner within the past eight weeks reported clinically measurable depression, and 12% displayed moderate to severe depression.[63,64] A meta-analysis of the brain regions activated when reliving a recent and unwanted breakup with a romantic partner indicated activation in brain areas involved in passionate love, rumination, emotion regulation, and self-awareness. Specifically, reliving rejection by a significant other revealed four main brain areas: the right caudate nucleus, the right anterior insula, the right anterior cingulate cortex, and the left inferior orbitofrontal cortex.[65]

When participants are asked to perform a computerized cyberball task with strangers,[66,67] the meta-analysis revealed that social rejection during the task was associated with activation in the anterior insula (bilaterally), the left anterior cingulate cortex, and the left inferior orbitofrontal cortex, but not the caudate nucleus.[65]

The dorsal part of the anterior cingulate, which has been often identified as a core region in both theoretical analyses and narrative reviews of this literature as a sign that social pain shares similar neural bases[68–70] with physical pain, was not related to rejection by a loved one or by strangers.[65] Instead, the region of the anterior cingulate associated with rejection by a loved one and by strangers has been associated with violation of expectancies rather than pain.[71] Rejection during a task by strangers, therefore, may be unexpected rather than physically painful and operate on attentional mechanisms.

In 2014, neuroscientist Choong-Wan Woo and colleagues also challenged the assumption that social rejection and physical pain share common neural mechanisms.[72] In their first neuroimaging study, Woo and colleagues recorded brain activity from participants who were exposed to (1) physical pain (via administration of painful heat ["Heat-pain" condition] or warm heat ["Warmth" condition] to the left volar forearm), and (2) social rejection (via the viewing of a head shot photograph of their ex-partner ["Ex-partner" condition] or a close friend ["Friend" condition]). All participants had recently experienced an unwanted breakup with a romantic partner and felt intensely rejected. Their results showed that social rejection was associated with activation in the right inferior frontal gyrus, ventromedial prefrontal cortex, perigenual anterior cingulate cortex, temporoparietal junction, dorsomedial prefrontal cortex, and precuneus—areas associated with self-related cognition and mentalizing—while physical pain was associated with activation in the periaqueductal gray (PAG), supramarginal gyrus, middle and dorsal posterior and ventral insula, amygdala, and thalamus—areas associated with sensory and emotional aspects of pain. These and related findings also indicated that physical pain and social pain are distinct types of sensation.[72]

9.6 Cognitive Benefits of Love

Love not only involves cognitive as well as affective processes, but there are also cognitive benefits of love. The popular notion that love diminishes a person's ability to focus generally may not be accurate. As discussed in chapter 2, being married and having salutary relationships is associated with better physical and mental health and more positive affective states. Stephanie Cacioppo and colleagues[73–75] showed that embodied cognition is improved in couples who are passionately in love and who have a strong sense of self–other overlap. Love makes it easier to connect with the beloved emotionally (e.g., rapport), cognitively (e.g., people you love are easier to read), and behaviorally (e.g., action coordination).

9.6.1 Benefits of Love on Embodied Cognition

In chapter 4, we discussed how expertise in a specific domain can facilitate embodied cognition in that domain. A similar facilitation effect can occur in a love relationship. For instance, the more members of a couple feel close to one another, the better and/or faster they anticipate one another's actions and intentions.[73–75] A significant other is also often easier to read and predict feelings, actions, desires, or intentions.[76,77] This increased facility makes it simpler to align interests, goals, and behaviors within the couple, thereby promoting cooperation and behaviors that promote mutual benefit (e.g., bi-parental care).[76,77] Moreover, an individual in a love relationship does not need to be physically in the presence of the partner to perform better at a task or to successfully anticipate the action and intentions of their partner vs. a stranger.[73–75] For instance, when individuals who are in love with a partner watch video clips of actions performed either by themselves, their beloved partner, or strangers (as controls), they are significantly faster to identify their own intentions and the intentions of their significant other than they are to identify the intentions of a stranger. Further, the higher a participant scores on the Passionate Love Scale[18] (see appendix B), the faster the individual is to understand their partner's actions and intentions.[73–75] These findings suggest that facilitation effects rely on shared action representations.

The working model of simulation mentalizing is that the anticipatory representations of a significant other's behaviors require internal predictive models of actions formed from pre-established, shared representations between the observer and the actor. This model suggests that observers can be better at predicting intentions performed by a significant other whom they feel close to, rather than a stranger. Although being in love is not a prerequisite for effective simulation mentalizing (see chapter 4 for facilitation effects among expert athletes), loving the person who is acting and seeing him or her as an extension of one's own self facilitates this automatic matching mechanism, even at a distance.[74,78] Indeed, theories on embodied cognition suggest that automatic reenactment of past shared motor experiences (rather than the agents' perceptual familiarity) facilitates action and intention understanding.[74,79] In line with this, people are better at recognizing a friend's walk than a stranger's walk, even in dark and without any familiarity cues.[79] In this example, kinematic cues from biological motion and a shared mental representation of a friend's walk are sufficient to provide implicit information about the person's identity despite the lack of existence of a definite recognition cue for individual identification.[79–81] In addition, the self-expansion theory of intention understanding suggests that the more people tend to see themselves in their significant other (or vice versa), the stronger their shared mental representation of one another is.

Neuroscientist Stephanie Cacioppo and colleagues recently tested the extent to which the mirror neuron system (discussed in chapter 4) is involved. Using fMRI, couples in love who had a strong self–other overlap (as gauged by the inclusion-of-the-other-in-the-self scale[7,8]) were scanned while they were completing a computerized behavioral intention task in which they predicted the intentions of different actors.[73] The results showed that participants were better at inferring intentions performed by themselves and their significant other than those performed by nonfamiliar actors. Interestingly, this better performance was associated with greater activation in the mirror neuron system. In addition, the more the participants reported being cognitively close to their partner on the inclusion-of-the-other-in-the-self scale,[8] the less the brain areas associated with action self–other comparison (e.g., inferior parietal lobule), attention (e.g., superior parietal lobule), recollection (hippocampus), and pair bond (ventral tegmental area) were recruited. That is, the more the observer and the actor share a common mental map (self–other overlap) of their actions, desires, or intentions, the smaller the differences in neural responses to one's own actions and the actions of the other person.

9.6.2 Benefits of Love on Social Cognition

Being passionately in love with someone promotes efficient cognitive processing in various situations, including situations that are not directly related to the significant other. Recently, psychologist Rafael Wlodarski and colleagues demonstrated that priming participants with the name and face of their beloved just before asking them to look at pictures of strangers and try to read the minds of these strangers just by looking at their eyes (see chapter 6 for an example[77]) extended the facilitation effect from the beloved to the stranger. Specifically, participants who are "truly, madly, deeply in love" with their current romantic partner were better at interpreting the emotional states of a stranger when the face and name of their beloved was presented for 45 seconds prior to the presentation of the face of the stranger.[77] Models of conceptual priming explain these effects by positing that the love prime

TABLE 9.2. Features of romantic love similar to some features of obsessive compulsive behaviors

Feature	Romantic love
Altered emotional state	++++
Idealization of the "SO"* as being perfect	++++
Longing for reciprocity	+++
Positive image of the future	+++
Worries of not being adequate or worthy	+++
Need to check SO is safe and secure	+++

Source: From Feygin, Swain, and Leckman[89]

* SO = Significant Other

reactivates a mental map or schema or memories that are directly or indirectly relevant to that stimulus, and in turn, these effects can apply to individuals beyond the romantic partner.[77,83,84] In other words, facilitation priming effects arise not only when the relation between prime and target is a perceptual one, but also when it is a purely conceptual one.[33,83–86] Conceptual priming plays a particular role in individuals in love, especially when they devote a significant amount of their time thinking about their significant other—a main feature of romantic love (see table 9.2).[33,84,87–90]

Similar facilitation priming effects and cognitive enhancement have been found when the love prime is subliminal (without the subjects being aware of it). For instance, when a word is presented immediately after the preconscious presentation of the name of a beloved, the decision maker may understand this word faster, even if the word is not directly related to their beloved.[53,84] This facilitation effect of love on cognition is associated with the level of activity in the angular gyrus, a brain area that is involved not only in self–other overlap but also in language, metaphors, spatial attention, and numbers, as well as the management of autobiographical data such as a person's self-image.

Passionate love is typically associated with positive affect, which has been found to enhance various aspects of cognition. For instance, social psychologist Alice Isen provided experimental evidence that positive affect facilitates problem solving, decision making, and innovation in various situations and populations.[91,92] Isen noted:

> as long as the situation is one that is either interesting or important to the decision maker, positive affect facilitates systematic, careful, cognitive processing, tending to make it both more efficient and more thorough, as well as more flexible and innovative.[91] (p. 75)

Isen went on to note that the cognitive effects of positive affect have various implications from customers' behavior and satisfaction, to employees' satisfaction, and even to doctor-patient interaction, medical decision making, and medical consumer satisfaction.[91] Consistent with the real world relevance of these effects, Isen's work on negotiations showed that positive affect facilitates the bargaining process and improves the outcomes of face-to-face negotiators trying to reach agreement. In another task involving decision making, people in whom positive affect had been introduced were faster and more efficient and thoughtful in the way they selected a car among six for purchase (based on information from nine dimensions).[92] Her body of work shows that the beneficial cognitive effect of positive affect, like love, may lead to altruism and interpersonal understanding, improved creativity, cognitive flexibility, and openness to information.[91]

9.7 Concluding Remarks

Throughout the centuries, passionate romantic love, which is defined here as "a state of intense longing for union with another,"[23,93] has often been regarded as little more than a "natural addiction"[87] with little redeeming value beyond reproduction. In favor of this love-addiction hypothesis, some individuals in the early stage of intense passionate love show behaviors and emotions that parallel symptoms of drug addiction (cocaine, heroin, alcohol, nicotine) or behavioral addiction (gambling, sex, food[94,95]), such as euphoria, craving, dependence, withdrawal and relapse, obsessive thinking, compulsive behaviors, and reduced inhibitory control in some conditions.[87,96 but see also 97]

At the heart of this love-addiction hypothesis lies the mesolimbic dopaminergic system—a brain network central to the "addiction brain network"[98–102] that is typically associated with increased arousal, motivation, reward, pleasure, and the detection of salient incentives.[98,100] Functional neuroimaging studies of drug intoxication or craving have shown that food, sex, and drugs increase activity in the mesolimbic system and its connected areas (e.g., nucleus accumbens, ventral tegmental area, orbitofrontal cortex, hippocampus) in a similar way as observed in response to a photograph of a loved one. Also in accord with the love-addiction hypothesis, neurochemical studies have shown that large and fast increases in the "feel-good" dopamine neurotransmitter are associated with the reinforcing effects of drugs of abuse[100] and in the formation of pair bonds, especially in the nucleus accumbens.[103]

However, recent developments in the neuroimaging of love and addiction have identified several important differences between the two.

For instance, while love is associated with increased functional connectivity within the dopaminergic reward/motivation system,[42] addiction correlates with a hypodopaminergic dysfunctional state within the reward circuitry of the brain.[104] For instance, Hong and colleagues showed that adolescents with Internet addiction have a reduced functional connectivity and impaired connections within cortico-subcortical circuits, including the prefrontal (~24%) and parietal (~27%) cortices.[105]

Moreover, the neuroimaging of love indicates that love recruits more than the putative addictive brain network.[29] Love activates three main systems: (1) the affect system, (2) the reward/motivation system, and (3) the cognitive system. The fact that these three complex systems work together in response to a beloved suggests that love is more than an instinct or emotional response. In addition, being passionately and romantically in love may serve multiple functions—not only to connect people emotionally (e.g., increasing motivation for bi-parental caregiving) but also cognitively (e.g., people you love are easier to read), and behaviorally (e.g., together, you and your beloved can coordinate actions better and more efficiently).

APPENDIX A

THE CACIOPPO EVOLUTIONARY THEORY OF LONELINESS (ETL)

Premise: An Evolutionary Classification of Social Behaviors. Evolutionary fitness refers to the probability that the line of descent from an individual with a specific trait will remain or increase in the population. As such, evolutionary fitness is a statistical rather than a psychological construct. A social behavior can be characterized according to the fitness consequences for the actor (and genetically related individuals) in terms of benefits (b_a) and costs (c_a), and according to the fitness consequences for the recipient in terms of benefits (b_r) and costs (c_r) (Gardner and West, 2006). The social behaviors expressed in an interaction can be categorized according to the fitness consequences for the actor and its social partners as one of the following: (1) *selfishness*—the actor benefits at a cost to the recipient, (2) *mutual benefit*—both the actor and the recipient benefit, (3) *altruism*—the recipient benefits at a cost to the actor, and (4) *spite*—both actor and recipient suffer a loss.

		Fitness Effect for Actor	
		$b_a > c_a$	$b_a < c_a$
Fitness Effect for Recipient	$b_r < c_r$	Selfishness $b_a > c_a$, $b_r < c_r$	Spite $b_a < c_a$, $b_r < c_r$
	$b_r > c_r$	Mutual Benefit $b_a > c_a$, $b_r > c_r$	Altruism $b_a < c_a$, $b_r > c_r$

Salutary Relationship Postulate. The Cacioppo Evolutionary Theory of Loneliness (ETL) posits that beneficial social interactions and reliable social relationships can contribute to the likelihood of survival, reproduction, and consequent genetic legacy. Not all social interactions or relationships are salutary, however; some can decrease the likelihood of survival or reproductive success, for instance, through increased exposure to violence or conflict, competition for limited resources including food and mates, exposure to infectious diseases, and risks for oppression and exploitation. Social interactions and relationships can range from hospitable to hostile, and the same objective relationship (e.g., sibling) can prove to be caring and protective or threatening and corrosive. Moreover, the nature of a given social relationship in primates can change quickly over time. Survival depends not only on the ability to curry the *presence* of conspecifics, but on the discrimination of foe from friend on a continuing basis. According to the salutary relationship postulate, the formation and maintenance of salutary social relationships increases evolutionary fitness, and the brain has evolved as the key organ for differentiating friend from foe and for the formation, maintenance, and repair or replacement of social relationships that satisfy the need for salutary social connection.

Self-Preservation Postulate. The perception by the brain that the organism lacks sufficient salutary social relationships (i.e., loneliness) has been associated over evolutionary time with a decreased likelihood of encountering social behaviors categorized in terms of evolutionary fitness as mutual benefit or altruism, and an increased likelihood of encountering social behaviors categorized as selfish or spiteful. These conditions increase attention to and motivation for self-preservation. Self-preservation is used here not in reference to an explicit (i.e., conscious) goal, but in reference to a probabilistic outcome of a behavioral predisposition orchestrated by the brain in part through nonconscious processes.

Conservation Postulate. The perception of a shift in the fitness consequences of behavior is posited to be evolutionarily old. Consequently, loneliness in the ETL is posited to operate in humans in part through nonconscious processes.

Aversive Signal Postulate. Among the processes triggered by loneliness is an aversive response. The evolutionary heritage of humans has shaped the brain to incline individuals toward certain ways of feeling, thinking, and acting. The aversion to physical pain is illustrative of a variety of conserved biological mechanisms that have evolved that capitalize on aversive signals to motivate behaviors that promote our chances of short-term survival. The aversion of loneliness is a part of the biological warning machinery that alerts us to threats to our social body and to our evolutionary fitness.

Repair/Replacement Postulate. The aversiveness of loneliness not only serves as a biological warning signal that alerts an individual to potential damage to the social body, it motivates the individual to repair or replace perceived deficiencies in one or more important salutary social relationships. The ETL, therefore, posits that loneliness increases the motivation to attend to and approach social stimuli to repair/replace the salutary social relationships that promote long-term evolutionary fitness.

Implicit Vigilance Postulate. As specified in the self-preservation postulate, loneliness has been associated over evolutionary time with environments or social behaviors in which the likelihood of mutual aid or altruism is relatively low and the likelihood and costs of betrayal are relatively high. The unfettered motivation to approach and interact with others could therefore prove fatal. Consequently, loneliness is posited to also trigger an increase in implicit vigilance for and avoidance of social threats. Moreover, in addition to the motivation to attend to and approach social stimuli posited by the repair/replacement postulate, the implicit vigilance postulate specifies an increase in motivation to avoid others, especially as interpersonal distance decreases and the potential for evolutionary costs increase. In sum, processes specified by the implicit vigilance postulate evolved to promote short-term survival, whereas those specified by the repair/replacement postulate evolved to promote survival over a longer term.

Selfishness Postulate. When the likelihood is low (or uncertainty is high) that social interactions are characterized by mutual benefit or altruism, evolutionary fitness favors an emphasis on selfishness or, in limited cases, spite (e.g., hostility). Consequently, loneliness is posited to increase responses that reflect concern for one's own interests and welfare. As is the case for the processes specified in the ETL postulates, the increased focus on self-centeredness is posited to be evolutionarily old and conserved, with the potential to operate through nonconscious processes. Consequently, people's conscious perceptions—such as being unselfish or having little influence or control over their interpersonal interactions or social relationships—cannot be assumed to be accurate.

Automatic Preparatory Adjustments Postulate. In the absence of reliable mutual aid or protection, the brain initiates a set of interrelated behavioral, neural, hormonal, cellular, and transcriptomic adjustments that promote short-term survival. Illustrative processes are addressed in section 2.4.1, "Theoretical pathways linking loneliness to mortality in the modern world."

Cumulative Deleterious Effects Postulate. These preparatory adjustments and responses may serve short-term self-preservation, but the long-term effects of these adjustments can have deleterious consequences for health and well-being across the life span.

Life span Postulate. These deleterious effects are especially evident in later life with the degradation of physiological resilience and compensatory physiological mechanisms. Consequently, the long-term costs of loneliness may be more apparent now than in any other period of human history due to factors including the rapidly rising proportion of older adults in industrialized nations and the relatively low biological resilience in older adults that results from the wear and tear of aging.

APPENDIX B

THE PASSIONATE LOVE SCALE

Elaine Hatfield and Susan Sprecher
University of Hawaii and Illinois State University

Purpose

Many classifications and typologies of love exist in the literature, but the most common distinction is between passionate love and companionate love. Hatfield and Walster (1978) described passionate love as:

> A state of intense longing for union with another. Reciprocated love (union with the other) is associated with fulfillment and ecstasy; unrequited love (separation) is associated with emptiness, anxiety, or despair. (p. 9)

In 1986, Hatfield and Sprecher published the Passionate Love Scale (PLS) for the purpose of promoting more research on this intense type of love. Although a companion scale to measure companionate love was not also developed by this team of researchers, other measures exist in the literature designed to assess this type of love (see, for example, Grote and Frieze's [1994] Friendship-Based Love Scale).

Description

The PLS scale was specifically designed to assess the cognitive, emotional, and behavioral components of passionate love. The *cognitive components* consist of intrusive thinking, preoccupation with the partner, idealization of the other or of the relationship, and desire to know the other and be known by him/her. *Emotional components* consist of attraction to the partner, especially sexual attraction; positive feelings when things go well; negative feelings when things go awry; longing for reciprocity (passionate lovers not only love but want to be loved in return); desire for complete and permanent union; and physiological (sexual) arousal. Finally, *behavioral components* consist of actions aimed at determining the other's feelings, studying the other person, service to the other, and maintaining physical closeness.

The most common form of the PLS is a 15-item scale, but an alternative 15-item version is also available. The two scales can be combined to form a 30-item scale. Although the scale was originally designed using North American young adults in pilot studies, the scale has subsequently been revised to be administered to children, and has been translated into many languages and administered to samples in other countries.

Response Mode and Timing

Participants are presented with statements such as: "I would feel deep despair if ____ left me" and are asked to indicate how true the statement is of them. Possible responses range from 1 = not at all true to 9 = definitely true. (The ____ in each statement refers to the partner.) The scale takes only a few minutes to complete, although often it is embedded in a larger questionnaire with other measures.

Scoring

The total score of the scale can be represented by either the mean of the scores for the items or by the sum of the ratings. Higher scores indicate greater passionate love. An average score for young adults across the items is approximately 7. Recently, for a popular press article, Hatfield and Sprecher (2004) provided for readers the following rubric to interpret their summed scores across 15 items:
106–135 points = Wildly, recklessly, in love.
86–105 points = Passionate but less intense.
66–85 points = Occasional bursts of passion.
45–65 points = Tepid, infrequent, passion.
15–44 points = The thrill is gone.

Reliability

Hatfield and Sprecher (1986) reported a coefficient alpha of 0.91 for the 15-item version and 0.94 for the 30-item version. Others have also reported high levels of reliability for the scale (e.g., Sprecher and Regan, 1998). The PLS appears to be primarily unidimensional, with one primary factor emerging from a principal components factoring.

Validity

The scale is uncontaminated by a social desirability bias, as indicated by a non-significant correlation between respondents' scores on the PLS and their scores on the 1964 Crowne and Marlowe *Social Desirability Scale* (Hatfield and Sprecher, 1986). There is some evidence for the construct validity of the PLS. For example, it has been found to be associated positively with conceptually similar scales and measures (Aron and Henkemeyer, 1995; Hatfield and Sprecher, 1986; Hendrick and Hendrick, 1989; Sprecher and Regan, 1998).

Other Information

Researchers have used the PLS in exploring many different topics, including cross-cultural differences in passionate love (Hatfield and Rapson, 2005; Hatfield, Rapson, and Martel, 2007; Landis and O'Shea, 2000), prototype approaches to love (Fehr, 2005), neural bases of passionate love (Aron et al., 2005; Bartels and Zeki, 2004), changes in passionate love over the family life cycle (Tucker and Aron, 1993), correlates of sexual desire (Beck, Bozman, and Qualtrough, 1991), the effects of an emotionally focused couples therapy (James, 2007), degree of bonding with an abusive partner (Graham et al., 1995), and the effects of having married couples engage in novel activities (Aron, Norman, Aron, McKenna, and Heyman, 2000).

The PLS is copyrighted by Hatfield and Sprecher (1986). Permission is given to all clinicians and researchers who wish to use the scale in their research (free of charge).

Passionate Love Scale (Version A)

We would like to know how you feel (or once felt) about the person you love, or have loved, most *passionately*. Some common terms for passionate love are romantic love, infatuation, love sickness, or obsessive love.

Please think of the person whom you love most passionately *right now*. If you are not in love, please think of the last person you loved. If you have never been in love, think of the person you came closest to caring for in that way.

Try to describe the way you felt when your feelings were most intense. Answers range from (1) Not at all true to (9) Definitely true.

Whom are you thinking of?

- Someone I love *right now*.
- Someone I *once* loved.
- I have never been in love.

	Not true								Definitely true
I would feel deep despair if _____ left me.	1	2	3	4	5	6	7	8	9
Sometimes I feel I can't control my thoughts; they are obsessively on _____.	1	2	3	4	5	6	7	8	9
I feel happy when I am doing something to make _____ happy.	1	2	3	4	5	6	7	8	9
I would rather be with _____ than anyone else.	1	2	3	4	5	6	7	8	9
I'd get jealous if I thought _____ were falling in love with someone else.	1	2	3	4	5	6	7	8	9
I yearn to know all about _____.	1	2	3	4	5	6	7	8	9
I want _____ physically, emotionally, mentally.	1	2	3	4	5	6	7	8	9
I have an endless appetite for affection from _____.	1	2	3	4	5	6	7	8	9
For me, _____ is the perfect romantic partner.	1	2	3	4	5	6	7	8	9

	Not true				Definitely true				
I sense my body responding when _____ touches me.	1	2	3	4	5	6	7	8	9
_____ always seems to be on my mind.	1	2	3	4	5	6	7	8	9
I want _____ to know me—my thoughts, my fears, and my hopes.	1	2	3	4	5	6	7	8	9
I eagerly look for signs indicating _____'s desire for me.	1	2	3	4	5	6	7	8	9
I possess a powerful attraction for _____.	1	2	3	4	5	6	7	8	9
I get extremely depressed when things don't go right in my relationship with _____.	1	2	3	4	5	6	7	8	9
	Total: _______								

Results:

- 106–135 points = Wildly, even recklessly, in love.
- 86–105 points = Passionate, but less intense.
- 66–85 points = Occasional bursts of passion.
- 45–65 points = Tepid, infrequent passion.
- 15–44 points = The thrill is gone.

The Passionate Love Scale (Version B)

We would like to know how you feel (or once felt) about the person you love, or have loved, most *passionately*. Some common terms for passionate love are romantic love, infatuation, love sickness, or obsessive love.

Please think of the person whom you love most passionately *right now*. If you are not in love, please think of the last person you loved. If you have never been in love, think of the person you came closest to caring for in that way.

Try to describe the way you felt when your feelings were most intense. Answers range from (1) Not at all true to (9) Definitely true.

Whom are you thinking of?

- Someone I love *right now*.
- Someone I *once* loved.
- I have never been in love.

	Not true Definitely true
Since I've been involved with ____, my emotions have been on a roller coaster.	1 2 3 4 5 6 7 8 9
Sometimes my body trembles with excitement at the sight of _____.	1 2 3 4 5 6 7 8 9
I take delight in studying the movements and angles of _____'s body.	1 2 3 4 5 6 7 8 9
No one else could love _____ like I do.	1 2 3 4 5 6 7 8 9
I will love _____ forever.	1 2 3 4 5 6 7 8 9
I melt when looking deeply into ____'s eyes.	1 2 3 4 5 6 7 8 9
_____ is the person who can make me feel happiest.	1 2 3 4 5 6 7 8 9
I feel tender toward _____.	1 2 3 4 5 6 7 8 9
If I were separated from ____ for a long time, I would feel intensely lonely.	1 2 3 4 5 6 7 8 9
I sometimes find it difficult to concentrate on work because thoughts of _____ occupy my mind.	1 2 3 4 5 6 7 8 9

	Not true Definitely true
Knowing that _____ cares about me makes me feel complete.	1 2 3 4 5 6 7 8 9
If _____ were going through a difficult time, I would put away my own concerns to help him/her out.	1 2 3 4 5 6 7 8 9
_____ can make me feel effervescent and bubbly.	1 2 3 4 5 6 7 8 9
In the presence of _____, I yearn to touch and be touched.	1 2 3 4 5 6 7 8 9
An existence without _____ would be dark and dismal.	1 2 3 4 5 6 7 8 9
	Total: ______

Results:

- 106–135 points = Wildly, even recklessly, in love.
- 86–105 points = Passionate, but less intense.
- 66–85 points = Occasional bursts of passion.
- 45–65 points = Tepid, infrequent passion.
- 15–44 points = The thrill is gone.

REFERENCES

References for Chapter 1

1. Frith CD, Wolpert DM. *The Neuroscience of Social Interaction: Decoding, Imitating, and Influencing the Actions of Others.* New York: Oxford University Press; 2004.
2. Gardner A, West SA. Spite. *Curr Biol.* 2006;16(17):662–4.
3. de Vladar HP, Szathmáry E. Beyond Hamilton's rule. *Science.* 2017;356(6337):485–6. doi:10.1126/science.aam6322.
4. Cacioppo JT, Amaral DG, Blanchard JJ, et al. Progress and implications for mental health. *Perspect Psychol Sci.* 2007;2(2):99–123.
5. Cacioppo S, Capitanio JP, Cacioppo JT. Toward a neurology of loneliness. *Psychol Bull.* 2014;140(6):1464–504. doi:10.1037/a0037618.
6. Ott SR, Rogers SM. Gregarious desert locusts have substantially larger brains with altered proportions compared with the olitarious phase. *Proc R Soc B Biol Sci.* 2010;277(1697):3087–96. doi:10.1098/rspb.2010.0694.
7. Dunbar R. Evolution of the social brain. *Science.* 2003;302(5648):1160–61.
8. Dunbar RI. Evolutionary basis of the social brain. In: Decety J, Cacioppo JT, eds. *The Oxford Handbook of Social Neuroscience.* Oxford, UK: Oxford University Press; 2011:28–38.
9. Dunbar RIM, Shultz S. Evolution in the social brain. *Science.* 2007;317(5843):1344–7. doi:10.1126/science.1145463.
10. Herrmann E, Call J, Hernandez-Lloreda MV, Hare B, Tomasello M. Humans have evolved specialized skills of social cognition: The cultural intelligence hypothesis. *Science.* 2007;317(5843):1360–6. doi:10.1126/science.1146282.
11. Cahill L. An issue whose time has come. *J Neurosci Res.* 2017;95(1–2):12–13. doi:10.1002/jnr.23972.
12. McEwen BS, Akil H. Introduction to social neuroscience: Gene, environment, brain, body. *Ann NY Acad Sci.* 2011;1231:vii–ix.
13. Kagan J. Why stress remains an ambiguous concept: Reply to McEwen & McEwen (2016) and Cohen et al. (2016). *Perspect Psychol Sci.* 2016;11(4):464–5. doi:10.1177/1745691616649952.
14. McEwen BS, Bowles NP, Gray JD, et al. Mechanisms of stress in the brain. *Nat Neurosci.* 2015;18(10):1353–63. doi:10.1038/nn.4086.
15. Halfon N, Larson K, Lu M, Tullis E, Russ S. Lifecourse health development: Past, present and future. *Matern Child Health J.* 2014;18(2):344–65. doi:10.1007/s10995-013-1346-2.
16. Hoekzema E, Barba-Müller E, Pozzobon C, et al. Pregnancy leads to long-lasting changes in human brain structure. *Nat Neurosci.* 2017;20(2):287–96. doi:10.1038/nn.4458.
17. McEwen BS, Gray JD, Nasca C. Redefining neuroendocrinology: Stress, sex and cognitive and emotional regulation. *J Endocrinol.* 2015;226(2):T67-T83. doi:10.1530/JOE-15-0121.
18. Cacioppo S, Cacioppo JT. Why may allopregnanolone help alleviate loneliness? *Med Hypotheses.* 2015;85(6). doi:10.1016/j.mehy.2015.09.004.

19. Cacioppo S, Grippo AJ, London S, Goossens L, Cacioppo JT. Loneliness: Clinical import and interventions. *Perspect Psychol Sci*. 2015;10(2). doi:10.1177/1745691615570616.
20. Cacioppo JT, Cacioppo S. Social neuroscience. *Perspect Psychol Sci*. 2013;8(6). doi:10.1177/1745691613507456.
21. Cacioppo JT, Berntson GG. Social psychological contributions to the decade of the brain. Doctrine of multilevel analysis. *Am Psychol*. 1992;47(8):1019–28. doi:10.1037/0003-066X.47.8.1019.
22. Gazzaniga MS. *Human: The Science behind What Makes Us Unique*. New York: Harper Collins; 2008.
23. Sousa AMM, Meyer KA, Santpere G, Gulden FO, Sestan N. Evolution of the human nervous system function, structure, and development. *Cell*. 2017;170:226–47.
24. He Z, Han D, Efimova O, et al. Comprehensive transcriptome analysis of neocortical layers in humans, chimpanzees and macaques. *Nat Neurosci*. 2017;20(6): 886–95.
25. Buckner RL, Krienen FM. The evolution of distributed association networks in the human brain. *Trends Cogn Sci*. 2013;17(12):648–65.
26. Robertson JM. Astrocytes and the evolution of the human brain. *Med Hypotheses*. 2014;82:236–9.
27. Semendeferi K, Lu A, Schenker N, Damasio H. Humans and great apes share a large frontal cortex. *Nat Neurosci*. 2002;5(3):272–6. doi:10.1038/nn814.
28. Semendeferi K, Teffer K, Buxhoeveden DP, et al. Spatial organization of neurons in the frontal pole sets humans apart from great apes. *Cereb Cortex*. 2011;21(7):1485–97. doi:10.1093/cercor/bhq191.
29. Hofman MA. Evolution of the human brain: When bigger is better. *Front Neuroanat*. 2014;8(Article 15):1–12.
30. Gazzaniga MS, LeDoux JE. *The Integrated Mind*. New York: Plenum Press; 1978.
31. Gazzaniga MS. Principles of human brain organization derived from split-brain studies. *Neuron*. 1995;14(2):217–28. http://www.ncbi.nlm.nih.gov/pubmed/7857634. Accessed August 31, 2010.
32. Nisbett RE, Wilson TD. Telling more than we know: Verbal reports on mental processes. *Psychol Rev*. 1977;84(3):231–59.
33. Cacioppo JT, Berntson GG, Semin GR. Scientific symbiosis represents the mutual benefit of iteratively adopting the perspective of realism and instrumentalism. *Am Psychol*. 2005;60(4):347–8. doi:10.1037/0003-066X.60.4.347.
34. O'Malley MAO, Brigandt I, Love AC, et al. Multilevel research strategies and biological systems. *Philos Sci*. 2014;81(5):811–28.
35. Kiecolt-Glaser JK, Goulin JP, Hantsoo LV. Close relationships, inflammation, and health. *Neurosci Biobehav Rev*. 2010;35(1):33–38.
36. Wu W, Yamaura T, Murakami K, et al. Social isolation stress enhanced liver metastasis of murine colon 26-L5 carcinoma cells by suppressing immune responses in mice. *Life Sci*. 2000;66(19):1827–38.
37. Haber SN, Barchas PR. The regulatory effect of social rank on behavior after amphetamine administration. In: Barchas PR, ed. *Social Hierarchies: Essays toward a Sociophysiological Perspective*. Westport, Connecticut: Greenwood Press; 1983:119–32.
38. Bernstein IS, Gordon TP, Rose RM. The interaction of hormones, behavior, and social context in nonhuman primates. Hormones and agressive behavior. In: Svare BB, ed. *Hormones and Aggressive Behavior*. New York: Plenum Press; 1983:535–61.
39. Rose RM, Gordon TP, Bernstein IS. Plasma testosterone levels in the male rhesus: Influences of sexual and social stimuli. *Science*. 1972;178:643–5.
40. Zhang T, Meaney MJ. Epigenetics and the environmental regulation of the genome and its function. *Annu Rev Psychol*. 2010;61:439–66.
41. Decety J, Cacioppo JT. Frontiers in human neuroscience: The golden triangle and beyond. *Perspect Psychol Sci*. 2010;5(6):767–71. doi:10.1177/1745691610388780.

42. Cacioppo S, Cacioppo JT. Cognizance of the neuroimaging methods for studying the social brain. Obhi SS, Cross ES, eds. *Shar Represent.* 2017;(May):86–106. doi:10.1017/CBO9781107279353.006.
43. Van Gerven M, Farquhar J, Schaefer R, Vlek R, Geuze J, Nijholt A, et al. The brain-computer interface cycle. *J Neural Eng.* 2009;6:041001. doi: 10.1088/1741-2560/6/4/041001.
44. Mogil JS. Animal models of pain: Progress and challenges. *Nat Rev Neurosci.* 2009;10:283–94.
45. Cacioppo JT, Berntson GG, Decety J. A history of social neuroscience. In: Van Lange P, Kruglanski AW, eds. *Handbook of the History of Social Psychology.* New York: Psychology Press; 2012:123–36.
46. Jaiswal MK. Toward a high-resolution neuroimaging biomarker for mild traumatic brain injury: From bench to bedside. *Front Neurol.* 2015;6(Jul):1–4. doi:10.3389/fneur.2015.00148.
47. Niven JE, Laughlin SB. Energy limitation as a selective pressure on the evolution of sensory systems. *J Exp Biol.* 2008;211(11):1792–804. doi:10.1242/jeb.017574.
48. Burrows M, Rogers SM, Ott SR. Epigenetic remodelling of brain, body and behaviour during phase change in locusts. *Neural Syst Circuits.* 2011;1(1–11):1–9. doi:10.1186/2042-1001-1-11.
49. Technau GM. Fiber number in the mushroom bodies of adult *Drosophila melanogaster* depends on age, sex and experience. *J Neurogenet.* 1984;1(2):113–26.
50. Heisenberg M. What do the mushroom bodies do for the insect brain? An introduction. *Learn Mem.* 1998;5(1–2):1–10. doi:10.1101/lm.5.1.1.
51. Krubitzer L, Kahn DM. Nature versus nurture revisited: An old idea with a new twist. *Prog Neurobiol.* 2003;70(1):33–52. doi:10.1016/S0301-0082(03)00088-1.
52. Northcutt RG. Understanding vertebrate brain evolution. *Integr Comp Biol.* 2002;42(4):743–56. doi:10.1093/icb/42.4.743.
53. Northcutt RG. Evolving large and complex brains. *Science.* 2011;331(6018):721–5. doi:10.1126/science.1201765.
54. Stanley DA, Adolphs R. Toward a neural basis for social behavior. *Neuron.* 2013;80(3):816–26. doi:10.1016/j.neuron.2013.10.038.
55. Donaldson ZR, Young LJ. Oxytocin, vasopressin, and the neurogenetics of sociality. *Science.* 2008;322(5903):900–904. doi:10.1126/science.1158668.
56. Ordikhani-Seyedlar M, Lebedev MA, Sorensen HB, Puthusserypady S. Neurofeedback therapy for enhancing visual attention: State-of-the-art and challenges. *Front Neurosci.* 2016 Aug 3;10:352. doi: 10.3389/fnins.2016.00352. eCollection 2016. Review. PubMed PMID: 27536212; PubMed Central PMCID: PMC4971093.

References for Chapter 2

1. Cacioppo JT, Cacioppo S. The growing problem of loneliness. *Lancet.* 2018; 391(10119):426.
2. Cacioppo JT, Cacioppo S. Loneliness in the modern age: An Evolutionary Theory of Loneliness (ETL). *Adv Exp Soc Psychol.* 2018;58:127–97. doi:10.1016/bs.aesp.2018.03.003
3. Cacioppo JT, Fowler JH, Christakis NA. Alone in the crowd: The structure and spread of loneliness in a large social network. *J Pers Soc Psychol.* 2009;97(6):977–91. doi:10.1037/a0016076.
4. Cacioppo JT, Hawkley LC, Ernst JM, et al. Loneliness within a nomological net: An evolutionary perspective. *J Res Pers.* 2006;40(6):1054–85. doi:10.1016/j.jrp.2005.11.007.
5. Cacioppo JT, Patrick W. *Loneliness: Human Nature and the Need for Social Connection.* New York: W. W. Norton Books; 2008.

6. Cacioppo JT, Cacioppo S, Boomsma DI. Evolutionary mechanisms for loneliness. *Cogn Emot*. 2014;28(1):3–21. doi:10.1080/02699931.2013.837379.
7. Goossens L, van Roekel E, Verhagen M, et al. The genetics of loneliness: Linking Evolutionary theory to genome-wide genetics, epigenetics, and social science. *Perspect Psychol Sci*. 2015;10(2). doi:10.1177/1745691614564878.
8. Cacioppo JT, Cacioppo S, Capitanio JP, Cole SW. The neuroendocrinology of social isolation. *Annu Rev Psychol*. 2015;66(1):733–67. doi:10.1146/annurev-psych-010814-015240.
9. Nunn CL, Craft ME, Gillespie TR, Schaller M, Kappeler PM. The sociality-health-fitness nexus: Synthesis, conclusions and future directions. *Philos Trans R Soc Lond B Biol Sci*. 2015;370(1669):20140115.
10. Cacioppo S. Neuroscience of salutary close relationships. In: Vangelist AL, Perlman D, eds. *Handbook of Personal Relationships*. Cambridge: Cambridge University Press; 2017.
11. Simon RW. Revisiting the relationships among gender, marital status, and mental health. *Am J Sociol*. 2002;107(4):1065–96.
12. Waite LJ, Lehrer EL. The benefits from marriage and religion in the United States: A comparative analysis. *Popul Dev Rev*. 2003;29(2):255–75.
13. Murphy M, Glaser K, Grundy E. Marital status and long-term illness in Great Britain. *J Marriage Fam*. 1997;59(1):156–64.
14. Goodwin JS, Hunt WC, Key CR, Samet JM. The effect of marital status on stage, treatment, and survival of cancer patients. *JAMA*. 1987;258(21):3125–30. http://www.ncbi.nlm.nih.gov/pubmed/3669259. Accessed September 10, 2017.
15. Li X, Liu Y, Wang Y, et al. The influence of marital status on survival of gallbladder cancer patients: A population-based study. *Sci Rep*. 2017;7(1):5322. doi:10.1038/s41598-017-05545-0.
16. Inverso G, Mahal BA, Aizer AA, Donoff RB, Chau NG, Haddad RI. Marital status and head and neck cancer outcomes. *Cancer*. 2015;121(8):1273–8. doi:10.1002/cncr.29171.
17. Merrill RM, Johnson E. Benefits of marriage on relative and conditional relative cancer survival differ between males and females in the USA. *J Cancer Surviv*. August 2017. doi:10.1007/s11764-017-0627-y.
18. Horwitz AV, White HR, Howell-White S. Becoming married and mental health: A longitudinal study of a cohort of young adults. *J Marriage Fam*. 1996;58(4):895–907. doi:10.2307/353978.
19. Waite LJ. Trends in men's and women's well-being in marriage. In: *The Ties That Bind: Perspectives on Marriage and Cohabitation*. New York: Walter de Gruyter Inc.; 2000:368–92.
20. Lillard LA, Waite LJ. 'Til death do us part: Marital disruption and mortality. *Am J Sociol*. 1995;100(5):1131–56.
21. Robles TF, Kiecolt-Glaser JK. The physiology of marriage: Pathways to health. *Physiol Behav*. 2003;79(3):409–16.
22. Holt-Lunstad J, Birmingham W, Jones BQ. Is there something unique about marriage? The relative impact of marital status, relationship quality, and network social support on ambulatory blood pressure and mental health. *Ann Behav Med*. 2008;35(2):239–44.
23. Holt-Lunstad J, Jones BQ, Birmingham W. The influence of close relationships on nocturnal blood pressure dipping. *Int J Psychophysiol*. 2009;71(3):211–17.
24. Cacioppo JT, Patrick B. *Loneliness: Human Nature and the Need for Social Connection*. First. New York: W. W. Norton & Company; 2008.
25. Cacioppo JT, Cacioppo S, Cole SW, Capitanio JP, Goossens L, Boomsma DI. Loneliness across phylogeny and a call for comparative studies and animal models. *Perspect Psychol Sci*. 2015;10(2). doi:10.1177/1745691614564876.

26. Cacioppo S, Cacioppo JT. Decoding the invisible forces of social connections. *Front Integr Neurosci.* 2012;6:51. doi:10.3389/fnint.2012.00051.
27. Silk JB, Beehner JC, Bergman TJ, et al. Strong and consistent social bonds enhance the longevity of female baboons. *Curr Biol.* 2010;20(15):1–3. doi:10.1016/j.cub.2010.05.067.
28. Holt-Lunstad J, Smith TB, Baker M, Harris T, Stephenson D. Loneliness and social isolation as risk factors for mortality: A meta-analytic review. *Perspect Psychol Sci.* 2015;10(2):227–37. doi:10.1177/1745691614568352.
29. Robles TF, Slatcher RB, Trombello JM, McGinn MM. Marital quality and health: A meta-analytic review. *Psychol Bull.* 2014;140(1):140–87. doi:10.1037/a0031859.
30. Hawkley LC, Cacioppo JT. Loneliness and pathways to disease. *Brain Behav Immun.* 2003;17 Suppl 1(1):S98–105. http://www.ncbi.nlm.nih.gov/pubmed/12615193.
31. Cacioppo JT, Cacioppo S, Capitanio JP, Cole SW. The neuroendocrinology of social isolation. *Annu Rev Psychol.* 2015;66(1):733–67. doi:10.1146/annurev-psych-010814-015240.
32. Cacioppo S, Capitanio JP, Cacioppo JT. Toward a neurology of loneliness. *Psychol Bull.* 2014;140(6):1464–504. doi:10.1037/a0037618.
33. Cacioppo JT, Cacioppo S, Boomsma DI. Evolutionary mechanisms for loneliness. *Cogn Emot.* 2014;28(1):3–21. doi:10.1080/02699931.2013.837379.
34. House JS, Landis KR, Umberson D. Social relationships and health. *Science.* 1988;241(4865):540–5. doi:10.1126/science.3399889.
35. Umberson D. Family status and health behaviors: Social control as a dimension of social integration. *J Health Soc Behav.* 1987;28(3). http://www.worldcat.org/title/family-status-and-health-behaviors-social-control-as-a-dimension-of-social-integration/oclc/480218621&referer=brief_results. Accessed February 2, 2011.
36. Cacioppo JT, Cacioppo S, Boomsma DI. Evolutionary mechanisms for loneliness. *Cogn Emot.* 2014;28(1):3–21. doi:10.1080/02699931.2013.837379.
37. Hawkley LC, Hughes ME, Waite LJ, Masi CM, Thisted RA, Cacioppo JT. From social structure factors to perceptions of relationship quality and loneliness: The Chicago Health, Aging, and Social Relations Study. *J Gerontol Soc Sci.* 2008; 63B(6):S375–S384.
38. Wheeler L, Reis H, Nezlek J. Loneliness, social interaction, and sex roles. *J Pers Soc Psychol.* 1983;45(4):943–93.
39. Boomsma DI, Willemsen G, Dolan CV, Hawkley LC, Cacioppo JT. Genetic and environmental contributions to loneliness in adults: The Netherlands twin register study. *Behav Genet.* 2005;35(6):745–52. doi:10.1007/s10519-005-6040-8.
40. Distel MA, Rebollo-Mesa I, Abdellaoui A, et al. Familial resemblance for loneliness. *Behav Genet.* 2010;40(4):480–94. doi:10.1007/s10519-010-9341-5.
41. Gao J, Davis LK, Hart AB, et al. Genome-wide association study of loneliness demonstrates a role for common variation. *Neuropsychopharmacology.* 2017;42(4):811–21. doi:10.1038/npp.2016.197.
42. Perlman D, Peplau LA. Toward a social psychology of loneliness. In: Duck S, Gilmour R, eds. *Personal Relationships in Disorder.* London: Academic Press; 1981:31–56.
43. Markham JA, Greenough WT. Experience-driven brain plasticity: Beyond the synapse. *Neuron Glia Biol.* 2004;1(4):351–63.
44. Rosenzweig MR, Bennett EL, Hebert M, Morimoto H. Social grouping cannot account for cerebral effects of enriched environments. *Brain Res.* 1978;153(3):563–76.
45. Nin MS, Martinez LA, Pibiri F, Nelson M, Pinna G. Neurosteroids reduce social isolation-induced behavioral deficits: A proposed link with neurosteroid-mediated upregulation of BDNF expression. *Front Endocrinol (Lausanne).* 2011;2:73. doi:10.3389/fendo.2011.00073.
46. Valzelli L. The "isolation syndrome" in mice. *Psychopharmacology (Berl).* 1973; 31(4):305–20.

47. Wallace DL, Han MH, Graham DL, et al. CREB regulation of nucleus accumbens excitability mediates social isolation-induced behavioral deficits. *Nat Neurosci*. 2009;12:200–209.
48. Steptoe A, Shankar A, Demakakos P, Wardle J. Social isolation, loneliness, and all-cause mortality in older men and women. *Proc Natl Acad Sci*. 2013;110(15):1–5. doi:10.1073/pnas.1219686110.
49. Weiss RS. *Loneliness: The Experience of Emotional and Social Isolation*. Cambridge, MA: MIT Press; 1973.
50. Russell DW. UCLA Loneliness Scale (version 3): Reliability, validity, and factor structure. *J Pers Assess*. 1996;66(1):20–40. http://www.ncbi.nlm.nih.gov/pubmed/8576833.
51. Russell D, Peplau LA, Cutrona CE. The revised UCLA Loneliness Scale: Concurrent and discriminant validity evidence. *J Pers Soc Psychol*. 1980;39(3):472–80. http://www.ncbi.nlm.nih.gov/pubmed/7431205.
52. Cacioppo S, Capitanio JP, Cacioppo JT. Toward a neurology of loneliness. *Psychol Bull*. 2014;140(6):1464–504. doi:10.1037/a0037618.
53. Capitanio JP, Hawkley LC, Cole SW, Cacioppo JT. A behavioral taxonomy of loneliness in humans and rhesus monkeys (*Macaca mulatta*). Bard KA, ed. *PLoS One*. 2014;9(10):e110307. doi:10.1371/journal.pone.0110307.
54. Martin AL, Brown RE. The lonely mouse: Verification of a separation-induced model of depression in female mice. *Behav Brain Res*. 2010;207(1):196–207. doi:10.1016/j.bbr.2009.10.006.
55. Ahern TH, Modi ME, Burkett JP, Young LJ. Evaluation of two automated metrics for analyzing partner preference tests. *J Neurosci Methods*. 2009;182(2):180–8. doi:10.1007/s11103-011-9767-z.Plastid.
56. Peplau L, Russell DW, Heim M. The experience of loneliness. In: Frieze IH, Bar-Tal D, Carroll JS, eds. *New Approaches to Social Problems: Applications of Attribution Theory*. San Fransisco: Jossey-Bass; 1979:53–78.
57. Anderson CA, Arnoult LH. Attributional models of depression, loneliness, and shyness. In: Harvey J and Weary G, eds. *Attribution: Basic issues and applications*. New York: Academic Press; 1985:235–79.
58. Karnick PM. Feeling lonely: Theoretical perspectives. *Nurs Sci Q*. 2005;18(1):7–12; discussion 6. doi:10.1177/0894318404272483.
59. Cacioppo JT, Cacioppo S, Cole SW, Capitanio JP, Goossens L, Boomsma DI. Loneliness across phylogeny and a call for comparative studies and animal models. *Perspect Psychol Sci*. 2015;10(2):202–12. doi:10.1177/1745691614564876.
60. Georgiev A V., Klimczuk AC, Traficonte DM, Maestripieri D. When violence pays: A cost-benefit analysis of aggressive behavior in animals and humans. *Evol Psychol*. 2013;11:678–99.
61. Gardner A, West SA. Spite. *Curr Biol*. 2006;16(17):662–4.
62. van Veelen M, Allen B, Hoffman M, Simon B, Veller C. Hamilton's rule. *J Theor Biol*. 2017;414:176–230.
63. de Vladar HP, Szathmáry E. Beyond Hamilton's rule. *Science*. 2017;356(6337):485–6. doi:10.1126/science.aam6322.
64. Cacioppo S, Bangee M, Balogh S, Cardenas-Iniguez C, Qualter P, Cacioppo JT. Loneliness and implicit attention to social threat: A high-performance electrical neuroimaging study. *Cogn Neurosci*. 2015;(August):1–22. doi:10.1080/17588928.2015.1070136.
65. Bangee M, Harris RA, Bridges N, Rotenberg KJ, Qualter P. Loneliness and attention to social threat in young adults: Findings from an eye tracker study. *Pers Individ Dif*. 2014;63:16–23.
66. Layden EA, Cacioppo JT, Cacioppo S, et al. Perceived social isolation is associated with altered functional connectivity in neural networks associated with tonic alert-

ness and executive control. *Neuroimage*. 2017;145(March 2016):58–73. doi:10.1016/j.neuroimage.2016.09.050.

67. Cacioppo JT, Chen HY, Cacioppo S. Reciprocal influences between loneliness and self-centeredness: A cross-lagged panel analysis in a population-based sample of African American, Hispanic, and Caucasian adults. *Personal Soc Psychol Bull*. 2017;43(8):1125–35. doi:10.1177/0146167217705120.
68. Worthman CM, Melby MK. Toward a comparative developmental ecology of human sleep. In: Carskadon MA, ed. *Adolescent Sleep Patterns: Biological, Social, and Psychological Influences*. Cambridge, England: Cambridge University Press; 2002:69–117.
69. Baglioni C, Nanovska S, Regen W, et al. Sleep and mental disorders: A meta-analysis of polysomnographic research. *Psychol Bull*. 2016;142(9):969–90. doi:10.1037/bul0000053.
70. Pace-Schott EF, Germain A, Milad MR. Effects of sleep on memory for conditioned fear and fear extinction. *Psychol Bull*. 2015;141(4):835–57. doi:10.1037/bul0000014.
71. Prather AA, Janicki-Deverts D, Hall MH, Cohen S. Behaviorally assessed sleep and susceptibility to the common cold. *Sleep*. 2015;38(9):1353–9. doi:10.5665/sleep.4968.
72. Luo Y, Hawkley LC, Waite LJ, Cacioppo JT. Loneliness, health, and mortality in old age: A national longitudinal study. *Soc Sci Med*. 2012;74(6):907–14. doi:10.1016/j.socscimed.2011.11.028.
73. Cacioppo JT, Hawkley LC, Berntson GG, et al. Do lonely days invade the nights? Potential social modulation of sleep efficiency. *Psychol Sci*. 2002;13(4):384–7. http://www.ncbi.nlm.nih.gov/pubmed/12137144. Accessed February 3, 2011.
74. Hawkley LC, Preacher K, Cacioppo JT. As we said, loneliness (not living alone) explains individual differences in sleep quality: Reply. *Health Psychol*. 2011;30(2):136. doi:10.1037/a0022366.
75. Aanes MM, Hetland J, Pallesen S, Mittelmark MB. Does loneliness mediate the stress-sleep quality relation? The Hordaland Health Study. *Int Psychogeriatrics*. 2011;23(6):994–1002. doi:10.1017/S1041610211000111.
76. Jacobs JM, Cohen A, Hammerman-Rozenberg R, Stessman J. Global sleep satisfaction of older people: The Jerusalem Cohort Study. *J Am Geriatr Soc*. 2006;54(2):325–9. doi:10.1111/j.1532-5415.2005.00579.x.
77. Kurina LM, Knutson KL, Hawkley LC, Cacioppo JT, Lauderdale DS, Ober C. Loneliness is associated with sleep fragmentation in a communal society. *US Natl Libr Med*. 2011;34(11):1519–26.
78. Matthews T, Danese A, Gregory AM, Caspi A, Moffitt TE, Arseneault L. Sleeping with one eye open: Loneliness and sleep quality in young adults. *Psychol Med*. 2017;47:2177–86.
79. Kaushal N, Nair D, Gozal D, Ramesh V. Socially isolated mice exhibit a blunted homeostatic sleep response to acute sleep deprivation compared to socially paired mice. *Brain Res*. 2012;1454:65–79. doi:10.1016/j.brainres.2012.03.019.
80. Rueggeberg R, Wrosch C, Miller GE, McDade TW. Associations between health-related self-protection, diurnal cortisol, and C-reactive protein in lonely older adults. *Psychosom Med*. October 2012:937–44. doi:10.1097/PSY.0b013e3182732dc6.
81. Bosch OJ, Nair HP, Ahern TH, Neumann ID, Young LJ. The CRF system mediates increased passive stress-coping behavior following the loss of a bonded partner in a monogamous rodent. *Neuropsychopharmacology*. 2009;34(6):1406–15. doi:10.1038/npp.2008.154.
82. Sun P, Smith AS, Lei K, Liu Y, Wang Z. Breaking bonds in male prairie vole: Long-term effects on emotional and social behavior, physiology, and neurochemistry. *Behav Brain Res*. 2014;265:22–31. doi:10.1016/j.bbr.2014.02.016.

83. Grippo AJ, Cushing BS, Carter CS. Depression-like behavior and stressor-induced neuroendocrine activation in female prairie voles exposed to chronic social isolation. *Psychosom Med*. 2007;69(2):149–57. doi:10.1097/PSY.0b013e31802f054b.
84. Grippo AJ, Gerena D, Huang J, et al. Social isolation induces behavioral and neuroendocrine disturbances relevant to depression in female and male prairie voles. *Psychoneuroendocrinology*. 2007;32(8):966–80. doi:10.1016/j.psyneuen.2007.07.004.
85. Young L, Alexander B. *The Chemistry between Us*. New York: Penguin Books; 2012.
86. Mendoza SP, Mason WA. Parental division of labour and differentiation of attachments in a monogamous primate (*Callicebus moloch*). *Anim Behav*. 1986;34(5): 1336–47.
87. Mendoza SP, Mason WA. Contrasting responses to intruders and to involuntary separation by monogamous and polygynous New World monkeys. *Physiol Behav*. 1986;38(6):795–801. http://www.ncbi.nlm.nih.gov/pubmed/3823197.
88. Mendoza SP, Hennessy MB, Lyons DM. Distinct immediate and prolonged effects of separation on plasma cortisol in adult female squirrel monkeys. *Psychobiology*. 1992;20(4):300–306. doi:10.3758/bf03332064.
89. Hennessy MB. Effects of social partners on pituitary-adrenal activity during novelty exposure in adult female squirrel monkeys. *Physiol Behav*. 1986;38(6):803–7. http://www.ncbi.nlm.nih.gov/pubmed/3823198. Accessed September 14, 2017.
90. Vogt JL, Levine S. Response of mother and infant squirrel monkeys to separation and disturbance. *Physiol Behav*. 1980;24(5):829–32. http://www.ncbi.nlm.nih.gov/pubmed/6773081. Accessed September 14, 2017.
91. Coe CL, Mendoza SP, Smotherman WP, Levine S. Mother-infant attachment in the squirrel monkey: Adrenal response to separation. *Behav Biol*. 1978;22(2):256–63. http://www.ncbi.nlm.nih.gov/pubmed/415729. Accessed September 14, 2017.
92. Mendoza SP, Smotherman WP, Miner MT, Kaplan J, Levine S. Pituitary-adrenal response to separation in mother and infant squirrel monkeys. *Dev Psychobiol*. 1978;11(2):169–75. doi:10.1002/dev.420110209.
93. Tuber DS, Sanders S, Hennessy MB, Miller JA. Behavioral and glucocorticoid responses of adult domestic dogs (*Canis familiaris*) to companionship and social separation. *J Comp Psychol*. 1996;110(1):103–8. http://www.ncbi.nlm.nih.gov/pubmed/8851558.
94. Cacioppo JT, Berntson GG, Malarkey WB, et al. Autonomic, neuroendocrine, and immune responses to psychological stress: The reactivity hypothesis. *Ann NY Acad Sci*. 1998;840:664–73.
95. Cohen S, Gianaros PJ, Manuck SB. A stage model of stress and disease. *Perspect Psychol Sci*. 2016;11(4):456–63. doi:10.1177/1745691616646305.
96. Mendes WB, Blascovich J, Lickel B, Hunter S. Challenge and threat during social interactions with White and Black men. *Personal Soc Psychol Bull*. 2002;28(7): 939–52.
97. Cacioppo JT, Hawkley LC, Crawford LE, et al. Loneliness and health: Potential mechanisms. *Psychosom Med*. 2002;64(3):407–17. http://www.ncbi.nlm.nih.gov/pubmed/12021415.
98. Hawkley LC, Burleson MH, Berntson GG, Cacioppo JT. Loneliness in everyday life: Cardiovascular activity, psychosocial context, and health behaviors. *J Pers Soc Psychol*. 2003;85(1):105–20. doi:10.1037/0022-3514.85.1.105.
99. Cacioppo JT, Hawkley LC, Berntson GG. The anatomy of loneliness. *Curr Dir Psychol Sci*. 2003;12(3):71–74. doi:10.1111/1467–8721.01232.
100. Mezuk B, DeSantis AS, Rapp SR, Roux AVD, Seeman T. Loneliness, depression, and inflammation: Evidence from the multi-ethnic study of atherosclerosis. *PLoS One*. 2016;11(7). doi:10.1371/journal.pone.0158056.
101. Steptoe A, Owen N, Kunz-Ebrecht SR, Brydon L. Loneliness and neuroendocrine, cardiovascular, and inflammatory stress responses in middle-aged men and

women. *Psychoneuroendocrinology*. 2004;29(5):593–611. doi:10.1016/S0306-4530(03)00086-6.
102. Shankar A, McMunn A, Banks J, Steptoe A. Loneliness, social isolation, and behavioral and biological health indicators in older adults. *Heal Psychol*. 2011;30(4):377–85. doi:10.1037/a0022826.
103. Cacioppo JT, Tassinary LG, Berntson GG. *Handbook of Psychophysiology*. 4th ed. New York: Cambridge University Press; 2017.
104. Parfitt DN: Loneliness and the paranoid syndrome. *J Neurol Psychopathol* 1937;17:318–21.
105. Lynch JJ, Convey WH. Loneliness, disease, and death: Altemative approaches. *Psychosomatics*. 1979;20(10):702–8. doi:10.1016/S0033-3182(79)73751-0.
106. Lynch J. *The Broken Heart: The Medical Consequences of Loneliness*. New York: Basic Books; 1977.
107. Lynch J. *A Cry Unheard: New Insights into the Medical Consequences of Loneliness*. Baltimore, MD: Bancroft Press; 2000.
108. Hawkley LC, Masi CM, Berry JD, Cacioppo JT. Loneliness is a unique predictor of age-related differences in systolic blood pressure. *Psychol Aging*. 2006;21(1):152–64. doi:10.1037/0882-7974.21.1.152.
109. Ong AD, Rothstein JD, Uchino BN. Loneliness accentuates age differences in cardiovascular responses to social evaluative threat. *Psychol Aging*. 2012;27(1):190–98.
110. Momtaz YA, Hamid TA, Yusoff S, et al. Loneliness as a risk factor for hypertension in later life. *J Aging Health*. 2012;24(4):696–710.
111. Hawkley LC, Thisted RA, Masi CM, Cacioppo JT. Loneliness predicts increased blood pressure: 5-year cross-lagged analyses in middle-aged and older adults. *Psychol Aging*. 2010;25(1):132–41. doi:10.1037/a0017805.
112. Caspi A, Harrington H, Moffitt TE, Milne BJ, Poulton R. Socially isolated children 20 years later: Risk of cardiovascular disease. *Arch Pediatr Adolesc Med*. 2006;160(8):805–11. doi:10.1001/archpedi.160.8.805.
113. Tomaka J, Thompson S, Palacios R. The relation of social isolation, loneliness, and social support to disease outcomes among the elderly. *J Aging Health*. 2006;18(3):359–84. doi:10.1177/0898264305280993.
114. Whisman MA. Loneliness and the metabolic syndrome in a population-based sample of middle-aged and older adults. *Health Psychol*. 2010;29(5):550–54.
115. Kunz-Ebrecht SR, Kirschbaum C, Marmot M, Steptoe A. Differences in cortisol awakening response on work days and weekends in women and men from the Whitehall II cohort. *Psychoneuroendocrinology*. 2004;29(4):516–28. doi:10.1016/S0306-4530(03)00072-6.
116. Geller J, Janson P, McGovern E, Valdini A. Loneliness as a predictor of hospital emergency department use. *J Fam Pract*. 1999;48(10):801–4.
117. Gerst-Emerson K, Jayawardhana J. Loneliness as a public health issue: The impact of loneliness on health care utilization among older adults. *Am J Public Health*. 2015;105:1013–19.
118. Musich S, Wang SS, Kraemer S, Hawkins K, Wicker E. Caregivers for older adults: Prevalence, characteristics, and health care utilization and expenditures. *Geriatr Nurs (Minneap)*. 2017;9–16.
119. Christiansen J, Larsen FB, Lasgaard M. Do stress, health behavior, and sleep mediate the association between loneliness and adverse health conditions among older people? *Soc Sci Med*. 2016;152:80–86. doi:10.1016/j.socscimed.2016.01.020.
120. Eaker ED, Pinsky J, Castelli WP. Myocardial infarction and coronary death among women: Psychosocial predictors from a 20-year follow-up of women in the Framingham study. *Am J Epidemiol*. 1992;135(8):854–64. http://www.ncbi.nlm.nih.gov/pubmed/1585898. Accessed September 22, 2010.

121. Olsen RB, Olsen J, Gunner-Svensson F, Waldstrom B. Social networks and longevity. A 14 year follow-up study among elderly in Denmark. *Soc Sci Med.* 1991;33(10):1189–95.
122. Sorkin D, Rook KS, Lu JL. Loneliness, lack of emotional support, lack of companionship, and the likelihood of having a heart condition in an elderly sample. *Ann Behav Med.* 2002;24(4):290–98.
123. Thurston RC, Kubzansky LD. Women, loneliness, and incident coronary heart disease. *Psychosom Med.* 2009;71(8):836–42. doi:10.1097/PSY.0b013e3181b40efc.
124. Valtorta NK, Kanaan M, Gilbody S, Ronzi S, Hanratty B. Loneliness and social isolation as risk factors for coronary heart disease and stroke: Systematic review and meta-analysis of longitudinal observational studies. *Heart.* 2016;102(13):1009–16. doi:10.1136/heartjnl-2015-308790.
125. Holt-Lunstad J, Smith TB. Loneliness and social isolation as risk factors for CVD: Implications for evidence-based patient care and scientific inquiry. *Hear Online First.* 2016;102(13):987–89. doi:10.1136/heartjnl-2015-309242.
126. Peuler JD, Scotti MAL, Phelps LE, McNeal N, Grippo AJ. Chronic social isolation in the prairie vole induces endothelial dysfunction: Implications for depression and cardiovascular disease. *Physiol Behav.* 2012;106(4):476–84.
127. Cruz FC, Duarte JO, Leão RM, Hummel LF, Planeta CS, Crestani CC. Adolescent vulnerability to cardiovascular consequences of chronic social stress: Immediate and long-term effects of social isolation during adolescence. *Dev Neurobiol.* 2016;76(1):34–46.
128. Maslova LN, Bulygina VV, Amstislavskaya TG. Prolonged social isolation and social instability in adolescence in rats: Immediate and long-term physiological and behavioral effects. *Neurosci Behav Physiol.* 2010;40(9):955–63.
129. Coelho AM, Carey KD, Shade RE. Assessing the effects of social environment on blood pressure and heart rates of baboons. *Am J Primatol.* 1991;23(4):257–67. doi:10.1002/ajp.1350230406.
130. Cole SW, Capitanio JP, Chun K, Arevalo JMG, Ma J, Cacioppo JT. Myeloid differentiation architecture of leukocyte transcriptome dynamics in perceived social isolation. *Proc Natl Acad Sci.* 2015;112(49):15142–7. doi:10.1073/pnas.1514249112.
131. Goossens L, van Roekel E, Verhagen M, et al. The genetics of loneliness. *Perspect Psychol Sci.* 2015;10(2):213–26. doi:10.1177/1745691614564878.
132. Slavich GM, Cole SW. The emerging field of human social genomics. *Clin Psychol Sci.* 2013;1(3):331–48. doi:10.1177/2167702613478594.
133. Cole SW, Hawkley LC, Arevalo JM, Sung CY, Rose RM, Cacioppo JT. Social regulation of gene expression in human leukocytes. *Genome Biol.* 2007;8(9):R189. doi:10.1186/gb-2007-8-9-r189.
134. Cole SW, Hawkley LC, Arevalo JMG, Cacioppo JT. Transcript origin analysis identifies antigen- presenting cells as primary targets of socially regulated gene expression in leukocytes. *Proc Natl Acad Sci.* 2011;108(7):3080–5. doi:10.1073/pnas.1014218108.
135. Creswell JD, Irwin MR, Burklund LJ, et al. Mindfulness-Based Stress Reduction training reduces loneliness and pro-inflammatory gene expression in older adults: A small randomized controlled trial. *Brain Behav Immun.* 2012;26(7):1095–101. doi:10.1016/j.bbi.2012.07.006.
136. Cole SW, Levine ME, Arevalo JMG, Ma J, Weir DR, Crimmins EM. Loneliness, eudaimonia, and the human conserved transcriptional response to adversity. *Psychoneuroendocrinology.* 2015;62:11–17. doi:10.1016/j.psyneuen.2015.07.001.
137. Cacioppo JT, Cacioppo S. The population-based longitudinal Chicago Health, Aging, and Social Relations Study (CHASRS): Study description and predictors of attrition in older adults. *Arch Sci Psychol.* 2018;6(1):21–31. doi:10.1037/arc0000036.

138. Mehl MR, Raison CL, Pace TWW, Arevalo JMG, Cole SW. Natural language indicators of differential gene regulation in the human immune system. 2017:1–6. doi:10.1073/pnas.1707373114.
139. Cacioppo JT, Hawkley LC, Kalil A, Hughes ME, Waite LJ, Thisted RA. Happiness and the invisible threads of social connection: The Chicago Health, Aging, and Social Relations Study. In: Eid M, Larson R, eds. *The Science of Well-Being*. New York: Guilford; 2008:195–219.
140. VanderWeele TJ, Hawkley LC, Cacioppo JT. On the reciprocal association between loneliness and subjective well-being. *Am J Epidemiol*. 2012;176(9):777–84. doi:10.1093/aje/kws173.
141. Canli T, Wen R, Wang X, et al. Differential transcriptome expression in human nucleus accumbens as a function of loneliness. *Mol Psychiatry*. 2017;22(7):1069–78. doi:10.1038/mp.2016.186.
142. Cacioppo JT, Norris CJ, Decety J, Monteleone G, Nusbaum H. In the eye of the beholder: Individual differences in perceived social isolation predict regional brain activation to social stimuli. *J Cogn Neurosci*. 2009;21(1):83–92. doi:10.1162/jocn.2009.21007.
143. Cacioppo JT, Ernst JM, Burleson MH, et al. Lonely traits and concomitant physiological processes: The MacArthur social neuroscience studies. *Int J Psychophysiol*. 2000;35(2):143–54. http://www.ncbi.nlm.nih.gov/pubmed/10677643.
144. Inagaki TK, Muscatell KA, Moieni M, et al. Yearning for connection? Loneliness is associated with increased ventral striatum activity to close others. *Soc Cogn Affect Neurosci*. 2016;11(7):1096–101. doi:10.1093/scan/nsv076.
145. D'Enes Z. Loneliness in old age. *Zeitschrift Alternsforsch*. 1980;35:475–80.
146. Kiecolt-Glaser JK, Ricker D, George J, et al. Urinary cortisol levels, cellular immunocompetency, and loneliness in psychiatric inpatients. *Psychosom Med*. 1984;46(1):15–23. http://www.ncbi.nlm.nih.gov/pubmed/6701251.
147. Kiecolt-Glaser JK, Garner W, Speicher C, Penn GM, Holliday J, Glaser R. Psychosocial modifiers of immunocompetence in medical students. *Psychosom Med*. 1984;46(1):7–14. http://www.ncbi.nlm.nih.gov/pubmed/6701256.
148. Kiecolt-Glaser JK, Speicher CE, Holliday JE, Glaser R. Stress and the transformation of lymphocytes by Epstein-Barr virus. *J Behav Med*. 1984;7:1–12.
149. Glaser R, Kiecolt-Glaser JK, Speicher CE, Holliday JE. Stress, loneliness, and changes in herpesvirus latency. *J Behav Med*. 1985;8(3):249–60. http://www.ncbi.nlm.nih.gov/pubmed/3003360.
150. Pressman SD, Cohen S, Miller GE, Barkin A, Rabin BS, Treanor JJ. Loneliness, social network size, and immune response to influenza vaccination in college freshmen. *Heal Psychol*. 2005;24(3):297–306. doi:10.1037/0278-6133.24.3.297.
151. Dixon D, Cruess S, Kilbourn K, Klimas N, Fletcher MA, Ironson G. Social support mediates loneliness and human herpesvirus type 6 (HHV-6) antibody titers. *J Appl Soc Psychol*. 2001;31(6):1111–32. doi:10.1111/j.1559-1816.2001.tb02665.x.
152. Straits-Troester K, Patterson TL, Semple SJ, Temoshok L. The relationship between loneliness, interpersonal competence, and immunologic status in HIV-infected men. *Psychol Heal*. 1994;9(3):205–19. http://www.worldcat.org/title/the-relationship-between-loneliness-interpersonal-competence-and-immunologic-status-in-hiv-infected-men/oclc/201691530&referer=brief_results. Accessed September 22, 2010.
153. Glasper ER, DeVries AC. Social structure influences effects of pair-housing on wound healing. *Brain Behav Immun*. 2005;19(1):61–68. doi:10.1016/j.bbi.2004.03.002.
154. Arenillas JF, Álvarez-Sabín J, Molina CA, et al. C-reactive protein predicts further ischemic events in first-ever transient ischemic attack or stroke patients with intracranial large-artery occlusive disease. *Stroke*. 2003;34(10):2463–8. doi:10.1161/01.STR.0000089920.93927.A7.

155. Heneka MT, O'Banion MK. Inflammatory processes in Alzheimer's disease. *J Neuroimmunol.* 2007;184(1–2):69–91. doi:10.1016/j.jneuroim.2006.11.017.
156. Schetter AJ, Heegaard NHH, Harris CC. Inflammation and cancer: Interweaving microRNA, free radical, cytokine and p53 pathways. *Carcinogenesis.* 2009; 31(1):37–49. doi:10.1093/carcin/bgp272.
157. Lavie CJ, Milani RV, Verma A, O'Keefe JH. C-reactive protein and cardiovascular diseases—Is it ready for primetime? *Am J Med Sci.* 2009;338(6):486–92. doi:10.1007/s00547-003-1018-y.
158. O'Luanaigh C, O'Connell H, Chin A, et al. Loneliness and vascular biomarkers: The Dublin healthy ageing study. *Int J Geriatr Psychiatry.* 2012;27(1):83–88.
159. Hackett RA, Hamer M, Endrighi R, Brydon L, Steptoe AA. Loneliness and stress-related inflammatory and neuroendocrine responses in older men and women. *Psychoneuroendocrinology.* 2012;37(11):1801–9. doi:10.1016/j.psyneuen.2012.03.016.
160. Jaremka LM, Fagundes CP, Peng J, et al. Loneliness promotes inflammation during acute stress. *Psychol Sci.* 2013;24(7):1089–97. doi:10.1177/0956797612464059.
161. Moieni M, Irwin MR, Jevtic I, et al. Trait sensitivity to social disconnection enhances pro-inflammatory responses to a randomized controlled trial of endotoxin. *Psychoneuroendocrinology.* 2015;62:336–42.
162. Karelina K, Norman GJ, Zhang N, Morris JS, Peng H, DeVries AC. Social isolation alters neuroinflammatory response to stroke. *Proc Natl Acad Sci.* 2009;106(14):5895–900. doi:10.1073/pnas.0810737106.
163. Norman GJ, Zhang N, Morris JS, Karelina K, Berntson GG, DeVries AC. Social interaction modulates autonomic, inflammatory, and depressive-like responses to cardiac arrest and cardiopulmonary resuscitation. *Proc Natl Acad Sci.* 2010;107(37): 16342–7. doi:10.1073/pnas.1007583107.
164. Karelina K, DeVries AC. Modeling social influences on human health. *Psychosom Med.* 2011;73(1):67–74. doi:10.1097/PSY.0b013e3182002116.
165. Cacioppo S, Grippo AJ, London S, Goossens L, Cacioppo JT. Loneliness: Clinical import and interventions. *Perspect Psychol Sci.* 2015;10(2):238–49. doi:10.1177/1745691615570616.
166. Heinrich LM, Gullone E. The clinical significance of loneliness: A literature review. *Clin Psychol Rev.* 2006;26(6):695–718. doi:10.1016/j.cpr.2006.04.002.
167. Booth R. Loneliness as a component of psychiatric disorders. *Medscape Gen Med.* 2000;2(2):1–7.
168. Vanderweele TJ, Hawkley LC, Thisted RA, Cacioppo JT. A marginal structural model analysis for loneliness: Implications for intervention trials and clinical practice. *J Consult Clin Psychol.* 2011;79(2):225–35. doi:10.1037/a0022610.
169. Cacioppo JT, Hawkley LC, Thisted RA. Perceived social isolation makes me sad: 5-year cross-lagged analyses of loneliness and depressive symptomatology in the Chicago Health, Aging, and Social Relations Study. *Psychol Aging.* 2010;25(2):453–63. doi:10.1037/a0017216.
170. Celano CM, Huffman JC. Depression and cardiac disease: A review. *Cardiol Rev.* 2011;19(3):130–42. doi:10.1097/CRD.0b013e31820e8106.
171. Laursen TM, Musliner KL, Benros ME, Vestergaard M, Munk-Olsen T. Mortality and life expectancy in persons with severe unipolar depression. *J Affect Disord.* 2016;193:203–7.
172. Lichtman JH, Bigger JT, Blumenthal JA, et al. Depression and coronary heart disease. *Circulation.* 2008;118(17):1768–75.
173. Schulz R, Beach SR, Ives DG, Martire LM, Ariyo AA, Kop WJ. Association between depression and mortality in older adults: The Cardiovascular Health Study. *Arch Intern Med.* 2000;160(12):1761–8.

174. Adam EK, Hawkley LC, Kudielka BM, Cacioppo JT. Day-to-day dynamics of experience—Cortisol associations in a population-based sample of older adults. *Proc Natl Acad Sci*. 2006;103(45):17058–63. doi:10.1073/pnas.0605053103.
175. Hawkley LC, Preacher KJ, Cacioppo JT. Loneliness impairs daytime functioning but not sleep duration. *Heal Psychol*. 2010;29(2):124–29. doi:10.1037/a0018646.
176. Allen NB, Badcock PBT. The social risk hypothesis of depressed mood: Evolutionary, psychosocial, and neurobiological perspectives. *Psychol Bull*. 2003;129(6): 887–913. doi:10.1037/0033-2909.129.6.887.
177. West DA, Kellner R, Moore-West M. The effects of loneliness: A review of the literature. *Compr Psychiatry*. 1986;27(4):351–63. http://www.ncbi.nlm.nih.gov/pubmed/3524985.
178. Ernst JM, Cacioppo JT. Lonely hearts: Psychological perspectives on loneliness. *Appl Prev Psychol*. 1999;8(1):1–22.
179. Adams K, Sanders S, Auth E. Loneliness and depression in independent living retirement communities: Risk and resilience factors. *Aging Ment Health*. 2004;8(6):475–85. doi:10.1080/13607860410001725054.
180. Anderson CA, Harvey RJ. Brief Report: Discriminating between problems in living: An examination of measures of depression, loneliness, shyness, and social anxiety. *J Soc Clin Psychol*. 1988;6(3–4):482–91.
181. Aylaz R, Aktürk Ü, Erci B, Öztürk H, Aslan H. Relationship between depression and loneliness in elderly and examination of influential factors. *Arch Gerontol Geriatr*. 2012;55(3):548–54. doi:10.1016/j.archger.2012.03.006.
182. Bodner E, Bergman YS. Loneliness and depressive symptoms among older adults: The moderating role of subjective life expectancy. *Psychiatry Res*. 2016;237:78–82. doi:10.1016/j.psychres.2016.01.074.
183. Cacioppo JT, Hughes ME, Waite LJ, Hawkley LC, Thisted RA. Loneliness as a specific risk factor for depressive symptoms: Cross-sectional and longitudinal analyses. *Psychol Aging*. 2006;21(1):140–51. doi:10.1037/0882-7974.21.1.140.
184. Goswick RA, Jones WH. Loneliness, self-concept, and adjustment. *J Psychol*. 1981;107(2):237–40.
185. Heikkinen R-L, Kauppinen M. Mental well-being: A 16-year follow-up among older residents in Jyväskylä. *Arch Gerontol Geriatr*. 2011;52(1):33–39. doi:10.1016/j.archger.2010.01.017.
186. Koenig LJ, Abrams RF. Adolescent loneliness and adjustment: A focus on gender differences. In: Rotenberg KJ, Hymel S, eds. *Loneliness in Childhood and Adolescence*. New York: Cambridge University Press; 1999:296–324.
187. Koenig LJ, Isaacs AM, Schwartz JA. Sex differences in adolescent depression and loneliness: Why are boys lonelier if girls are more depressed? *J Res Pers*. 1994;28(1):27–43.
188. Mahon NE, Yarcheski A, Yarcheski TJ. Mental health variables and positive health practices in early adolescents. *Psychol Rep*. 2001;88(3_suppl):1023–30.
189. Moore D, Schultz NR. Correlates, attributions, and coping. *J Youth Adolesc*. 1983;12(2):95–100.
190. Nolen-Hoeksema S, Ahrens C. Age differences and similarities in the correlates of depressive symptoms. *Psychol Aging*. 2002;17(1):116–24. doi:10.1037//0882-7974.17.1.116.
191. Rook KS. Social support versus companionship: Effects on life stress, loneliness, and evaluations by others. *J Pers Soc Psychol*. 1987;52(6):1132–47.
192. van Beljouw IM, van Exel E, De Jong Gierveld J, et al. "Being all alone makes me sad": Loneliness in older adults with depressive symptoms. *Int Psychogeriatrics*. 2014;26(9):1541–51.
193. Weeks DG, Michela JL, Peplau L, Bragg ME. Relation between loneliness and depression: A structural equation analysis. *J Pers Soc Psychol*. 1980;39:1238–44.

194. Zawadzki MJ, Graham JE, Gerin W. Rumination and anxiety mediate the effect of loneliness on depressed mood and sleep quality in college students. *Heal Psychol.* 2012;32(2):212–22. doi:10.1037/a0029007.
195. Jackson J, Cochran SD. Loneliness and psychological distress. *J Psychol.* 1991;125(3):257–62.
196. Schinka KC, van Dulmen MHM, Mata AD, Bossarte R, Swahn M. Psychosocial predictors and outcomes of loneliness trajectories from childhood to early adolescence. *J Adolesc.* 2013;36(6):1251–60. doi:10.1016/j.adolescence.2013.08.002.
197. Young JE. Loneliness, depression, and cognitive therapy: Theory and application. In: Peplau LA, Perlman D, eds. *Loneliness: A Sourcebook of Current Theory, Research, and Therapy.* New York: Wiley; 1982:379–405.
198. Brage D, Meredith W, Woodward J. Correlates of loneliness among Midwestern adolescents. *Adolescence.* 1993;28(111):685–93.
199. Green BH, Copeland JR, Dewey ME, et al. Risk factors for depression in elderly people: A prospective study. *Acta Psychiatr Scand.* 1992;86(3):213–17. doi:10.1111/j.1600-0447.1992.tb03254.x.
200. Hagerty BM, Williams AR. The effects of sense of belonging, social support, conflict, and loneliness on depression. *Nurs Res.* 1999;48(4):215–19.
201. Heikkinen R-L, Kauppinen M. Depressive symptoms in late life: A 10-year follow-up. *Arch Gerontol Geriatr.* 2004;38(3):239–50.
202. Santini ZI, Fiori KL, Feeney J, Tyrovolas S, Haro JM, Koyanagi A. Social relationships, loneliness, and mental health among older men and women in Ireland: A prospective community-based study. *J Affect Disord.* 2016;204:59–69.
203. Wei M, Russell DW, Zakalik RA. Adult attachment, social self-efficacy, self-disclosure, loneliness, and subsequent depression for freshman college students: A longitudinal study. *J Couns Psychol.* 2005;52(4):602–14.
204. Walker DL, Davis M. Role of the extended amygdala in short-duration versus sustained fear: A tribute to Dr. Lennart Heimer. *Brain Struct Funct.* 2008;213(1–2):29–42.
205. Matthews GA, Nieh EH, Vander Weele CM, et al. Dorsal raphe dopamine neurons represent the experience of social isolation. *Cell.* 2016;164(4):617–31. doi:10.1016/j.cell.2015.12.040.
206. Makinodan M, Rosen KM, Ito S, Corfas G. A critical period for social experience-dependent oligodendrocyte maturation and myelination. *Science.* 2012;337(6100):1357–60. doi:10.1126/science.1220845.
207. Liu J, Dietz K, DeLoyht JM, et al. Impaired adult myelination in the prefrontal cortex of socially isolated mice. *Nat Neurosci.* 2012;15(12):1621–3. doi:10.1038/nn.3263.
208. Nin MS, Martinez LA, Pibiri F, Nelson M, Pinna G. Neurosteroids reduce social isolation-induced behavioral deficits: A proposed link with neurosteroid-mediated upregulation of BDNF expression. *Front Endocrinol (Lausanne).* 2011;2(November):1–12. doi:10.3389/fendo.2011.00073.
209. Nelson M, Pinna G. S-norfluoxetine microinfused into the basolateral amygdala increases allopregnanolone levels and reduces aggression in socially isolated mice. *Neuropharmacology.* 2011;60:1154–9.
210. Wilson RS, Krueger KR, Arnold SE, et al. Loneliness and risk of Alzheimer disease. *Arch Gen Psychiatry.* 2007;64(2):234–40. doi:10.1001/archpsyc.64.2.234.
211. Karelina K, Norman GJ, Zhang N, DeVries AC. Social contact influences histological and behavioral outcomes following cerebral ischemia. *Exp Neurol.* 2009;220(2):276–82. doi:10.1016/j.expneurol.2009.08.022.
212. Venna VR, Weston G, Benashski SE, et al. NF-κB contributes to the detrimental effects of social isolation after experimental stroke. *Acta Neuropathol.* 2012;124(3):425–38. doi:10.1007/s00401-012-0990-8.

213. Cole SW, Hawkley LC, Arevalo JMG, Cacioppo JT. Transcript origin analysis identifies antigen-presenting cells as primary targets of socially regulated gene expression in leukocytes. *Proc Natl Acad Sci.* 2011;108(7):3080–5. doi:10.1073/pnas.1014218108.
214. Heijne A, Rossi F, Sanfey AG. Why we stay with our social partners: Neural mechanisms of stay/leave decision-making. *Soc Neurosci.* 2018;13(6)667–79. doi:10.1080/17470919.2017.1370010.
215. Ioannou CC, Guttal V, Couzin ID. Predatory fish select for coordinated collective motion in virtual prey. *Science.* 2012;337(6099):1212–15. doi:10.1126/science.1218919.
216. Grippo AJ, Ihm E, Wardwell J, et al. The effects of environmental enrichment on depressive and anxiety-relevant behaviors in socially isolated prairie voles. *Psychosom Med.* 2014;76(4):277–84. doi:10.1097/PSY.0000000000000052.
217. Hughes ME, Waite LJ, Hawkley LC, Cacioppo JT. A short scale for measuring loneliness in large surveys: Results from two population-based studies. *Res Aging.* 2004;26(6):655–72. doi:10.1177/0164027504268574.
218. Allaert FA, Urbinelli R. Sociodemographic profile of insomniac patients across national surveys. *CNS Drugs.* 2004;18(1):3–7.
219. Hayley AC, Downey LA, Stough C, Sivertsen B, Knapstad M, Øverland S. Social and emotional loneliness and self-reported difficulty initiating and maintaining sleep (DIMS) in a sample of Norwegian university students. *Scand J Psychol.* 2017;58(1):91–99.
220. Hawkley LC, Preacher KJ, Cacioppo JT. Loneliness impairs daytime functioning but not sleep duration. *Heal Psychol.* 2010;29(2):124–29. doi:10.1037/a0018646.
221. Segrin C, Burke TJ. Loneliness and sleep quality: Dyadic effects and stress effects. *Behav Sleep Med.* 2015;13(3):241–54.
222. Segrin C, Domschke T. Social support, loneliness, recuperative processes, and their direct and indirect effects on health. *Health Commun.* 2011;26(3):1–12. doi:10.1080/10410236.2010.546771.
223. Segrin C, Passalacqua SA. Functions of loneliness, social support, health behaviors, and stress in association with poor health. *Health Commun.* 2010;25(4):312–22. doi:10.1080/10410231003773334.
224. Stickley A, Koyanagi A, Leinsalu M, Ferlander S, Sabawoon W, McKee M. Loneliness and health in Eastern Europe: Findings from Moscow, Russia. *Public Health.* 2015;129(4):403–10.
225. Harris RA, Qualter P, Robinson SJ. Loneliness trajectories from middle childhood to pre-adolescence: Impact on perceived health and sleep disturbance. *J Adolesc.* 2013;36(6):1295–304.
226. Zilioli S, Slatcher RB, Chi P, Li X, Zhao J, Zhao G. The impact of daily and trait loneliness on diurnal cortisol and sleep among children affected by parental HIV/AIDS. *Psychoneuroendocrinology.* 2017;75:64–71.
227. Okamura H, Tsuda A, Matsuishi T. The relationship between perceived loneliness and cortisol awakening responses on work days and weekends. *Jpn Psychol Res.* 2011;53(2):113–20.
228. Doane LD, Adam EK. Loneliness and cortisol: Momentary, day-to-day, and trait associations. *Psychoneuroendocrinology.* 2010;35(3):430–41. doi:10.1016/j.psyneuen.2009.08.005.
229. McNeal N, Scotti MAL, Wardwell J, et al. Disruption of social bonds induces behavioral and physiological dysregulation in male and female prairie voles. *Auton Neurosci.* 2014;180:9–16.
230. Caspi A, Harrington H, Moffitt TE, Milne BJ, Poulton R. Socially isolated children 20 years later: Risk of cardiovascular disease. *Arch Pediatr Adolesc Med.* 2006;160(8):805–11. doi:10.1001/archpedi.160.8.805.

231. Thurston RC, Kubzansky LD. Women, loneliness, and incident coronary heart disease. *Psychosom Med*. 2009;71(8):836–42.
232. Sorkin D, Rook KS, Lu JL. Loneliness, lack of emotional support, lack of companionship, and the likelihood of having a heart condition in an elderly sample. *Ann Behav Med*. 2002;24(4):290–98.
233. Cole SW. Social regulation of leukocyte homeostasis: The role of glucocorticoid sensitivity. *Brain Behav Immun*. 2008;22(7):1049–55. doi:10.1016/j.bbi.2008.02.006.
234. Norman GJ, Zhang N, Morris JS, Karelina K, Berntson GG, DeVries AC. Social interaction modulates autonomic, inflammatory, and depressive-like responses to cardiac arrest and cardiopulmonary resuscitation. *Proc Natl Acad Sci*. 2010;107: 16342–7.
235. Weil ZM, Norman GJ, Barker JM, Su AJ, Nelson RJ, DeVries AC. Social isolation potentiates cell death and inflammatory responses after global ischemia. *Mol Psychiatry*. 2008;13(10):913–15. doi:10.1038/mp.2008.70.
236. Hawkley LC, Thisted RA, Cacioppo JT. Loneliness predicts reduced physical activity: Cross-sectional & longitudinal analyses. *Heal Psychol*. 2009;28(3):354–63. doi:10.1037/a0014400.
237. Baumeister RF, DeWall CN. *The Inner Dimensions of Social Exclusion: Intelligent Thought and Self-Regulation among Rejected Persons*. Sydney Sym. (Williams KD, Forgas JP, von Hippel W, eds.). New York: Psychology Press; 2005.
238. Campbell WK, Krusemark EA, Dyckman KA, et al. A magnetoencephalography investigation of neural correlates for social exclusion and self-control. *Soc Neurosci*. 2006;1(2):124–34. doi:10.1080/17470910601035160.
239. Zeeb FD, Wong AC, Winstanley CA. Differential effects of environmental enrichment, social-housing, and isolation-rearing on a rat gambling task: Dissociations between impulsive action and risky decision-making. *Psychopharmacology (Berl)*. 2013;225:381–95.
240. Barrot M, Wallace DL, Bolaños CA, et al. Regulation of anxiety and initiation of sexual behavior by CREB in the nucleus accumbens. *Proc Natl Acad Sci*. 2005;102(23):8357–62. doi:10.1073/pnas.0500587102.
241. Evans J, Sun Y, McGregor A, Connor B. Allopregnanolone regulates neurogenesis and depressive/anxiety-like behavior in a social isolation rodent model of chronic stress. *Neuropharmacology*. 2012;63:1315–26.
242. Suomi SJ, Eisele CD, Grady SA, Harlow HF. Depressive behavior in adult monkeys following separation from family environment. *J Abnorm Psychol*. 1975; 84:576–78.
243. Wilson S. *Evolution for Everyone*. New York: Delacorte Press; 2007.
244. Cacioppo JT, Cacioppo S, Capitanio JP, Cole SW. *The Neuroendocrinology of Social Isolation*. Vol 66.; 2015. doi:10.1146/annurev-psych-010814-015240.

References for Chapter 3

1. Northcutt RG. Understanding vertebrate brain evolution. *Integr Comp Biol*. 2002;42(4):743–56. doi:10.1093/icb/42.4.743.
2. Northcutt RG. Evolving large and complex brains. *Science*. 2011;331(6018):721–25. doi:10.1126/science.1201765.
3. Cacioppo S, Cacioppo JT. Cognizance of the neuroimaging methods for studying the social brain. Obhi SS, Cross ES, eds. *Shar Represent*. 2017;(May):86–106. doi:10.1017/CBO9781107279353.006.
4. Sandrone S, Bacigaluppi M. Learning from default mode network: The predictive value of resting state in traumatic brain injury. *J Neurosci*. 2012;32(6):1915–17. doi:10.1523/JNEUROSCI.5637-11.2012.

5. Cacioppo JT, Berntson GG, Lorig TS, Norris CJ, Rickett E, Nusbaum H. Just because you're imaging the brain doesn't mean you can stop using your head: A primer and set of first principles. *J Pers Soc Psychol.* 2003;85(4):650–61. doi:10.1037/0022-3514.85.4.650.
6. Bullmore E, Sporns O. Complex brain networks: Graph theoretical analysis of structural and functional systems. *Nat Rev Neurosci.* 2009;10(3):186–98. doi:10.1038/nrn2575.
7. Cacioppo S, Frum C, Asp E, Weiss RM, Lewis JW, Cacioppo JT. A quantitative meta-analysis of functional imaging studies of social rejection. *Sci Rep.* 2013;3. doi:10.1038/srep02027.
8. Wager TD, Lindquist MA, Nichols TE, Kober H, Van Snellenberg JX. Evaluating the consistency and specificity of neuroimaging data using meta-analysis. *Neuroimage.* 2009;45(1 Suppl):S210–21. doi:10.1016/j.neuroimage.2008.10.061.
9. Cacioppo JT, Tassinary LG, Berntson GG. *Handbook of Psychophysiology.* 4th ed. New York: Cambridge University Press; 2017.
10. Cacioppo JT, Cacioppo S, Capitanio JP, Cole SW. The neuroendocrinology of social isolation. *Annu Rev Psychol.* 2015;66(1):733–67. doi:10.1146/annurev-psych-010814–015240.
11. Stanley DA, Adolphs R. Toward a neural basis for social behavior. *Neuron.* 2013;80(3):816–26. doi:10.1016/j.neuron.2013.10.038.
12. Kennedy DP, Adolphs R. The social brain in psychiatric and neurological disorders. *Trends Cogn Sci.* 2012;16(11):559–72. doi:10.1016/j.tics.2012.09.006.
13. Saxe R, Kanwisher N. People thinking about thinking people. The role of the temporo-parietal junction in "theory of mind." *Neuroimage.* 2003;19(4):1835–42. doi:10.1016/S1053-8119(03)00230-1.
14. Downing PE, Jiang Y, Shuman M, Kanwisher N. A cortical area selective for visual processing of the human body. *Science.* 2001;293(5539):2470–3.
15. Sabatinelli D, Fortune EE, Li Q, et al. Emotional perception: Meta-analyses of face and natural scene processing. *Neuroimage.* 2011;54(3):2524–33. doi:10.1016/j.neuroimage.2010.10.011.
16. Mather M, Cacioppo JT, Kanwisher N. Introduction to the special section: 20 years of fMRI—What has it done for understanding cognition? *Perspect Psychol Sci.* 2013;8(1):41–43. doi:10.1177/1745691612469036.
17. Mather M, Cacioppo JT, Kanwisher N. How fMRI can inform cognitive theories. *Perspect Psychol Sci.* 2013;8(1):108–13. doi:10.1177/1745691612469037.
18. Raichle ME, MacLeod AM, Snyder AZ, Powers WJ, Gusnard DA, Shulman GL. A default mode of brain function. *Proc Natl Acad Sci.* 2001;98(2):676–82. doi:10.1073/pnas.98.2.676.
19. Seghier ML. The angular gyrus: Multiple functions and multiple subdivisions. *Neuroscientist.* 2013;19(1):43–61. doi:10.1177/1073858412440596.
20. Cacioppo JT, Tassinary LG. Inferring psychological significance from physiological signals. *Am Psychol.* 1990;45(1):16–28. doi:10.1037/0003-066X.45.1.16.
21. Button KS, Ioannidis JPA, Mokrysz C, et al. Power failure: Why small sample size undermines the reliability of neuroscience. *Nat Rev Neurosci.* 2013;14(5):365–76. doi:10.1038/nrn3475.
22. Van Overwalle F, Baetens K. Understanding others' actions and goals by mirror and mentalizing systems: A meta-analysis. *Neuroimage.* 2009;48(3):564–84. doi:10.1016/j.neuroimage.2009.06.009.
23. Astolfi L, Toppi J, De Vico Fallani F, et al. Imaging the social brain by simultaneous hyperscanning during subject interaction. *IEEE Intell Syst.* 2011;26(5):38–45. doi:10.1109/MIS.2011.61.
24. Astolfi L, Toppi J, Borghini G, et al. Study of the functional hyperconnectivity between couples of pilots during flight simulation: An EEG hyperscanning study. *Conf Proc IEEE Eng Med Biol So.* 2011;2011:2338–41. doi:10.1109/IEMBS.2011.6090654.

25. Babiloni F, Cincotti F, Mattia D, et al. Hypermethods for EEG hyperscanning. *Conf Proc IEEE Eng Med Biol Soc*. 2006;1:3666–69. doi:10.1109/IEMBS.2006.260754.
26. Montague PR, Berns GS, Cohen JD, et al. Hyperscanning: Simultaneous fMRI during linked social interactions. *Neuroimage*. 2002;16(4):1159–64. http://www.ncbi.nlm.nih.gov/pubmed/12202103. Accessed September 10, 2017.
27. Goossens L, van Roekel E, Verhagen M, et al. The genetics of loneliness. *Perspect Psychol Sci*. 2015;10(2):213–26. doi:10.1177/1745691614564878.
28. Sweatt JD, Meaney MJ, Nestler E, Akbarian S. *Epigenetic Regulation in the Nervous System: Basic Mechanisms and Clinical Impact*. Waltham, MA: Academic Press; 2013.
29. Kumsta R, Heinrichs M. Oxytocin, stress and social behavior: Neurogenetics of the human oxytocin system. *Curr Opin Neurobiol*. 2013;23(1):11–16. doi:10.1016/j.conb.2012.09.004.
30. Cacioppo JT, Cacioppo S, Cole SW. Social neuroscience and social genomics: The emergence of multi-level integrative analyses. *Int J Psychol Res*. 2013;6(SPEC. ISSUE):1–6.
31. Cole SW, Hawkley LC, Arevalo JM, Sung CY, Rose RM, Cacioppo JT. Social regulation of gene expression in human leukocytes. *Genome Biol*. 2007;8(9):R189. doi:10.1186/gb-2007-8-9-r189.
32. Cole SW, Hawkley LC, Arevalo JMG, Cacioppo JT. Transcript origin analysis identifies antigen-presenting cells as primary targets of socially regulated gene expression in leukocytes. *Proc Natl Acad Sci*. 2011;108(7):3080–5. doi:10.1073/pnas.1014218108.
33. Harari YN. *Sapiens: A Brief History of Humankind*. New York: Harper Collins; 2015.
34. Hublin J-J, Ben-Ncer A, Bailey SE, et al. New fossils from Jebel Irhoud (Morocco) and the pan-African origin of *Homo sapiens*. *Nature*. 2017;546(7657):289–92. doi:10.1038/nature22336.
35. Pearce E, Stringer C, Dunbar RIM. New insights into differences in brain organization between Neanderthals and anatomically modern humans. *Proc R Soc B Biol Sci*. 2013;280(1758):20130168. doi:10.1098/rspb.2013.0168.
36. Gibbons A. Neandertals mated early with modern humans. *Science*. 2017;357(6346):14. doi:10.1126/science.357.6346.14.
37. Whiten A, Van Schaik CP. The evolution of animal "cultures" and social intelligence. *Philos Trans R Soc Lond B Biol Sci*. 2007;362(1480):603–20.
38. Henrich J. *The Secret of Our Success: How Culture Is Driving Human Evolution, Domesticating Our Species, and Making Us Smarter*. Princeton, NJ: Princeton University Press; 2016.
39. Coolidge FL, Wynn T. An introduction to cognitive archaeology. *Curr Dir Psychol Sci*. 2016;25(6):386–92. doi:10.1177/0963721416657085.
40. Sousa AMM, Meyer KA, Santpere G, Gulden FO, Sestan N. Evolution of the human nervous system function, structure, and development. *Cell*. 2017;170(2):226–47. doi:10.1016/j.cell.2017.06.036.
41. Gamble C, Gowlett J, Dunbar R. *Thinking Big: How the Evolution of Social Life Shaped the Human Mind*. London, UK: Thames & Hudson; 2014.
42. Silk JB. Social components of fitness in primate groups. *Science*. 2007;317(5843):1347–51. doi:10.1126/science.1140734.
43. Jerison HJ. *Gross Brain Indices and the Meaning of Brain Size*. New York: Academic Press; 1973.
44. Barton RA. Neocortex size and behavioural ecology in primates. *Proc R Soc London Ser B*. 1996;263:173–7.
45. Harvey PH, Krebs JR. Comparing brains. *Science*. 1990;249(4965):140–46. doi:10.1126/science.2196673.
46. DeCasien AR, Williams SA, Higham JP. Primate brain size is predicted by diet but not sociality. *Nat Ecol Evol*. 2017;1(5):112. doi:10.1038/s41559-017-0112.

47. Dunbar RI. Evolutionary basis of the social brain. In: Decety J, Cacioppo JT, eds. *The Oxford Handbook of Social Neuroscience*. Oxford, UK: Oxford University Press; 2011:28–38.
48. Ross MD, Owren MJ, Zimmermann E. The evolution of laughter in great apes and humans. *Commun Integr Biol*. 2010;3(2):191–4. doi:10.4161/cib.3.2.10944.
49. Blakemore S-J, Wolpert DM, Frith CD. Central cancellation of self-produced tickle sensation. *Nat Neurosci*. 1998;1(7):635–40. doi:10.1038/2870.
50. Wattendorf E, Westermann B, Fiedler K, Kaza E, Lotze M, Celio MR. Exploration of the neural correlates of ticklish laughter by functional magnetic resonance imaging. *Cereb Cortex*. 2013;23(6):1280–9. doi:10.1093/cercor/bhs094.
51. Darwin C. The expression of emotion in animals and man. *London Methuen(1877), A Biogr sketch an*. 1872. https://scholar.google.com/scholar?hl=en&q=darwin+1872+expression&btnG=&as_sdt=1%2C14&as_sdtp=&oq=darwin+1872+e. Accessed September 16, 2017.
52. Panksepp JB, Lahvis GP. Rodent empathy and affective neuroscience. *Neurosci Biobehav Rev*. 2011;35(9):1864–75. doi:10.1016/j.neubiorev.2011.05.013.
53. Barger N, Hanson KL, Teffer K, Schenker-Ahmed NM, Semendeferi K. Evidence for evolutionary specialization in human limbic structures. *Front Hum Neurosci*. 2014;8:277. doi:10.3389/fnhum.2014.00277.
54. Semendeferi K, Teffer K, Buxhoeveden DP, et al. Spatial organization of neurons in the frontal pole sets humans apart from great apes. *Cereb Cortex*. 2011;21(7):1485–97. doi:10.1093/cercor/bhq191.
55. Semendeferi K, Lu A, Schenker N, Damasio H. Humans and great apes share a large frontal cortex. *Nat Neurosci*. 2002;5(3):272–76. doi:10.1038/nn814.
56. Semendeferi K, Damasio H, Frank R, Van Hoesen GW. The evolution of the frontal lobes: A volumetric analysis based on three-dimensional reconstructions of magnetic resonance scans of human and ape brains. *J Hum Evol*. 1997;32(4):375–88. doi:10.1006/jhev.1996.0099.
57. Dunbar RIM, Shultz S. Evolution in the social brain. *Science*. 2007;317(5843):1344–7. doi:10.1126/science.1145463.
58. Dunbar RIM. The social brain: Mind, language, and society in evolutionary perspective. *Annu Rev Anthropol*. 2003;32(1):163–81. doi:10.1146/annurev.anthro.32.061002.093158.
59. Dunbar RIM. The social brain hypothesis and its implications for social evolution. *Ann Hum Biol*. 2009;36(5):562–72.
60. Opie C, Atkinson QD, Dunbar RIM, Shultz S. Male infanticide leads to social monogamy in primates. *Proc Natl Acad Sci*. 2013;110(33):13328–32. doi:10.1073/pnas.1307903110.
61. Gavrilets S. Human origins and the transition from promiscuity to pair-bonding. *Proc Natl Acad Sci*. 2012;109(25):9923–8. doi:10.1073/pnas.1200717109.
62. Nelson CA, Furtado EA, Fox NA, Zeanah CH. The deprived human brain. *Am Sci*. 2009;97:222–29. http://www.americanscientist.org/issues/id.6380,y.2009,no.3,css.print/issue.aspx.
63. Grossmann T. The development of social brain functions in infancy. *Psychol Bull*. 2015;141(6):1266–87. doi:10.1037/bul0000002.
64. Fausey C, Jayaraman S, Smith L. From faces to hands: Changing visual input in the first two years. *Cognition*. 2016;152:101–7.
65. Jayaraman S, Fausey C, Smith L. The faces in infant-perspective scenes change over the first year of life. *PLoS One*. 2015;10(5):e0123780.
66. Jayaraman S, Fausey C, Smith L. Why are faces denser in the visual experiences of younger than older infants? *Dev Psychol*. 2017;53:38–49.
67. Kuhl PK. Social mechanisms in early language acquisition: Understanding integrated brain systems supporting language. In: Decety J, Cacioppo JT, eds. *The*

Oxford Handbook of Social Neuroscience. New York: Oxford University Press; 2011:649–67.
68. Kuhl PK, Tsao F-M, Liu H-M. Foreign-language experience in infancy: Effects of short-term exposure and social interaction on phonetic learning. *Proc Natl Acad Sci*. 2003;100(15):9096–101. doi:10.1073/pnas.1532872100.
69. Kuhl PK. Is speech learning "gated" by the social brain? *Dev Sci*. 2007;10(1):110–20. doi:10.1111/j.1467-7687.2007.00572.x.
70. Gogtay N, Giedd JN, Lusk L, et al. Dynamic mapping of human cortical development during childhood through early adulthood. *Proc Natl Acad Sci*. 2004;101(21):8174–9. doi:10.1073/pnas.0402680101.
71. Vasung L, Lepage C, Radoš M, et al. Quantitative and qualitative analysis of transient fetal compartments during prenatal human brain development. *Front Neuroanat*. 2016;10:11. doi:10.3389/fnana.2016.00011.
72. Petrican R, Burris CT, Bielak T, Schimmack U, Moscovitch M. For my eyes only: Gaze control, enmeshment, and relationship quality. *J Pers Soc Psychol*. 2010;100(6):1111–23. doi:10.1037/a0021714.
73. Thompson PM, Vidal C, Giedd JN, et al. Mapping adolescent brain change reveals dynamic wave of accelerated gray matter loss in very early-onset schizophrenia. *Proc Natl Acad Sci*. 2001;98(20):11650–5. doi:10.1073/pnas.201243998.
74. Baron-Cohen S. *Mindblindness: An Essay on Autism and Theory of Mind*. Cambridge, MA: Bradford; 1995.
75. Baron-Cohen S. *Theory of Mind*. Hove: Psychology Press; 2007. http://www.worldcat.org/title/theory-of-mind/oclc/488257120&referer=brief_results. Accessed October 18, 2010.
76. Perrett DII, Emery NJJ. Understanding the intentions of others from visual signals: Neurophysiological evidence. *Curr Psychol Cogn*. 1994;13(5):683–94.
77. Williams KD. *Ostracism: The Power of Silence*. New York: Guilford Press; 2001.
78. Petrican R, Schimmack U. The role of dorsolateral prefrontal function in relationship commitment. *J Res Pers*. 2008;42(4):1130–35. doi:10.1016/j.jrp.2008.03.001.
79. Insel TR. The challenge of translation in social neuroscience: A review of oxytocin, vasopressin, and affiliative behavior. *Neuron*. 2010;65(6):768–79. doi:10.1016/j.neuron.2010.03.005.
80. Carter CS, Grippo AJ, Pournajafi-Nazarloo H, Ruscio MG, Porges SW. Oxytocin, vasopressin and sociality. *Prog Brain Res*. 2008;170:331–36. doi:10.1016/S0079-6123(08)00427-5.
81. Williams JR, Insel TR, Harbaugh CR, Carter CS. Oxytocin administered centrally facilitates formation of a partner preference in female prairie voles (*Microtus ochrogaster*). *J Neuroendocrinol*. 1994;6(3):247–50. doi:10.1111/j.1365-2826.1994.tb00579.x.
82. Carter CS, Williams JR, Witt DM, Insel TR. Oxytocin and social bonding. *Ann NY Acad Sci*. 1992;652(1):204–11. doi:10.1111/j.1749-6632.1992.tb34356.x.
83. Donaldson ZR, Young LJ. Oxytocin, vasopressin, and the neurogenetics of sociality. *Science*. 2008;322(5903):900–904. doi:10.1126/science.1158668.
84. Young LJ, Wang Z. The neurobiology of pair bonding. *Nat Neurosci*. 2004;7(10):1048–54. doi:10.1038/nn1327.
85. Nagasawa M, Mitsui S, En S, et al. Oxytocin-gaze positive loop and the co-evolution of human-dog bond. *Science*. 2015;348(6232):333–36. doi:10.1126/science.1261022.
86. Scheele D, Striepens N, Gunturkun O, et al. Oxytocin modulates social distance between males and females. *J Neurosci*. 2012;32(46):16074–9. doi:10.1523/JNEUROSCI.2755-12.2012.
87. Baek EC, Scholz C, O'Donnell MB, Falk EB. The value of sharing information: A neural account of information transmission. *Psychol Sci*. 2017;28(7):851–61. doi:10.1177/0956797617695073.

88. Bartra O, McGuire JT, Kable JW. The valuation system: A coordinate-based meta-analysis of BOLD fMRI experiments examining neural correlates of subjective value. *Neuroimage*. 2013;76:412–27. doi:10.1016/j.neuroimage.2013.02.063.
89. Murray RJ, Schaer M, Debbané M. Degrees of separation: A quantitative neuro-imaging meta-analysis investigating self-specificity and shared neural activation between self- and other-reflection. *Neurosci Biobehav Rev*. 2012;36(3):1043–59. doi:10.1016/j.neubiorev.2011.12.013.
90. Dufour N, Redcay E, Young L, et al. Similar brain activation during false belief tasks in a large sample of adults with and without autism. *PLoS One*. 2013;8(9). e75468 doi:10.1371/journal.pone.0075468.
91. Niven JE, Laughlin SB. Energy limitation as a selective pressure on the evolution of sensory systems. *J Exp Biol*. 2008;211(11):1792–804. doi:10.1242/jeb.017574.
92. Burrows M, Rogers SM, Ott SR. Epigenetic remodelling of brain, body and behaviour during phase change in locusts. *Neural Syst Circuits*. 2011;1(1):11. doi:10.1186/2042-1001-1-11.
93. Ott SR, Rogers SM. Gregarious desert locusts have substantially larger brains with altered proportions compared with the solitarious phase. *Proc R Soc B Biol Sci*. 2010;277(1697):3087–96. doi:10.1098/rspb.2010.0694.
94. Cacioppo S, Capitanio JP, Cacioppo JT. Toward a neurology of loneliness. *Psychol Bull*. 2014;140(6):1464–504. doi:10.1037/a0037618.
95. Gheusi G, Ortega-Perez I, Murray K, Lledo P. A niche for adult neurogenesis in social behavior. *Behav Brain Res*. 2009;200(2):315–22. doi:10.1016/j.bbr.2009.02.006.
96. Makinodan M, Rosen KM, Ito S, Corfas G. A critical period for social experience-dependent oligodendrocyte maturation and myelination. *Science*. 2012;337(6100): 1357–60. doi:10.1126/science.1220845.
97. Liu J, Dietz K, DeLoyht JM, et al. Impaired adult myelination in the prefrontal cortex of socially isolated mice. *Nat Neurosci*. 2012;15(12):1621–3. doi:10.1038/nn.3263.
98. Kanai R, Bahrami B, Duchaine B, Janik A, Banissy MJ, Rees G. Brain structure links loneliness to social perception. *Curr Biol*. 2012;22(20):1975–9. doi:10.1016/j.cub.2012.08.045.
99. Kanai R, Bahrami B, Roylance R, Rees G. Online social network size is reflected in human brain structure. *Proc Biol Sci*. 2012;279(1732):1327–34. doi:10.1098/rspb.2011.1959.
100. Cacioppo JT, Cacioppo S, Capitanio JP, Cole SW. The neuroendocrinology of social isolation. *Annu Rev Psychol*. 2015;66(1):733–67. doi:10.1146/annurev-psych-010814-015240.
101. Nakagawa S, Takeuchi H, Taki Y, et al. White matter structures associated with loneliness in young adults. *Sci Rep*. 2015;5(1):17001. doi:10.1038/srep17001.
102. Layden EA, Cacioppo JT, Cacioppo S, et al. Perceived social isolation is associated with altered functional connectivity in neural networks associated with tonic alertness and executive control. *Neuroimage*. 2017;145(March 2016):58–73. doi:10.1016/j.neuroimage.2016.09.050.
103. Cacioppo JT, Cacioppo S, Boomsma DI. Evolutionary mechanisms for loneliness. *Cogn Emot*. 2014;28(1):3–21. doi:10.1080/02699931.2013.837379.
104. Cacioppo JT, Cacioppo S, Cole SW, Capitanio JP, Goossens L, Boomsma DI. Loneliness across phylogeny and a call for comparative studies and animal models. *Perspect Psychol Sci*. 2015;10(2):202–12. doi:10.1177/1745691614564876.
105. Cacioppo S, Cacioppo JT. Research in social neuroscience: How perceived social isolation, ostracism, and romantic rejection affect our brain. In: Riva P, Eck J, eds. *The Many Faces of Social Exclusion*. New York: Springer International Publishing; 2016:73–87. doi:10.1007/978-3-319-33033-4.
106. Cacioppo JT, Norris CJ, Decety J, Monteleone G, Nusbaum H. In the eye of the beholder: Individual differences in perceived social isolation predict regional brain

activation to social stimuli. *J Cogn Neurosci*. 2009;21(1):83–92. doi:10.1162/jocn.2009.21007.
107. Cacioppo JT, Chen HY, Cacioppo S. Reciprocal influences between loneliness and self-centeredness: A cross-lagged panel analysis in a population-based sample of African American, Hispanic, and Caucasian adults. *Personal Soc Psychol Bull*. 2017;43(8):1125–35. doi:10.1177/0146167217705120.
108. Beadle JN, Keady B, Brown V, Tranel D, Paradiso S. Trait empathy as a predictor of individual differences in perceived loneliness. *Psychol Rep*. 2012;110(1):3–15. doi:10.2466/07.09.20.PR0.110.1.3-15.
109. Rilling J, Gutman D, Zeh T, Pagnoni G, Berns G, Kilts C. A neural basis for social cooperation. *Neuron*. 2002;35(2):395–405. http://www.ncbi.nlm.nih.gov/pubmed/12160756.
110. Fliessbach K, Weber B, Trautner P, et al. Social comparison affects reward-related brain activity in the human ventral striatum. *Science*. 2007;318(5854):1305–8. doi:10.1126/science.1145876.
111. de Quervain DJ-F, Fischbacher U, Treyer V, et al. The neural basis of altruistic punishment. *Science*. 2004;305(5688):1254–8. doi:10.1126/science.1100735.
112. Ortigue S, Bianchi-Demicheli F, Patel N, Frum C, Lewis JW. Neuroimaging of love: fMRI meta-analysis evidence toward new perspectives in sexual medicine. *J Sex Med*. 2010;7(11):3541–52. doi:10.1111/j.1743-6109.2010.01999.x.
113. Cacioppo S, Bianchi-Demicheli F, Frum C, Pfaus JG, Lewis JW. The common neural bases between sexual desire and love: A multilevel kernel density fMRI analysis. *J Sex Med*. 2012;9(4):1048–54. doi:10.1111/j.1743-6109.2012.02651.x.
114. Cacioppo JT, Ernst JM, Burleson MH, et al. Lonely traits and concomitant physiological processes: The MacArthur social neuroscience studies. *Int J Psychophysiol*. 2000;35(2):143–54. http://www.ncbi.nlm.nih.gov/pubmed/10677643.
115. Inagaki TK, Muscatell KA, Moieni M, et al. Yearning for connection? Loneliness is associated with increased ventral striatum activity to close others. *Soc Cogn Affect Neurosci*. 2016;11(7):1096–101. doi:10.1093/scan/nsv076.
116. Powers KE, Wagner DD, Norris CJ, Heatherton TF. Socially excluded individuals fail to recruit medial prefrontal cortex for negative social scenes. *Soc Cogn Affect Neurosci*. 2013;8(2):151–57. doi:10.1093/scan/nsr079.
117. Cacioppo S, Cacioppo JT. Do you feel lonely? You are not alone: Lessons from social neuroscience. *Front Young Minds*. 2013;1(November):1–6. doi:10.3389/frym.2013.00009.

References for Chapter 4

1. Erle TM, Topolinski S. The grounded nature of psychological perspective-taking. *J Pers Soc Psychol*. 2017;112(5):683–95. doi:10.1037/pspa0000081.
2. Hillis AE. Inability to empathize: Brain lesions that disrupt sharing and understanding another's emotions. *Brain*. 2014;137(4):981–97.
3. Hatfield E, Cacioppo JT, Rapson RL. Emotional contagion: Cambridge studies in emotion and social interaction. *Stat Med*. 1994;21:1089–101.
4. Hatfield E, Cacioppo JT, Rapson RL. *Emotional Contagion*. New York: Cambridge University Press; 1994.
5. Prochazkova E, Kret ME. Connecting minds and sharing emotions through mimicry: A neurocognitive model of emotional contagion. *Neurosci Biobehav Rev*. 2017;80:99–114. doi:10.1016/j.neubiorev.2017.05.013.
6. Cacioppo JT, Berntson GG, Larsen JT, Poehlmann KM, Ito TA. The psychophysiology of emotion. *Handb Emot*. 2000;2:173–91.
7. Nyström P. The infant mirror neuron system studied with high density EEG. *Soc Neurosci*. 2008;3(3–4):334–47.

8. Cacioppo JT, Tassinary LG, Fridlund AJ. The skeletomotor system. In: Cacioppo JT, Tassinary LG, eds. *Principles of Psychophysiology: Physical, Social, and Inferential Elements*. New York: Cambridge University Press; 1990:325–84.
9. Cacioppo JT, Petty RE. Electromyograms as measures of extent and affectivity of information processing. *Am Psychol*. 1981;36(5):441–56. doi:10.1037/0003-066X.36.5.441.
10. Carr L, Iacoboni M, Dubeau M-C, Mazziotta JC, Lenzi GL. Neural mechanisms of empathy in humans: A relay from neural systems for imitation to limbic areas. *Proc Natl Acad Sci*. 2003;100(9):5497–502. doi:10.1073/pnas.0935845100.
11. Decety J, Jackson PL. A social-neuroscience perspective on empathy. *Curr Dir Psychol Sci*. 2006;15(2):54–58.
12. Decety J, Cacioppo JT. Dissecting the neural mechanisms mediating empathy. *Emot Rev*. 2011;3(1):92–108. doi:10.1177/1754073910374662.
13. Lanzetta JT, Englis BG. Expectations of cooperation and competition and their effects on observers' vicarious emotional responses. *J Pers Soc Psychol*. 1989;56(4):543–54. doi:10.1037/0022-3514.56.4.543.
14. Bernhardt BC, Singer T. The neural basis of empathy. *Annu Rev Neurosci*. 2012;35(1):1–23. doi:10.1146/annurev-neuro-062111-150536.
15. Molenberghs P, Johnson H, Henry JD, Mattingley JB. Understanding the minds of others: A neuroimaging meta-analysis. *Neurosci Biobehav Rev*. 2016;65:276–91.
16. Lamm C, Decety J, Singer T. Meta-analytic evidence for common and distinct neural networks associated with directly experienced pain and empathy for pain. *Neuroimage*. 2011;54(3):2492–502. doi:10.1016/j.neuroimage.2010.10.014.
17. Decety J, Cacioppo S. The speed of morality: A high-density electrical neuroimaging study. *J Neurophysiol*. 2012;108(11):3068–72. doi:10.1152/jn.00473.2012.
18. Stanley DA, Adolphs R. Toward a neural basis for social behavior. *Neuron*. 2013;80(3):816–26. doi:10.1016/j.neuron.2013.10.038.
19. Zahn R, Oliveira-Souza R de, Moll J. The neuroscience of moral cognition and emotion. In: Decety J, Cacioppo JT, eds. *The Handbook of Social Neuroscience*. New York: Oxford University Press; 2011:477–90.
20. Moll J, Zahn R, Krueger F, Grafman J. The neural basis of human moral cognition. *Nat Rev*. 2005;6:799–809. doi:10.1038/nrn1768.
21. Moll J, Schulkin J. Social attachment and aversion in human moral cognition. *Neurosci Biobehav Rev*. 2009;33(3):456–65. doi:10.1016/j.neubiorev.2008.12.001.
22. Decety J, Chen C, Harenski C, Kiehl KA. An fMRI study of affective perspective taking in individuals with psychopathy: Imagining another in pain does not evoke empathy. *Front Hum Neurosci*. 2013;7(September):489. doi:10.3389/fnhum.2013.00489.
23. Anderson NE, Kiehl KA. The psychopath magnetized: Insights from brain imaging. *Trends Cogn Sci*. 2012;16(1):52–60. doi:10.1016/j.tics.2011.11.008.
24. Kiehl KA, Buckholtz JW. Inside the mind of a psychopath. *Sci Am Mind*. 2010;21(4):22–29. doi:10.1038/scientificamericanmind0910-22.
25. Langford DJ, Crager SE, Shehzad Z, et al. Social modulation of pain as evidence for empathy in mice. *Science*. 2006;312(5782):1967–70. doi:10.1126/science.1128322.
26. Martin LJ, Hathaway G, Isbester K, et al. Reducing social stress elicits emotional contagion of pain in mouse and human strangers. *Curr Biol*. 2015;25(3):326–32.
27. Ben-Ami Bartal I, Decety J, Mason P. Empathy and pro-social behavior in rats. *Science*. 2011;334(6061):1427–30. doi:10.1126/science.1210789.
28. Cialdini RB, Schaller M, Houlihan D, Arps K, Fultz J, Beaman AL. Empathy-based helping: Is it selflessly or selfishly motivated? *J Pers Soc Psychol*. 1987;52(4):749–58. doi:10.1037/0022-3514.52.4.749.
29. Bartal IBA, Shan H, Molasky N, et al. Pro-social behavior in rats requires an affective motivation. *bioRxiv*. 2016;44180.

30. Byrne RW, Russon AE. Learning by imitation: A hierarchical approach. *Behav Brain Sci*. 1998;21(5):667–84.
31. Moody E, McIntosh D. Imitation in autism findings and controversies. In: Rogers SJ, Williams JHG, eds. *Imitation and the Social Mind: Autism and Typical Development*. New York: Guilford Press; 2006:71–88.
32. Subiaul F, Renner E, Krajkowski E. The comparative study of imitation mechanisms in non-human primates. Obhi SS, Cross ES, eds. *Shar Represent*. 2017;(May): 109–35. doi:10.1017/CBO9781107279353.007.
33. Subiaul F. What's special about human imitation? A comparison with enculturated apes. *Behav Sci (Basel)*. 2016;6(3):13. doi:10.3390/bs6030013.
34. Caspers S, Zilles K, Laird AR, Eickhoff SB. ALE meta-analysis of action observation and imitation in the human brain. *Neuroimage*. 2010;50(3):1148–67. doi:10.1016/j.neuroimage.2009.12.112.
35. Iacoboni M. Neurobiology of imitation. *Curr Opin Neurobiol*. 2009;19(6):661–65.
36. Grafton ST. Embodied cognition and the simulation of action to understand others. *Ann NY Acad Sci*. 2009;1156(1):97–117.
37. Legare CH, Nielsen M. Imitation and innovation: The dual engines of cultural learning. *Trends Cogn Sci*. 2015;19(11):688–99.
38. Buckner RL, Carroll DC. Self-projection and the brain. *Trends Cogn Sci*. 2007;11(2): 49–57. doi:10.1016/j.tics.2006.11.004.
39. Mitchell JP, Banaji MR, Macrae CN. The link between social cognition and self-referential thought in the medial prefrontal cortex. *J Cogn Neurosci*. 2005;17(8):1306–15. doi:10.1162/0898929055002418.
40. Krueger F, Barbey AK, Grrafman J, Grafman J. The medial prefrontal cortex mediates social event knowledge. *Trends Cogn Sci*. 2009;13(3):103–9. doi:10.1016/j.tics.2008.12.005.
41. Epley N, Converse BA, Delbosc A, Monteleone GA, Cacioppo JT. Believers' estimates of God's beliefs are more egocentric than estimates of other people's beliefs. *Proc Natl Acad Sci*. 2009;106(51):21533–8. doi:10.1073/pnas.0908374106.
42. Todd AR, Forstmann M, Burgmer P, Brooks AW, Galinsky AD. Anxious and egocentric: How specific emotions influence perspective taking. *J Exp Psychol Gen*. 2015;144(2):374–91. doi:10.1037/xge0000048.
43. Snyder M, Tanke ED, Berscheid E. Social perception and interpersonal behavior: On the self-fulfilling nature of social stereotypes. *J Pers Soc Psychol*. 1977;35(9): 656–66.
44. Mitchell JP, Macrae CN, Banaji MR. Dissociable medial prefrontal contributions to judgments of similar and dissimilar others. *Neuron*. 2006;50(4):655–63. doi:10.1016/j.neuron.2006.03.040.
45. Gallese V, Keysers C, Rizzolatti G. A unifying view of the basis of social cognition. *Trends Cogn Sci*. 2004;8(9):396–403. doi:10.1016/j.tics.2004.07.002.
46. Schurz M, Kogler C, Scherndl T, Kronbichler M, Kühberger A. Differentiating self-projection from simulation during mentalizing: Evidence from fMRI. *PLoS One*. 2015;10(3):e0121405. doi:10.1371/journal.pone.0121405.
47. Rizzolatti G, Fabbri-Destro M. The mirror neuron system. Volume 2. In: Berntson GG, Cacioppo JT, eds. *Handbook of Neuroscience for the Behavioral Sciences*. New York: John Wiley & Sons; 2009:337–60.
48. Rizzolatti G, Craighero L. The mirror-neuron system. *Annu Rev Neurosci*. 2004;27:169–92. doi:10.1146/annurev.neuro.27.070203.144230.
49. Cacioppo S, Fontang F, Patel N, Decety J, Monteleone G, Cacioppo JT. Intention understanding over T: A neuroimaging study on shared representations and tennis return predictions. *Front Hum Neurosci*. 2014;8(Oct):781. doi:10.3389/fnhum.2014.00781.

50. Aglioti SM, Cesari P, Romani M, Urgesi C. Action anticipation and motor resonance in elite basketball players. *Nat Neurosci*. 2008;11(9):1109–16. doi:10.1038/nn.2182.
51. Cross ES, Calvo-Merino B. The impact of action expertise on shared representations. In: Obhi SS, Cross ES, eds. *Shared Representations*. Cambridge: Cambridge University Press; 2017:541–62. doi:10.1017/CBO9781107279353.027.
52. Ortigue S, Patel N, Bianchi-Demicheli F, Grafton ST. Implicit priming of embodied cognition on human motor intention understanding in dyads in love. *J Soc Pers Relat*. September 2010:1–15. doi:10.1177/0265407510378861.
53. Ortigue S, Bianchi-Demicheli F. Why is your spouse so predictable? Connecting mirror neuron system and self-expansion model of love. *Med Hypotheses*. 2008;71(6):941–44. http://www.ncbi.nlm.nih.gov/pubmed/18722062.
54. Cacioppo S, Juan E, Monteleone G. Predicting intentions of a familiar significant other beyond the mirror neuron system. *Front Behav Neurosci*. 2017;11(August):1–12. doi:10.3389/fnbeh.2017.00155.
55. Ortigue S, Sinigaglia C, Rizzolatti G, Grafton ST. Understanding actions of others: The electrodynamics of the left and right hemispheres. A high-density EEG neuroimaging study. Tractenberg RE, ed. *PLoS One*. 2010;5(8):13. doi:10.1371/journal.pone.0012160.
56. Ortigue S, Thompson JC, Parasuraman R, et al. Spatio-temporal dynamics of human intention understanding in temporo-parietal cortex: A combined EEG/fMRI repetition suppression paradigm. *PLoS One*. 2009;4(9):e6962. doi:10.1371/journal.pone.0006962.
57. Mahy CE, Moses LJ, Pfeifer JH. How and where: Theory-of-mind in the brain. *Dev Cogn Neurosci*. 2014;9:68–81.
58. Wimmer H, Perner J. Beliefs about beliefs: Representation and constraining function of wrong beliefs in young children's understanding of deception. *Cognition*. 1983;13(1):103–28.
59. Slaughter V. Theory of mind in infants and young children: A review. *Aust Psychol*. 2015;50(3):169–72.
60. Schurz M, Radua J, Aichhorn M, Richlan F, Perner J. Fractionating theory of mind: A meta-analysis of functional brain imaging studies. *Neurosci Biobehav Rev*. 2014;42:9–34. doi:10.1016/j.neubiorev.2014.01.009.
61. Amodio DM, Frith CD. Meeting of minds: The medial frontal cortex and social cognition. *Nat Rev Neurosci*. 2006;7(4):268–77. http://discovery.ucl.ac.uk/70261/.
62. Aichhorn M, Perner J, Kronbichler M, Staffen W, Ladurner G. Do visual perspective tasks need theory of mind? *Neuroimage*. 2006;30(3):1059–68.
63. Krienen FM, Tu PC, Buckner RL. Clan mentality: Evidence that the medical prefrontal cortex responds to close others. *J Neurosci*. 2010;30(41):13906–15.
64. Perner J, Roessler J. From infants' to children's appreciation of belief. *Trends Cogn Sci*. 2012;16(10):519–25.
65. Apperly IA, Samson D, Humphreys GW. Domain-specificity and theory of mind: Evaluating neuropsychological evidence. *Trends Cogn Sci*. 2005;9(12):572–77. doi:10.1016/j.tics.2005.10.004.
66. Young L, Camprodon JA, Hauser M, Pascual-Leone A, Saxe R. Disruption of the right temporoparietal junction with transcranial magnetic stimulation reduces the role of beliefs in moral judgments. *Proc Natl Acad Sci*. 2010;107(15):6753–8. doi:10.1073/pnas.0914826107.
67. Caruso EM, Burns ZC, Converse BA. Slow motion increases perceived intent. *Proc Natl Acad Sci*. 2016;113(33):9250–5.
68. Muscatell KA, Morelli SA, Falk EB, et al. Social status modulates neural activity in the mentalizing network. *Neuroimage*. 2012;60(3):1771–7.

69. de Lange FP, Spronk M, Willems RM, Toni I, Bekkering H. Complementary systems for understanding action intentions. *Curr Biol.* 2008;18(6):454–57.
70. Ma N, Vandekerckhove M, Van Overwalle F, Seurinck R, Fias W. Spontaneous and intentional trait inferences recruit a common mentalizing network to a different degree: Spontaneous inferences activate only its core areas. *Soc Neurosci.* 2011;6(2):123–38. doi:10.1080/17470919.2010.485884.
71. Suzuki S, Adachi R, Dunne S, Bossaerts P, O'Doherty JP. Neural mechanisms underlying human consensus decision-making. *Neuron.* 2015;86(2):591–602. doi:10.1016/j.neuron.2015.03.019.
72. Zentall TR. Perspectives on observational learning in animals. *J Comp Psychol.* 2012;126(2):114–28. doi:10.1037/a0025381.
73. Gariepy J-F, Watson KK, Du E, et al. Social learning in humans and other animals. *Front Neurosci.* 2014;8(March):1–13. doi:10.3389/fnins.2014.00058.
74. Penn DC, Povinelli DJ. On the lack of evidence that non-human animals possess anything remotely resembling a "theory of mind." *Philos Trans R Soc Lond B Biol Sci.* 2007;362(1480):731–44.
75. Kelley WM, Wagner DD, Heatherton TF. In search of a human self-regulation system. *Annu Rev Neurosci.* 2015;38:389–411. doi:10.1146/annurev-neuro-071013-014243.
76. Sliwa J, Freiwald WA. A dedicated network for social interaction processing in the primate brain. *Science.* 2017;356(6339):745–49. doi:10.1126/science.aam6383.
77. Krupenye C, Kano F, Hirata S, Call J, Tomasello M. Great apes anticipate that other individuals will act according to false beliefs. *Science.* 2016;354(6308):110–14. doi:10.1126/science.aaf8110.
78. de Waal FBM. Apes know what others believe. *Science.* 2016;354(6308):39–40. doi:10.1126/science.aai8851.
79. Hamilton AFDC. Reflecting on the mirror neuron system in autism: A systematic review of current theories. *Dev Cogn Neurosci.* 2013;3(1):91–105. doi:10.1016/j.dcn.2012.09.008.
80. Kim K, Johnson MK. Activity in ventromedial prefrontal cortex during self-related processing: Positive subjective value or personal significance? *Soc Cogn Affect Neurosci.* 2015;10(4):494–500.
81. Gillihan SJ, Farah MJ. Is self special? A critical review of evidence from experimental psychology and cognitive neuroscience. *Psychol Bull.* 2005;131(1):76–97. doi:10.1037/0033-2909.131.1.76.
82. Munevar G, Cole ML, Ye Y, et al. fMRI study of self vs. others' attributions of traits consistent with evolutionary understanding of the self. *Neurosci Discov.* 2014;2:3. doi:10.7243/2052-6946–2-3.
83. McGuire WJ, Padawer-Singer A. Trait salience in the spontaneous self-concept. *J Pers Soc Psychol.* 1976;33(6):743–54.
84. Fliessbach K, Weber B, Trautner P, et al. Social comparison affects reward-related brain activity in the human ventral striatum. *Science.* 2007;318(5854):1305–8. doi:10.1126/science.1145876.
85. Cacioppo JT, Berntson GG. *Social Neuroscience: Key Readings.* New York: Psychology Press; 2005.
86. Spelke ES, Kinzler KD. Core knowledge. *Dev Sci.* 2007;10(1):89–96. doi:10.1111/j.1467-7687.2007.00569.x.

References for Chapter 5

1. Cacioppo JT, Tassinary LG, Fridlund AJ. The skeletomotor system. In: Cacioppo JT, Tassinary LG, eds. *Principles of Psychophysiology: Physical, Social, and Inferential Elements.* New York: Cambridge University Press; 1990:325–84.

2. Prendergast PM. Anatomy of the face and neck. In: *Cosmetic Surgery: Art and Techniques*. New York: Springer; 2013:29–45. doi:10.1007/978-3-642-21837-8.
3. Farah MJ, Wilson KD, Drain M, Tanaka JN. What is "special" about face perception? *Psychol Rev*. 1998;105(3):482.
4. Calder AJ, Young AW, Keane J, Dean M. Configural information in facial expression perception. *J Exp Psychol Hum Percept Perform*. 2000;26(2):527–51. doi:10.1037//0096-1523.26.2.527.
5. Parr LA, Dove T, Hopkins WD. Why faces may be special: Evidence of the inversion effect in chimpanzees. *J Cogn Neurosci*. 1998;10(5):615–22. doi:10.1162/089892998563013.
6. Cross ES, Ramsey R, Liepelt R, Prinz W, Hamilton AFDC. The shaping of social perception by stimulus and knowledge cues to human animacy. *Philos Trans R Soc B Biol Sci*. 2016;371(1686):20150075. doi:10.1098/rstb.2015.0075.
7. Kanwisher N. The quest for the FFA and where it led. *J Neurosci*. 2017;37(5):1056–61. doi:10.1523/JNEUROSCI.1706-16.2016.
8. Kanwisher N, McDermott J, Chun MM. The fusiform face area: A module in human extrastriate cortex specialized for face perception. *J Neurosci*. 1997;17(11):4302–11. doi:10.1523/JNEUROSCI.17-11-04302.1997.
9. McCarthy G, Puce A, Gore JC, Allison T. Face-specific processing in the human fusiform gyrus. *J Cogn Neurosci*. 1997;9(5):605–10.
10. Downing PE, Chan AY, Peelen MV, Dodds CM, Kanwisher N. Domain specificity in visual cortex. *Cereb Cortex*. 2006;16(10):1453–61.
11. Behrmann M, Avidan G. Congenital prosopagnosia: Face-blind from birth. *Trends Cogn Sci*. 2005;9(4):180–87.
12. Gainotti G, Marra C. Differential contribution of right and left temporo-occipital and anterior temporal lesions to face recognition disorders. *Front Hum Neurosci*. 2011;5:55. doi:10.3389/fnhum.2011.00055.
13. Epstein R, Kanwisher N. A cortical representation of the local visual environment. *Nature*. 1998;392(6676):598–601.
14. Downing PE, Jiang Y, Shuman M, Kanwisher N. A cortical area selective for visual processing of the human body. *Science*. 2001;293(5539):2470–3.
15. Cohen L, Dehaene S, Naccache L, et al. The visual word form area. *Brain*. 2000;123(2):291–307.
16. Moeller S, Freiwald WA, Tsao DY. Patches with links: A unified system for processing faces in the macaque temporal lobe. *Science*. 2008;320(2008):1355–60. doi:10.1126/science.1157436.
17. Moeller S, Crapse T, Chang L, Tsao DY. The effect of face patch microstimulation on perception of faces and objects. *Nat Neurosci*. 2017;20(5):743–52. doi:10.1038/nn.4527.
18. Bushneil IWR, Sai F, Mullin JT. Neonatal recognition of the mother's face. *Br J Dev Psychol*. 1989;7(1):3–15. doi:10.1111/j.2044-835X.1989.tb00784.x.
19. Asch SE. Forming impressions of personalities. *J Abnorm Soc Psychol*. 1946;(41):258–90. doi:10.1037/h0060423.
20. Todorov A, Said CP, Engell AD, Oosterhof NN. Understanding evaluation of faces on social dimensions. *Trends Cogn Sci*. 2008;12(12):455–60. doi:10.1016/j.tics.2008.10.001.
21. Todorov A, Olivola CY, Dotsch R, Mende-Siedlecki P. Social attributions from faces: Determinants, consequences, accuracy, and functional significance. *Annu Rev Psychol*. 2015;66:519–45.
22. Oosterhof NN, Todorov A. The functional basis of face evaluation. *Proc Natl Acad Sci*. 2008;105(32):11087–92.
23. Rezlescu C, Duchaine B, Olivola CY, Chater N. Unfakeable facial configurations affect strategic choices in trust games with or without information about past behavior. *PLoS ONE* 2012;7(3):e34293. doi:10.1371/journal.pone.0034293.

24. Zebrowitz LA. *Reading Faces: Window to the Soul?* Boulder, CO: Westview Press; 1997.
25. Bar M, Neta M, Linz H. Very first impressions. *Emotion*. 2006;6(2):269–78.
26. Willis J, Todorov A. First impressions making up your mind after a 100-ms exposure to a face. *Psychol Sci*. 2006;17(7):592–98.
27. Todorov A, Pakrashi M, Oosterhof NN. Evaluating faces on trustworthiness after minimal time exposure. *Soc Cogn*. 2009;27(6):813–33.
28. Knutson B. Facial expressions of emotion influence interpersonal trait inferences. *J Nonverbal Behav*. 1996;20(3):165–82.
29. Said C, Sebe N, Todorov A. Structural resemblance to emotional expressions predicts evaluation of emotionally neutral faces. *Emotion*. 2009;9(2):260.
30. Zebrowitz LA, Collins MA. Accurate social perception at zero acquaintance: The affordances of a Gibsonian approach. *Personal Soc Psychol Rev*. 1997;1(3): 204–23.
31. Oosterhof NN, Todorov A. Shared perceptual basis of emotional expressions and trustworthiness impressions from faces. *Emotion*. 2009;9(1):128–33.
32. Mende-Siedlecki P, Said CP, Todorov A. The social evaluation of faces: A meta-analysis of functional neuroimaging studies. *Soc Cogn Affect Neurosci*. 2013;8(3):285–99. doi:10.1093/scan/nsr090.
33. Darwin C. The expression of emotion in animals and man. *London Methuen(1877), A Biogr sketch an*. 1872. https://scholar.google.com/scholar?hl=en&q=darwin+1872+expression&btnG=&as_sdt=1%2C14&as_sdtp=&oq=darwin+1872+e. Accessed September 16, 2017.
34. Wollmer MA, de Boer C, Kalak N, et al. Facing depression with botulinum toxin: A randomized controlled trial. *J Psychiatr Res*. 2012;46(5):574–81.
35. Finzi E, Rosenthal NE. Treatment of depression with onabotulinumtoxinA: A randomized, double-blind, placebo controlled trial. *J Psychiatr Res*. 2014;52:1–6.
36. Magid M, Reichenberg JS, Poth PE, et al. Treatment of major depressive disorder using botulinum toxin A: A 24-week randomized, double-blind, placebo-controlled study. *J Clin Psychiatry*. 2014;75(8):837–44.
37. Carr L, Iacoboni M, Dubeau M-C, Mazziotta JC, Lenzi GL. Neural mechanisms of empathy in humans: A relay from neural systems for imitation to limbic areas. *Proc Natl Acad Sci* 2003;100(9):5497–502. doi:10.1073/pnas.0935845100.
38. Dapretto M, Davies MS, Pfeifer JH, et al. Understanding emotions in others: Mirror neuron dysfunction in children with autism spectrum disorders. *Nat Neurosci*. 2006;9(1):28–30. doi:10.1038/nn1611.
39. Hennenlotter A, Dresel C, Castrop F, Ceballos-Baumann AO, Wohlschläger AM, Haslinger B. The link between facial feedback and neural activity within central circuitries of emotion—New insights from botulinum toxin-induced denervation of frown muscles. *Cereb Cortex*. 2009;19(3):537–42.
40. LeDoux JE. Emotion circuits in the brain. *Annu Rev Neurosci*. 2000;23:155–84.
41. Snyder M, Tanke ED, Berscheid E. Social perception and interpersonal behavior: On the self-fulfilling nature of social stereotypes. *J Pers Soc Psychol*. 1977; 35(9):656–66.
42. Izard CE. *The Face of Emotion*. New York: Appleton-Century-Crofts; 1971.
43. Du S, Tao Y, Martinez AM. Compound facial expressions of emotion. *Proc Natl Acad Sci*. 2014;111(15):E1454–E1462.
44. Courville J. The nucleus of the facial nerve: The relation between cellular groups and peripheral branches of the nerve. *Brain Res*. 1966;1(4):338–54.
45. Rinn WE. The neuropsychology of facial expression: A review of the neurological and psychological mehanisms for producing facial expressions. *Psychol Bull*. 1984;95(1):52–77.

46. Sherwood CC, Hof PR, Holloway RL, et al. Evolution of the brainstem orofacial motor system in primates: A comparative study of trigeminal, facial, and hypoglossal nuclei. *J Hum Evol.* 2005;48(1):45–84. doi:10.1016/j.jhevol.2004.10.003.
47. Ekman P. Basic emotions. In: Dalgleish T, Power M, eds. *Handbook of Cognition and Emotion.* New York: John Wiley & Sons; 1999:45–60.
48. Matsumoto D, Keltner D, Shiota MN, O'Sullivan M, Frank M. Facial expressions of emotion. *Handb Emot.* 2008;3:211–34.
49. Lien JJ, Cohn JF, Kanade T, Li C-C. Automated facial expression recognition based on FACS action units. In: *IEEE Proceedings of FG'98.* 1998:390–95.
50. Haxby JV, Hoffman EA, Gobbini MI. The distributed human neural system for face perception. *Trends Cogn Sci.* 2000;4(6):223–33.
51. Kanwisher N, Stanley D, Harris A. The fusiform face area is selective for faces not animals. *Neuroreport.* 1999;10(1):183–87.
52. Puce A, Allison T, Asgari M, Gore JC, McCarthy G. Differential sensitivity of human visual cortex to faces, letterstrings, and textures: A functional magnetic resonance imaging study. *J Neurosci.* 1996;16(16):5205–15.
53. Breiter HC, Etcoff NL, Whalen PJ, et al. Response and habituation of the human amygdala during visual processing of facial expession. *Neuron.* 1996;17(5): 875–87.
54. Morris JS, Friston KJ, Büchel C, et al. A neuromodulatory role for the human amygdala in processing emotional facial expressions. *Brain.* 1998;121(1):47–57.
55. Said CP, Haxby JV, Todorov A. Brain systems for assessing the affective value of faces. *Philos Trans R Soc Lond B Biol Sci.* 2011;366(1571):1660–70.
56. Sabatinelli D, Fortune EE, Li Q, et al. Emotional perception: Meta-analyses of face and natural scene processing. *Neuroimage.* 2011;54(3):2524–33. doi:10.1016/j.neuroimage.2010.10.011.
57. Adolphs R, Tranel D, Damasio H, Damasio AR. Fear and the human amygdala. *J Neurosci.* 1995;15(9):5879–91.
58. Adolphs R, Tranel D. Intact recognition of emotional prosody following amygdala damage. *Neuropsychologia.* 1999;37(11):1285–92.
59. Adolphs R, Gosselin F, Buchanan TW, Tranel D, Schyns P, Damasio AR. A mechanism for impaired fear recognition after amygdala damage. *Nature.* 2005;433 (7021):68–72.
60. Adolphs R. What does the amygdala contribute to social cognition? *Ann NY Acad Sci.* 2010;1191(1):42–61.
61. Parr LA, Waller BM, Heintz M. Facial expression categorization by chimpanzees using standardized stimuli. *Emotion.* 2008;8(2):216–31.
62. Parr LA, Waller BM, Vick SJ. New developments in understanding emotional facial signals in chimpanzees. *Curr Dir Psychol Sci.* 2007;16(3):117–22.
63. Parr LA, Waller BM, Vick SJ, Bard KA. Classifying chimpanzee facial expressions using muscle action. *Emotion.* 2007;7(1):172–81.
64. Vick SJ, Waller BM, Parr LA, Pasqualini MCS, Bard KA. A cross-species comparison of facial morphology and movement in humans and chimpanzees using the facial action coding system (FACS). *J Nonverbal Behav.* 2007;31(1):1–20.
65. Parr LA, Waller BM. Understanding chimpanzee facial expression: Insights into the evolution of communication. *Soc Cogn Affect Neurosci.* 2006;1(3):221–28.
66. Mühlberger A, Wieser MJ, Gerdes ABM, Frey MCM, Weyers P, Pauli P. Stop looking angry and smile, please: Start and stop of the very same facial expression differentially activate threat- and reward-related brain networks. *Soc Cogn Affect Neurosci.* 2011;6(3):321–29. doi:10.1093/scan/nsq039.
67. Preuschoft S, van Hooff JA. The social function of "smile" and "laughter": Variations across primate species and societies. In: Segerstrale UC, Molnar P, eds. *Nonverbal*

Communication: Where Nature Meets Culture. Mahwah, NJ: Lawrence Erlbaum; 1997:171–90.

68. de Waal F, Luttrell LM. The formal hierarchy of rhesus macaques: An investigation of the bared-teeth display. *Am J Primatol*. 1985;9(2):73–85.
69. Waller BM, Dunbar RI. Differential behavioural effects of silent bared teeth display and relaxed open mouth display in chimpanzees (*Pan troglodytes*). *Ethology*. 2005;111(2):129–42.
70. Parr LA, Waller BM, Fugate J. Emotional communication in primates: Implications for neurobiology. *Curr Opin Neurobiol*. 2005;15(6):716–20.
71. Perrett DI, Mistlin AJ. Perception of facial characteristics by monkeys. In: Stebbins WC, Berkley MA, eds. *Comparative Perception, Vol. 2: Complex Signals*. Oxford, UK: John Wiley & Sons; 1990:187–215.
72. Hasselmo ME, Rolls ET, Baylis GC, Nalwa V. Object-centered encoding by face-selective neurons in the cortex in the superior temporal sulcus of the monkey. *Exp Brain Res*. 1989;75(2):417–29.
73. Sugase Y, Yamane S, Ueno S, Kawano K. Global and fine information coded by single neurons in the temporal visual cortex. *Nature*. 1999;400(6747):869–73.
74. Barraclough NE, Perrett DI. From single cells to social perception. *Philos Trans R Soc Lond B Biol Sci*. 2011;366(1571):1739–52.
75. Gothard KM, Battaglia FP, Erickson CA, Spitler KM, Amaral DG. Neural responses to facial expression and face identity in the monkey amygdala. *J Neurophysiol*. 2007;97(2):1671–83.
76. Keltner D, Buswell BN. Evidence for the distinctness of embarrassment, shame, and guilt: A study of recalled antecedents and facial expressions of emotion. *Cogn Emot*. 1996;10(2):155–72.
77. Ekman P. Darwin, deception, and facial expression. *Ann NY Acad Sci*. 2003; 1000(1):205–21.
78. Cacioppo JT, Bush LK, Tassinary LG. Microexpressive facial actions as a function of affective stimuli: Replication and extension. *Personal Soc Psychol Bull*. 1992;18(5):515–26.
79. Porter S, Ten Brinke L. Reading between the lies: Identifying concealed and falsified emotions in universal facial expressions. *Psychol Sci*. 2008;19(5):508–14.
80. Takahashi H, Yahata N, Koeda M, Matsuda T, Asai K, Okubo Y. Brain activation associated with evaluative processes of guilt and embarrassment: An fMRI study. *Neuroimage*. 2004;23(3):967–74.
81. Beer JS, Knight RT, D'Esposito M. Controlling the integration of emotion and cognition: The role of frontal cortex in distinguishing helpful from hurtful emotional information. *Psychol Sci*. 2006;17(5):448–53. doi:10.1111/j.1467-9280.2006.01726.x.
82. Bastin C, Harrison BJ, Davey CG, Moll J, Whittle S. Feelings of shame, embarrassment and guilt and their neural correlates: A systematic review. *Neurosci Biobehav Rev*. 2016;71:455–71.
83. Ekman P, Friesen WV. Detecting deception from the body or face. *J Pers Soc Psychol*. 1974;29(3):288–98.
84. Matsumoto D, Hwang HS. Evidence for training the ability to read microexpressions of emotion. *Motiv Emot*. 2011;35(2):181–91.
85. Cacioppo JT, Martzke JS, Petty RE, Tassinary LG. Specific forms of facial EMG response index emotions during an interview: From Darwin to the continuous flow hypothesis of affect-laden information processing. *J Pers Soc Psychol*. 1988; 54(4):592.
86. Grens K. A face to remember. *TheScientist*. 2014;28(11).
87. Todorov A, Oosterhof NN. Modeling social perception of faces. *IEEE Signal Process Mag*. 2011;(March):117–22.

88. Landi SM, Freiwald WA. Two areas for familiar face recognition in the primate brain. *Science*. 2017;357(6351):591–95. doi:10.1126/science.aan1139.
89. Gobbini MA, Haxby JV. Neural systems for recognition of familiar faces. *Neuropsychologia*. 2007;45(1):32–41.

References for Chapter 6

1. Ekman P, Friesen WV. Felt, false, and miserable smiles. *J Nonverbal Behav.* 1982;6(4):238–52.
2. Hager JC, Ekman P. The asymmetry of facial actions is inconsistent with models of hemispheric specialization. *Psychophysiology*. 1985;22:307–18.
3. Frank MG, Ekman P, Friesen WV. Behavioral markers and recognizability of the smile of enjoyment. *J Pers Soc Psychol*. 1993;64:83–93.
4. Ekman P. Basic emotions. In: Dalgleish T, Power M, eds. *Handbook of Cognition and Emotion*. New York: John Wiley & Sons; 1999:45–60.
5. Ekman P, Davidson RJ, Friesen WV. The Duchenne smile: Emotional expression and brain physiology. II. *J Pers Soc Psychol*. 1990;58(2):342–53. http://www.ncbi.nlm.nih.gov/pubmed/2319446. Accessed September 10, 2017.
6. Rinn WE. The neuropsychology of facial expression: A review of the neurological and psychological mehanisms for producing facial expressions. *Psychol Bull*. 1984;95(1):52–77.
7. Calvo MG, Gutiérrez-García A, Del Líbano M. What makes a smiling face look happy? Visual saliency, distinctiveness, and affect. *Psychol Res*. 2016:1–14.
8. Williams LM, Senior C, David AS, Loughland CM, Gordon E. In search of the "Duchenne Smile": Evidence from eye movements. *J Psychophysiol*. 2001;15:122–27.
9. Baron-Cohen S, Wheelwright S, Hill J, Raste Y, Plumb I. The "Leading the Mind in the Eyes" test revised version: A study with normal adults, and adults with Asperger syndrome or high-functioning autism. *J Child Psychol Psychiatry*. 2001;42(2):241–51. doi:10.1111/1469-7610.00715.
10. Baron-Cohen S, Jolliffe T, Mortimore C, Robertson M. Another advanced test of theory of mind: Evidence from very high functioning adults with autism or asperger syndrome. *J Child Psychol Psychiatry*. 1997;38(7):813–22. http://www.ncbi.nlm.nih.gov/pubmed/9363580.
11. Ekman P, Friesen W V. *Unmasking the Face: A Guide to Recognizing Emotions from Facial Cues*. Englewood Cliffs, NJ: Prentice Hall; 1975.
12. Baron-Cohen S. *Mindblindness: An Essay on Autism and Theory of Mind*. Cambridge, MA: Bradford; 1995.
13. Land M, Mennie N, Rusted J. The roles of vision and eye movements in the control of activities of daily living. *Perception*. 1999;28(11):1311–28.
14. Hart D, Sussman RW. *Man the Hunted: Primates, Predators, and Human Evolution*. Boulder, CO: Westview Press; 2008.
15. Isbell LA. Snakes as agents of evolutionary change in primate brains. *J Hum Evol*. 2006;51(1):1–35. doi:10.1016/j.jhevol.2005.12.012.
16. Yorzinski JL, Penkunas MJ, Platt ML, Coss RG. Dangerous animals capture and maintain attention in humans. *Evol Psychol*. 2014;12(3):534–48. doi:10.1177/147470491401200304.
17. Caro T. *Antipredator Defenses in Birds and Mammals*. Chicago: University of Chicago Press; 2005.
18. LeDoux JE. Emotion: Clues from the brain. *Annu Rev Psychol*. 1995;46(1):209–35.
19. LeDoux JE. Emotion circuits in the brain. *Annu Rev Neurosci*. 2000;23:155–84.

20. LeDoux JE. The emotional brain, fear, and the amygdala. *Cell Mol Neurobiol.* 2003;23:727–38. doi:10.1023/A.
21. Norman GJ, Cacioppo JT, Berntson GG. Social neuroscience of evaluative motivation. In: Decety, J, Cacioppo JT, eds. *The Oxford Handbook of Social Neuroscience.* New York: Oxford University Press; 2011:164–77.
22. Argyle M, Cook M. *Gaze and Mutual Gaze.* New York: Cambridge University Press; 1976.
23. Böckler A, van der Wel RP, Welsh TN. Catching eyes: Effects of social and nonsocial cues on attention capture. *Psychol Sci.* 2014;25(3):720–27.
24. Calder AJ, Lawrence AD, Keane J, et al. Reading the mind from eye gaze. *Neuropsychologia.* 2002;40(8):1129–38.
25. Tomasello M, Hare B, Lehmann H, Call J. Reliance on head versus eyes in the gaze following of great apes and human infants: The cooperative eye hypothesis. *J Hum Evol.* 2007;52(3):314–20. doi:10.1016/j.jhevol.2006.10.001.
26. Mundy P, Newell L. Attention, joint attention, and social cognition. 2007;16(5):269–74.
27. Kobayashi H, Kohshima S. Unique morphology of the human eye and its adaptive meaning: Comparative studies on external morphology of the primate eye. *J Hum Evol.* 2001;40(5):419–35. doi:10.1006/jhev.2001.0468.
28. Kobayashi H, Kohshima S. Unique morphlogy of the human eye. *Nature.* 1997;387:767–68.
29. Whalen PJ, Rauch SL, Etcoff NL, McInerney SC, Lee MB, Jenike MA. Masked presentations of emotional facial expressions modulate amygdala activity without explicit knowledge. *J Neurosci.* 1998;18(1):411–18. doi:9412517.
30. Amaral DG. The amygdala, social behavior, and danger detection. *Ann NY Acad Sci.* 2003;1000:337–47. doi:10.1196/annals.1280.015.
31. Whalen PJ. Human amygdala responsivity to masked fearful eye whites. *Science.* 2004;306(5704):2061. doi:10.1126/science.1103617.
32. Sato W, Okada T, Toichi M. Attentional shift by gaze is triggered without awareness. *Exp Brain Res.* 2007;183(1):87–94. doi:10.1007/s00221-007-1025-x.
33. Hostetter AB, Russell JL, Freeman H, Hopkins WD. Now you see me, now you don't: Evidence that chimpanzees understand the role of the eyes in attention. *Anim Cogn.* 2007;10(1):55–62. doi:10.1007/s10071-006-0031-x.
34. Friesen CK, Kingstone A. The eyes have it! Reflexive orienting is triggered by nonpredictive gaze. *Psychon Bull Rev.* 1998;5(3):490–95.
35. Langton SR, Bruce V. Reflexive visual orienting in response to the social attention of others. *Vis Cogn.* 1999;6(5):541–67.
36. Macrae CN, Hood BM, Milne AB, Rowe AC, Mason MF. Are you looking at me? Eye gaze and person perception. *Psychol Sci.* 2002;13(5):460–64.
37. Haxby JV, Hoffman EA, Gobbini MI. The distributed human neural system for face perception. *Trends Cogn Sci.* 2000;4(6):223–33.
38. Langston SR, Watt RJ, Bruce V. Do the eyes have it? Cues to the direction of social attention. *Trends Cogn Sci.* 2000;4(2):50–59.
39. Emery NJ. The eyes have it: The neuroethology, function and evolution of social gaze. *Neurosci Biobehav Rev.* 2000;24(6):581–604.
40. Bolmont M, Cacioppo JT, Cacioppo S. Love is in the gaze: An eye-tracking study of love and sexual desire. *Psychol Sci.* 2014;25(9):1748–56. doi:10.1177/0956797614539706.
41. Brosch T, Sander D, Scherer KR. That baby caught my eye . . . : Attention capture by infant faces. *Emotion.* 2007;7(3):685–89. doi:10.1037/1528-3542.7.3.685.
42. Yorzinski JL, Patricelli GL, Babcock JS, Pearson JM, Platt ML. Through their eyes: Selective attention in peahens during courtship. *J Exp Biol.* 2013;216(Pt 16):3035–46. doi:10.1242/jeb.087338.

43. Perrett DI, Hietanen JK, Oram MW, Benson PJ. Organisation and functions of cells responsive to faces in the temporal cortex. *Philos Trans R Soc B Biol Sci.* 1992;335:23–30. http://www.st-andrews.ac.uk/~mwo/Papers/Perrett_etal_PhilosTransRSocB_1992.pdf.
44. Perrett DI, Smith PA, Potter DD, et al. Visual cells in the temporal cortex sensitive to face view and gaze direction. *Proc R Soc London Ser B.* 1985;223:293–317. doi:10.1098/rspb.1985.0003.
45. Barraclough NE, Perrett DI. From single cells to social perception. *Philos Trans R Soc Lond B Biol Sci.* 2011;366(1571):1739–52.
46. Itier RJ, Batty M. Neural bases of eye and gaze processing: The core of social cognition. *Neurosci Biobehav Rev.* 2009;33(6):843–63.
47. Puce A, Perrett DI. Electrophysiology and brain imaging of biological motion. *Philos Trans R Soc Lond B Biol Sci.* 2003;358:435–45. doi:10.1098/rstb.2002.1221.
48. Grosbras M-H, Laird AR, Paus T. Cortical regions involved in eye movements, shifts of attention, and gaze perception. *Hum Brain Mapp.* 2005;25(1):140–54.
49. Adolphs R, Gosselin F, Buchanan TW, Tranel D, Schyns P, Damasio AR. A mechanism for impaired fear recognition after amygdala damage. *Nature.* 2005; 433(7021):68–72.
50. Kim H, Somerville LH, Johnstone T, Alexander AL, Whalen PJ. Inverse amygdala and medial prefrontal cortex responses to surprised faces. *Neuroreport.* 2003;14(18):2317–22. doi:10.1097/01.wnr.0000101520.44335.20.
51. Kim H, Somerville LH, Johnstone T, et al. Contextual modulation of amygdala responsivity to surprised faces. *J Cogn Neurosci.* 2004;16(10):1730–45. doi:10.1162/0898929042947865.
52. Sato W, Kochiyama T, Uono S, Yoshikawa S. Amygdala integrates emotional expression and gaze direction in response to dynamic facial expressions. *Neuroimage.* 2010;50(4):1658–65. doi:10.1016/j.neuroimage.2010.01.049.
53. Sato W, Yoshikawa S, Kochiyama T, Matsumura M. The amygdala processes the emotional significance of facial expressions: An fMRI investigation using the interaction between expression and face direction. *Neuroimage.* 2004;22(2):1006–13. doi:10.1016/j.neuroimage.2004.02.030.
54. Berntson GG, Boysen ST, Cacioppo JT. Neurobehavioral organization and the cardinal principle of evaluative bivalence. *Ann NY Acad Sci.* 1993;702(1):75–102.
55. Berridge KC. Motivation concepts in behavioral neuroscience. *Physiol Behav.* 2004;81(2):179–209.
56. Cacioppo JT, Berntson GG, Norris CJ, Gollan J. The evaluative space model. In: VanLange P, Kruglanski AW, Higgins ET, eds. *Handbook of Theories of Social Psychology.* Thousand Oaks, CA: SAGE Publications; 2012:50–72.
57. Berntson GG, Cacioppo JT. The neuroevolution of motivation. In: Shah J, Gardner W, eds. *Handbook of Motivation Science.* New York: Guilford Press; 2008:188–200.
58. Rathelot JA, Strick PL. Subdivisions of primary motor cortex based on corticomotoneuronal cells. *Proc Natl Acad Sci.* 2009;106(3):918–23.
59. Smith ER, Mackie DM, Claypool HM. *Social Psychology,* 4th ed. New York: Psychology Press; 2015.
60. Benjamin LT, Simpson JA. The power of the situation: The impact of Milgram's obedience studies on personality and social psychology. *Am Psychol.* 2009;64(1): 12–19. doi:10.1037/a0014077.
61. Cacioppo JT, Cacioppo S. Minimal replicability, generalizability, and scientific advances in psychological science. *Eur J Pers.* 2013;27(2):121–22. doi:10.1002/per.
62. Monteleone GT, Phan KL, Nusbaum HC, et al. Detection of deception using fMRI: Better than chance, but well below perfection. *Soc Neurosci.* 2009;4(6):528–38. doi:10.1080/17470910801903530.

63. Winston JS, Strange BA, O'Doherty J, Dolan RJ. Automatic and intentional brain responses during evaluation of trustworthiness of faces. *Nat Neurosci.* 2002;5(3):277–83. doi:10.1038/nn816.
64. Engell AD, Haxby JV, Todorov A. Implicit trustworthiness decisions: Automatic coding of face properties in the human amygdala. *J Cogn Neurosci.* 2007;19(9):1508–19. doi:10.1162/jocn.2007.19.9.1508.
65. Berntson GG, Cacioppo JT. Psychobiology and social psychology: Past, present, and future. *Personal Soc Psychol Rev.* 2000;4(1):3–15.
66. Slepian ML, Young SG, Rule NO, Weisbuch M, Ambady N. Embodied impression formation: Social judgments and motor cues to approach or avoidance. *Soc Cogn.* 2012;30(2):232–40. doi:10.1521/soco.2012.30.2.232.
67. Willis ML, Palermo R, Burke D. Judging approachability on the face of it: The influence of face and body expressions on the perception of approachability. *Emotion.* 2011;11(3):514–23. doi:10.1037/a0022571.
68. Young SG, Slepian ML, Sacco DF. Sensitivity to perceived facial trustworthiness is increased by activating self-protection motives. *Soc Psychol Personal Sci.* 2015;6(6):607–13. doi:10.1177/1948550615573329.
69. Slepian ML, Ames DR. Internalized impressions. *Psychol Sci.* 2016;27(2):282–88. doi:10.1177/0956797615594897.
70. Kim MP, Rosenberg S. Comparison of two structural models of implicit personality theory. *J Pers Soc Psychol.* 1980;38(3):375.
71. Fiske ST, Cuddy AJC, Glick P. Universal dimensions of social cognition: Warmth and competence. *Trends Cogn Sci.* 2007;11(2):77–83. doi:10.1016/j.tics.2006.11.005.
72. Zebrowitz LA, Montepare JM. Social psychological face perception: Why appearance matters. *Soc Personal Psychol Compass.* 2008;23(10):1497–517. doi:10.1111/j.1751-9004.2008.00109.x.
73. Forgas JP. Why do highly visible people appear more important? Affect mediates visual fluency effects in impression formation. *J Exp Soc Psychol.* 2015;58:136–41. doi:10.1016/j.jesp.2015.01.007.
74. Forgas JP, Laham SM. Halo effects. In: Pohl RF, ed. *Cognitive Illusions.* New York: Routledge; 2017:276–92.
75. Fiske ST, Taylor SE. *Social Cognition: From Brains to Culture*, 3rd ed. Thousand Oaks, CA: SAGE Publications; 2017.
76. Taylor SE, Fiske ST. Point of view and perceptions of causality. *J Pers Soc Psychol.* 1975;32(3):439–45. doi:10.1037/h0077095.
77. Robinson J, McArthur LZ. Impact of salient vocal qualities on causal attribution for a speaker's behavior. *J Pers Soc Psychol.* 1982;43(2):236–47. doi:10.1037/0022-3514.43.2.236.
78. McArthur L, Post DL. Figural emphasis and person perception. *J Exp Soc Psychol.* 1977;13:520–35.
79. Petty RE, Cacioppo JT. *Attitudes and Persuasion: Classic and Contemporary Approaches.* Dubuque, IA: W. C. Brown Co. Publishers; 1981.
80. Petty RE, Cacioppo JT. *Communication and Persuasion: Central and Peripheral Routes to Attitude Change.* New York: Springer-Verlag; 1986.
81. Evans JSB. Dual-processing accounts of reasoning, judgment, and social cognition. *Annu Rev Psychol.* 2008;59:255–78.
82. Dore BP, Zerubavel N, Ochsner KN. Social cognitive neuroscience: A review of core processes. *Annu Rev Psychol.* 2007;58(1):259–89. doi:10.1146/annurev.psych.58.110405.085654.
83. Boysen ST, Bernston GG, Hannan MB, et al. Quantity-based interference and symbolic representations in chimpanzees (*Pan troglodytes*). *J Exp Psychol Anim Behav Process.* 1996;22(1):76–86. http://www.ncbi.nlm.nih.gov/pubmed/8568498.

References for Chapter 7

1. Okasha S. The relation between kin and multilevel selection: An approach using causal graphs. *Br J Philos Sci*. 2016;67(2):435–70. doi:10.1093/bjps/axu047.
2. van Veelen M, Allen B, Hoffman M, Simon B, Veller C. Hamilton's rule. *J Theor Biol*. 2017;414:176–230.
3. Griffin AS, West SA, Buckling A. Cooperation and competition in pathogenic bacteria. *Nature*. 2004;430:1024–7. doi:10.1038/nature02802.1.
4. Birch J, Okasha S. Kin selection and its critics. *Bioscience*. 2015;65(1):22–32.
5. de Waal FBM, Polllick AS. The biology of family values: Reproductive strategies of our fellow primates. In: Tipton SM, Witte J, eds. *Family Transformed: Religion, Values, and Society in American Life*. Washington, DC: Georgetown University Press; 2005:34–51.
6. Lovejoy CO. Reexamining human origins in light of *Ardipithecus ramidus*. *Science*. 2009;326(5949):74, 74e1–74e8. doi:10.1126/science.1175834.
7. Gibbons A. A new kind of ancestor: *Ardipithecus* unveiled. *Science*. 2009;326 (October 2009):36–40.
8. Opie C, Atkinson QD, Dunbar RIM, Shultz S. Reply to Lukas and Clutton-Brock: Infanticide still drives primate monogamy. *Proc Natl Acad Sci*. 2014;111(17):E1675. http://www.ncbi.nlm.nih.gov/pubmed/24895760. Accessed September 10, 2017.
9. Gavrilets S. Human origins and the transition from promiscuity to pair-bonding. *Proc Natl Acad Sci*. 2012;109(25):9923–8. doi:10.1073/pnas.1200717109.
10. Sherif M, Harvey OJ, White BJ, Hood WR, Sherif GW. *Intergroup Conflict and Cooperation: The Robbers Cave Experiment*. Norman, OK: University Book Exchange; 1961.
11. Bowles S. Did warfare among ancestral hunter-gatherers affect the evolution of human social behaviors? *Science*. 2009;324(5932):12938. doi:10.1126/science.1168112.
12. LeVine RA, Campbell DT. *Ethnocentrism: Theories of Conflict, Ethnic Attitudes, and Group Behavior*. New York: Wiley; 1972.
13. Gilpin W, Feldman MW, Aoki K. An ecocultural model predicts Neanderthal extinction through competition with modern humans. *Proc Natl Acad Sci*. 2016; 113(8):2134–9. doi:10.1073/pnas.1524861113.
14. Harari YN. *Sapiens: A Brief History of Humankind*. New York: Harper Collins; 2015.
15. Henrich J. *The Secret of Our Success: How Culture Is Driving Human Evolution, Domesticating Our Species, and Making Us Smarter*. Princeton, NJ: Princeton University Press; 2016.
16. Laland K. *Darwin's Unfinished Symphony: How Culture Made the Human Mind*. Princeton, NJ: Princeton University Press; 2017.
17. Nowak MA. Five rules for the evolution of cooperation. *Science*. 2006;314(5805):1560–3. doi:10.1126/science.1133755.
18. Imhof LA, Fudenberg D, Nowak MA. Tit-for-tat or win-stay, lose-shift? *J Theor Biol*. 2007;247(3):574–80. doi:10.1016/j.jtbi.2007.03.027.
19. Nowak M, Sigmund K. A strategy of win-stay, lose-shift that outperforms tit-for-tat in the Prisoner's Dilemma game. *Nature*. 1993;364:56–58.
20. Brewer MB. The psychology of prejudice: Ingroup love and outgroup hate? *J Soc Issues*. 1999;55(3):429–44. doi:10.1111/0022-4537.00126.
21. Brewer MB. Taking the social origins of human nature seriously: Toward a more imperialist social psychology. *Personal Soc Psychol Rev*. 2004;8(2):107–13.
22. Efferson D, LaLive R, Fehr E. The coevolution of cultural groups and ingroup favoritism. *Science*. 2008;321(September):1844–9. doi:10.1126/science.1155805.
23. Durrheim K, Quayle M, Tredoux CG, Titlestad K, Tooke L. Investigating the evolution of ingroup favoritism using a minimal group interaction paradigm: The effects of inter- and intragroup interdependence. *PLoS One*. 2016;11(11):1–26. doi:10.1371/journal.pone.0165974.

24. Takagi E. The generalized exchange perspective on the evolution of altruism. In: Liebrand W, Messick D, eds. *Frontiers in Social Dilemmas Research*. Berlin: Springer-Verlag; 1996:311–36.
25. Tajfel H. Experiments in intergroup discrimination. *Sci Am*. 1970;223:96–102.
26. Brewer MB. In-group bias in the minimal intergroup situation: A cognitive-motivational analysis. *Psychol Bull*. 1979;86(2):307–24. doi:10.1037/0033-2909.86.2.307.
27. Caporael LR. Evolutionary theory for social and cultural psychology. In: Kruglanski AW, Higgins ET, eds. *Social Psychology: Handbook of Basic Principles*, 2nd ed. New York: Guilford Press; 2007:3–18.
28. Xu X, Zuo X, Wang X, Han S. Do you feel my pain? Racial group membership modulates empathic neural responses. *J Neurosci*. 2009;29(26):8525–9. doi:10.1523/JNEUROSCI.2418-09.2009.
29. Tajfel H, Turner JC. An integrative theory of intergroup conflict. In: Austin WG, Worchel S, eds. *The Social Psychology of Intergroup Relations*. Monterey, CA.: Brooks/Cole; 1979:33–47.
30. Everett JAC, Faber NS, Crockett M. Preferences and beliefs in ingroup favoritism. *Front Behav Neurosci*. 2015;9(February):1–21. doi:10.3389/fnbeh.2015.00015.
31. Balliet D, Wu J, De Dreu CKW. Ingroup favoritism in cooperation: A meta-analysis. *Psychol Bull*. 2014;140(6):1556–81. doi:10.1037/a0037737.
32. De Dreu CKW, Aaldering H, Saygi O. Conflict and negotiation within and between groups. In: Mikulincer M, Shaver PR, Dovidio JF, Simpson JA, eds. *APA Handbook of Personality and Social Psychology: Volume 2: Group Processes*. Washington, DC: American Psychological Association; 2014:151–76.
33. Rosch E. Principles of categorization. In: Rosch E, Lloyd BB, eds. *Cognition and Categorization*. Hillsdale, NJ: Erlbaum; 1978:27–48.
34. Rips LJ. Inductive judgements about natural categories. *J Verbal Learning Verbal Behav*. 1975;681:665–81.
35. Meissner CA, Brigham JC. Thirty years of investigating the own-race bias in memory for faces: A meta-analytic review. *Psychol Public Policy, Law*. 2001;7(1):3–35. doi:10.1037/1076-8971.7.1.3.
36. Amodio DM. The neuroscience of prejudice and stereotyping. *Nat Rev Neurosci*. 2014;15(10):670–82. doi:10.1038/nrn3800.
37. Contreras JM, Banaji MR, Mitchell JP. Dissociable neural correlates of stereotypes and other forms of semantic knowledge. *Soc Cogn Affect Neurosci*. 2012;7(7):764–70. doi:10.1093/scan/nsr053.
38. Gilbert SJ, Swencionis JK, Amodio DM. Evaluative vs. trait representation in intergroup social judgments: Distinct roles of anterior temporal lobe and prefrontal cortex. *Neuropsychologia*. 2012;50(14):3600–11. doi:10.1016/j.neuropsychologia.2012.09.002.
39. Gallate J, Wong C, Ellwood S, Chi R, Snyder A. Noninvasive brain stimulation reduces prejudice scores on an implicit association test. *Neuropsychology*. 2011;25(2):185–92. doi:10.1037/a0021102.
40. LeDoux JE. Emotion circuits in the brain. *Annu Rev Neurosci*. 2000;23:155–84.
41. Cacioppo JT, Berntson GG. The affect system and racial prejudice. In: Bargh J, Apsley DK, eds. *Unravelling the Complexities of Social Life: A Festschrift in Honor of Robert B. Zajonc*. Washington, DC: American Psychological Association; 2001:95–110.
42. Richeson JA, Trawalter S. The threat of appearing prejudiced and race-based attentional biases. 2008;19(2):98–102. doi:10.1111/j.1467-9280.2008.02052.x.
43. Van Bavel JJ, Packer DJ, Cunningham WA. The neural substrates of in-group bias: A functional magnetic resonance imaging investigation. *Psychol Sci*. 2008;19(11):1131–9. doi: 10.1111/j.1467-9280.2008.02214.x.

44. Knutson B, Adams CM, Fong GW, Hommer D. Anticipation of increasing monetary reward selectively recruits nucleus accumbens. *J Neurosci.* 2001;21(16):RC159. doi:10.1523/JNEUROSCI.21-16-j0002.2001.
45. O'Doherty J. Dissociable roles of ventral and dorsal striatum in instrumental conditioning. *Science.* 2004;304(5669):452–54. doi:10.1126/science.1094285.
46. Cikara M, Eberhardt JL, Fiske ST. From agents to objects: Sexist attitudes and neural responses to sexualized targets. *J Cogn Neurosci.* 2011;23(3):540–51. doi:10.1162/jocn.2010.21497.
47. Amodio DM, Kubota JT, Harmon-Jones E, Devine PG. Alternative mechanisms for regulating racial responses according to internal vs external cues. *Soc Cogn Affect Neurosci.* 2006;1(1):26–36. doi:10.1093/scan/nsl002.
48. Gozzi M, Raymont V, Solomon J, Koenigs M, Grafman J. Dissociable effects of prefrontal and anterior temporal cortical lesions on stereotypical gender attitudes. *Neuropsychologia.* 2009;47(10):2125–32. doi:10.1016/j.neuropsychologia.2009.04.002.
49. Amodio DM, Devine PG, Harmon-Jones E. A dynamic model of guilt. *Psychol Sci.* 2007;18(6):524–30. doi:10.1111/j.1467-9280.2007.01933.x.
50. Babiloni F, Astolfi L. Social neuroscience and hyperscanning techniques: past, present and future. *Neurosci Biobehav Rev.* 2014;44:76–93. doi: 10.1016/j.neubiorev.2012.07.006.
51. Gürerk O, Irlenbusch B, Rockenbach B. The competitive advantage of sanctioning institutions. *Science.* 2006;312(5770):108–11. doi:10.1126/science.1123633.
52. Sanfey AG, Rilling JK, Aronson JA, Nystrom LE, Cohen JD. The neural basis of economic decision-making in the Ultimatum Game. *Science.* 2003;300(5626):1755–8. doi:10.1126/science.1082976.
53. Knoch D, Pascual-Leone A, Meyer K, Treyer V, Fehr E. Diminishing reciprocal fairness by disrupting the right prefrontal cortex. *Science.* 2006;314(5800):829–32. doi:10.1126/science.1129156.
54. Rilling J, Gutman D, Zeh T, Pagnoni G, Berns G, Kilts C. A neural basis for social cooperation. *Neuron.* 2002;35(2):395–405. http://www.ncbi.nlm.nih.gov/pubmed/12160756.
55. Tabibnia G, Lieberman MD. Fairness and cooperation are rewarding: Evidence from social cognitive neuroscience. *Ann NY Acad Sci.* 2007;1118:90–101. doi:10.1196/annals.1412.001.
56. Fehr E, Gächter S. Altruistic punishment in humans. *Nature.* 2002;415(6868):137–40. doi:10.1038/415137a.
57. de Quervain DJ-F, Fischbacher U, Treyer V, et al. The neural basis of altruistic punishment. *Science.* 2004;305(5688):1254–8. doi:10.1126/science.1100735.
58. Rilling JK, Sanfey AG, Aronson JA, Nystrom LE, Cohen JD. The neural correlates of theory of mind within interpersonal interactions. *Neuroimage.* 2004;22(4):1694–703. doi:10.1016/j.neuroimage.2004.04.015.
59. Nisbett RE, Cohen D. *Culture of Honor: The Psychology of Violence in the South.* New York: Westview Press; 1996.
60. Cohen D, Nisbett RE, Bowdle BF, Schwarz N. Insult, aggression, and the Southern culture of honor: An "experimental ethnography." *J Personal Soc.* 1996;70(5):945–60. doi:10.1037/0022-3514.70.5.945.
61. Trivers RL. The evolution of reciprocal altruism. *Q Rev Biol.* 1971;46(1):35–57. doi:10.1086/406755.
62. Ernst F, Fischbacher U, Simon G. Strong reciprocity, human cooperation and the enforcement of social norms. *Hum Nat.* 2002;13(1):1–25.
63. Seymour B, Singer T, Dolan R. The neurobiology of punishment. *Nat Rev Neurosci.* 2007;8(4):300–311. doi:10.1038/nrn2119.
64. Bowles S, Gintis H. The evolution of strong reciprocity: Cooperation in heterogeneous populations. *Theor Popul Biol.* 2004;65(1):17–28. doi:10.1016/j.tpb.2003.07.001.

65. Gardner A, West SA. Spite. *Curr Biol*. 2006;16(17):662–64.
66. Forber P, Smead R. The evolution of fairness through spite. *Proc R Soc B Biol Sci*. 2014;281(1780):20132439.

References for Chapter 8

1. van de Waal E, Borgeaud C, Whiten A. Potent social learning and conformity shape a wild primate's foraging decisions. *Science*. 2013;340(6131):483–85.
2. van de Waal E, Bshary R, Whiten A. Wild vervet monkey infants acquire the food-processing variants of their mothers. *Anim Behav*. 2014;90:41–45.
3. Whiten A, van de Waal E. Identifying and dissecting conformity in animals in the wild: Further analysis of primate data. *Anim Behav*. 2016;122:e1–e4. doi:10.1016/j.anbehav.2016.04.002.
4. Ioannou CC, Guttal V, Couzin ID. Predatory fish select for coordinated collective motion in virtual prey. *Science*. 2012;337(6099):1212–15. doi:10.1126/science.1218919.
5. Galef BG, Giraldeau L-A. Social influences on foraging in vertebrates: Causal mechanisms and adaptive functions. *Anim Behav*. 2001;61(1):3–15. doi:10.1006/anbe.2000.1557.
6. Rapaport LG, Brown GR. Social influences on foraging behavior in young nonhuman primates: Learning what, where, and how to eat. *Evol Anthropol*. 2008;17(4):189–201. doi:10.1002/evan.20180.
7. Britton NF, Franks NR, Pratt SC, Seeley TD. Deciding on a new home: How do honeybees agree? *Proc R Soc B Biol Sci*. 2002;269(1498):1383–8.
8. Kendal RL, Coolen I, Laland KN. The role of conformity in foraging when personal and social information conflict. *Behav Ecol*. 2004;15(2):269–77.
9. Webster MM, Laland KN. Social information, conformity and the opportunity costs paid by foraging fish. *Behav Ecol Sociobiol*. 2012;66(5):797–809.
10. Aplin LM, Farine DR, Morand-Ferron J, Cockburn A, Thornton A, Sheldon BC. Counting conformity: Evaluating the units of information in frequency-dependent social learning. *Anim Behav*. 2015;110:e5–e8. doi:10.1016/j.anbehav.2015.09.015.
11. Galef BG, Whiskin EE. "Conformity" in Norway rats? *Anim Behav*. 2008;75(6):2035–9.
12. Whiten A, Van Schaik CP. The evolution of animal "cultures" and social intelligence. *Philos Trans R Soc Lond B Biol Sci*. 2007;362(1480):603–20.
13. Deutsch M, Gerard HB. A study of normative and informational social influences upon individual judgment. *J Abnorm Soc Psychol*. 1955;51(3):629–36. doi:10.1037/h0046408.
14. Campbell JD, Fairey PJ. Informational and normative routes to conformity: The effect of faction size as a function of norm extremity and attention to the stimulus. *J Pers Soc Psychol*. 1989;57(3):457–68. doi:10.1037/0022-3514.57.3.457.
15. Claidière N, Whiten A. Integrating the study of conformity and culture in humans and nonhuman animals. *Psychol Bull*. 2011;138(1):126–45. doi:10.1037/a0025868.
16. Van Schaik CP. Animal culture: Chimpanzee conformity? *Curr Biol*. 2012;22(10):R402–4. doi:10.1016/j.cub.2012.04.001.
17. Konopasky R, Telegdy G. Conformity in the rat: A leader's selection of door color versus a learned door-color discrimination. *Percept Mot Skills*. 1977;44(1):31–37. doi:10.2466/pms.1977.44.1.31.
18. Asch SE. Effects of group pressure upon the modification and distortion of judgments. In: Harold G, ed. *Groups, Leadership, and Men: Research in Human Relations*. Pittsburgh: Carnegie Press; 1951:177–90.
19. Asch SE. Opinions and social pressure. In: Aronson E, ed. *Readings about the Social Animal*. New York: Worth Publishers; 1955:17–26.

20. Klucharev V, Hytönen K, Rijpkema M, Smidts A, Fernández G. Reinforcement learning signal predicts social conformity. *Neuron*. 2009;61(1):140–51. doi:10.1016/j.neuron.2008.11.027.
21. Izuma K. The neural basis of social influence and attitude change. *Curr Opin Neurobiol*. 2013;23(3):456–62. doi:10.1016/j.conb.2013.03.009.
22. Berns GS, Capra CM, Moore S, Noussair C. Neural mechanisms of the influence of popularity on adolescent ratings of music. *Neuroimage*. 2010;49(3):2687–96.
23. Campbell-Meiklejohn DK, Bach DR, Roepstorff A, Dolan RJ, Frith CD. How the opinion of others affects our valuation of objects. *Curr Biol*. 2010;20(13):1165–70.
24. Chen J, Wu Y, Tong G, Guan X, Zhou X. ERP correlates of social conformity in a line judgment task. *BMC Neurosci*. 2012;13(1):43.
25. Kim BR, Liss A, Rao M, Singer Z, Compton RJ. Social deviance activated the brain's error-monitoring system. *Cogn Affect Behav Neurosci*. 2012;12(1):65–73.
26. Klucharev V, Munneke MA, M, Smidts A, Fernandez G. Downregulation of the posterior medial frontal cortex prevents social conformity. *J Neurosci*. 2011; 31(33):11934–40. doi:10.1523/JNEUROSCI.1869-11.2011.
27. Schnuerch R, Gibbons H. A review of neurocognitive mechanisms of social conformity. *Soc Psychol (Gott)*. 2014;45(6):466–78.
28. Izuma K, Adolphs R. Social manipulation of preference in the human brain. *Neuron*. 2013;78(3):563–73.
29. Milgram S. Behavioral study of obedience. *J Abnorm Soc Psychol*. 1963;67(4):371–78.
30. Milgram S. Liberating effects of group pressure. *J Pers Soc Psychol*. 1965;1(2):127–34.
31. Milgram S. Some conditions of obedience and disobedience to authority. *Hum Relations*. 1965;18(1):57–76.
32. de Waal F. *The Bonobo and the Atheist: In Search of Humanism among the Primates*. New York: W. W. Norton & Company; 2013.
33. Chapais B. Competence and the evolutionary origins of status and power in humans. *Hum Nat*. 2015;26(2):161–83.
34. Boehm C. The moral consequences of social selection. *Behaviour*. 2014;151 (2–3):167–83.
35. Cialdini RB, Goldstein NJ. Social influence: Compliance and conformity. *Annu Rev Psychol*. 2004;55:591–621.
36. Caspar EA, Christensen JF, Cleeremans A, Haggard P. Coercion changes the sense of agency in the human brain. *Curr Biol*. 2016;26(5):585–92.
37. Cheetham M, Pedroni A, Antley A, Slater M, Jäncke L. Virtual milgram: Empathic concern or personal distress? Evidence from functional MRI and dispositional measures. *Front Hum Neurosci*. 2009;3:29. doi:10.3389/neuro.09.029.2009.
38. Lehmann J, Boesch C. Social influences on ranging patterns among chimpanzees (*Pan troglodytes verus*) in the Taï National Park, Côte d'Ivoire. *Behav Ecol*. 2003;14(5):642–49.
39. Gintis H, Van Schaik C, Boehm C, et al. Zoon Politikon: The evolutionary origins of human political systems. *Curr Anthropol*. 2015;56(3):327–53.
40. McGuire WJ. The nature of attitudes and attitude change. *Handb Soc Psychol*. 1969;3(2):136–314.
41. Petty RE, Cacioppo JT. *Attitudes and Persuasion—Classic and Contemporary Approaches*. Dubuque, IA: W. C. Brown Co. Publishers; 1981.
42. Fazio RH, Lenn TM, Effrein EA. Spontaneous attitude formation. *Soc Cogn*. 1984;2(3):217–34.
43. Ito TA, Cacioppo JT. Electrophysiological evidence of implicit and explicit categorization processes. *J Exp Soc Psychol*. 2000;36(6):660–76. doi:10.1006/jesp.2000.1430.
44. Zimbardo PG, Leippe MR. *The Psychology of Attitude Change and Social Influence*. New York: Mcgraw-Hill Book Company; 1991.

45. Jowett GS, O'Donnell V. *Propaganda & Persuasion*, 6th ed. Thousand Oaks, CA: SAGE Publications; 2015.
46. Nemeth CJ. Differential contributions of majority and minority influence. *Psychol Rev.* 986;93(1):23–32.
47. Nemeth CJ. Minority influence theory. In: Van Lange PAM, Kruglanski AW, Higgins ET, eds. *The Handbook of Theories of Social Psychology*, Vol. 2. Thousand Oaks, CA: SAGE Publications, 2012;362–78.
48. Cacioppo JT, Berntson GG, Petty RE. Persuasion. In: Dulbecco R, ed. *Encyclopedia of Human Biology*, 2nd ed. Orlando: Academic Press; 1997:679–90.
49. Cacioppo JT, Berntson GG. Relationship between attitudes and evaluative space: A critical review, with emphasis on the separability of positive and negative substrates. *Psychol Bull.* 1994;115(3):401–23.
50. Fazio RH. Attitudes as object-evaluation associations: Determinants, consequences, and correlates of attitude accessibility. *Attitude Strength: Antecedents and consequences.* 1995;4:247–82.
51. Kruglanski AW, Jasko K, Chernikova M, et al. The rocky road from attitudes to behaviors: Charting the goal systemic course of actions. *Psychol Rev.* 2015;122(4):598–620. doi:10.1037/a0039541.
52. Petty RE, Wegener DT, Fabrigar LR. Attitudes and attitude change. *Annu Rev Psychol.* 1997;48:609–47. doi:10.1146/annurev.psych.48.1.609.
53. McCloskey D, Klamer A. One quarter of GDP is persuasion. *Am Econ Rev.* 1995;85(2):191–95.
54. APWG. *Phishing Attack Trends Reports.* https://www.antiphishing.org/resources/apwg-reports/. Published 2017. Accessed January 1, 2017.
55. Hovland CI, Janis IL, Kelley HH. *Communication and Persuasion: Psychological Studies of Opinion Change.* New Haven, CT: Yale University Press Communications; 1953.
56. Sheeran P, Maki A, Montanaro E, et al. The impact of changing attitudes, norms, and self-efficacy on health-related intentions and behavior: A meta-analysis. *Heal Psychol.* 2016. doi:10.1037/hea0000387.
57. Proust J. The evolution of primate communication and metacommunication. *Mind Lang.* 2016;31(2):177–203.
58. Krebs JR, Dawkins R. Animal signals: Mind-reading and manipulation. In: Krebs JR, Davies NB, eds. *Behavioural Ecology: An Evolutionary Approach.* Oxford: Blackwell Scientific Publications; 1984:380–402.
59. Petty RE, Cacioppo JT. *Communication and Persuasion: Central and Peripheral Routes to Attitude Change.* New York: Springer-Verlag; 1986.
60. Petty RE, Cacioppo JT. The elaboration likelihood model of persuasion. *Adv Exp Soc Psychol.* 1986;19:123–205.
61. Cacioppo JT, Petty RE, Kao CF, Rodriguez R. Central and peripheral routes to persuasion: An individual differences perspective. *J Pers Soc Psychol.* 1986;51:1032–43.
62. Cacioppo JT, Petty RE. Effects of message repetition on argument processing, recall, and persuasion. *Basic Appl Soc Psych.* 1989;10(1):3–12.
63. Eagly AH, Kulesa P, Chen S, Chaiken S. Do attitudes affect memory? Tests of the congeniality hypothesis. *Curr Dir Psychol Sci.* 2001;10(1):5–9.
64. Petty RE, Cacioppo JT. The effects of involvement on responses to argument quantity and quality: Central and peripheral routes to persuasion. *J Pers Soc Psychol.* 1984;46:69–81.
65. Lieberman MD, Ochsner KN, Gilbert DT, Schacter DL. Do amnesics exhibit cognitive dissonance reduction? The role of explicit memory and attention in attitude change. *Psychol Sci.* 2001;12(2):135–40.
66. Johnson MK, Kim JK, Risse G. Do alcoholic Korsakoff's syndrome patients acquire affective reactions? *J Exp Psychol Learn Mem Cogn.* 1985;11(1):22–36.

67. Cacioppo JT, Cacioppo S, Petty RE. The neuroscience of persuasion: A review with an emphasis on issues and opportunities. *Soc Neurosci.* 2017;0(0):1–44. doi:10.1080/17470919.2016.1273851.
68. Carpenter CJ. A meta-analysis of the ELM's argument quality processing type predictions. *Hum Commun Res.* 2015;41(4):501–34.
69. Petty RE, Cacioppo JT, Distinction I, Johnson I. Involvement and persuasion: Tradition versus integration. *Psychol Bull.* 1990;107(3):367–64.
70. Chua HF, Liberzon I, Welsh RC, Strecher VJ. Neural correlates of message tailoring and self-relatedness in smoking cessation programming. *Biol Psychiatry.* 2009;65(2):165–68. doi:10.1016/j.biopsych.2008.08.030.
71. Chua HF, Ho SS, Jasinska AJ, et al. Self-related neural response to tailored smoking-cessation messages predicts quitting. *Nat Neurosci.* 2011;14(4):426.
72. Cooper N, Tompson S, O'Donnell MB, Falk EB. Brain activity in self- and value-related regions in response to online antismoking messages predicts behavior change. *J Media Psychol.* 2015;27(3):93–108. doi:10.1027/1864-1105/a000146.
73. Falk EB, Berkman ET, Whalen D, Lieberman MD. Neural activity during health messaging predicts reductions in smoking above and beyond self-report. *Heal Psychol.* 2011;30(2):177.
74. Vezich IS, Katzman PL, Ames DL, Falk EB, Lieberman MD. Modulating the neural bases of persuasion: Why/how, gain/loss, and users/non-users. *Soc Cogn Affect Neurosci.* 2016;12(2):283–97. doi:10.1093/scan/nsw113.
75. Riddle PJ, Newman-Norlund RD, Baer J, Thrasher JF. Neural response to pictorial health warning labels can predict smoking behavioral change. *Soc Cogn Affect Neurosci.* 2016;11(11):1802–11. doi:10.1093/scan/nsw087.
76. Wang A-L, Ruparel K, Loughead JW, et al. Content matters: Neuroimaging investigation of brain and behavioral impact of televised anti-tobacco public service announcements. *J Neurosci.* 2013;33(17):7420–7. doi:10.1523/JNEUROSCI.3840-12.2013.
77. Stallen M, Smidts A, Rijpkema M, Smit G, Klucharev V, Fernández G. Celebrities and shoes on the female brain: The neural correlates of product evaluation in the context of fame. *J Econ Psychol.* 2010;31(5):802–11.
78. Hamilton WD. The genetical evolution of social behaviour. II. *J Theor Biol.* 1964;7(1):17–52.
79. Nowak MA, Tarnita CE, Wilson EO. The evolution of eusociality. *Nature.* 2010;466(7310):1057–62. doi:10.1038/nature09205.
80. Birch J, Okasha S. Kin selection and its critics. *Bioscience.* 2015;65(1):22–32.
81. van Veelen M, Allen B, Hoffman M, Simon B, Veller C. Hamilton's rule. *J Theor Biol.* 2017;414:176–230.
82. Cialdini RB. Harnessing the science of persuasion. *Harv Bus Rev.* 2001;79(9):72–81.

References for Chapter 9

1. Cacioppo S, Cacioppo JT. Decoding the invisible forces of social connections. *Front Integr Neurosci.* 2012;6(July):51. doi:10.3389/fnint.2012.00051.
2. Cacioppo JT, Cacioppo S, Boomsma DI. Evolutionary mechanisms for loneliness. *Cogn Emot.* 2014;28(1):3–21. doi:10.1080/02699931.2013.837379.
3. Dunbar RI. Evolutionary basis of the social brain. In: Decety J, Cacioppo JT, eds. *The Oxford Handbook of Social Neuroscience.* Oxford, UK: Oxford University Press; 2011:28–38.
4. Opie C, Atkinson QD, Dunbar RIM, Shultz S. Male infanticide leads to social monogamy in primates. *Proc Natl Acad Sci.* 2013;110(33):13328–32. doi:10.1073/pnas.1307903110.
5. Neisser U. Five kinds of self knowledge. *Philos Psychol.* 1988;1:35–39.

6. Neisser U. The roots of self-knowledge: Perceiving self, it, and thou. *Ann NY Acad Sci.* 1997;818(1):19–33. http://onlinelibrary.wiley.com/doi/10.1111/j.1749–6632.1997.tb48243.x/full. Accessed September 13, 2017.
7. Aron A, Aron EN. Love and expansion of the self: The state of the model. *Pers Relatsh.* 1996;3:45–58.
8. Aron A, Aron EN, Smollan D. Inclusion of other in the self scale and the structure of interpersonal closeness. *J Pers Soc Psychol.* 1992;63(4):596–612.
9. Wegner DM, Erber R, Raymond P. Transactive memory in close relationships. *J Pers Soc Psychol.* 1991;61(6):923–29. http://www.ncbi.nlm.nih.gov/pubmed/1774630. Accessed September 10, 2017.
10. Cross ES, Calvo-Merino B. The impact of action expertise on shared representations. In: Obhi SS, Cross ES, eds. *Shared Representations.* Cambridge: Cambridge University Press; 2017:541–62. doi:10.1017/CBO9781107279353.027.
11. Cacioppo S, Fontang F, Patel N, Decety J, Monteleone G, Cacioppo JT. Intention understanding over T: A neuroimaging study on shared representations and tennis return predictions. *Front Hum Neurosci.* 2014;8:781. doi:10.3389/fnhum.2014.00781.
12. Vesper C, Sebanz N. Acting together: Representations and coordination processes. In: Obhi SS, Cross ES, eds. *Shared Representations: Sensorimotor Foundations of Social Life.* Cambridge: Cambridge University Press; 2017:216–35. doi:10.1017/CBO9781107279353.012.
13. Wang G, Mao L, Ma Y, et al. Neural representations of close others in collectivistic brains. *Soc Cogn Affect Neurosci.* March 2011:222–29. doi:10.1093/scan/nsr002.
14. Decety J, Sommerville JA. Shared representations between self and other: A social cognitive neuroscience view. *Trends Cogn Sci.* 2003;7(12):527–33. doi:10.1016/j.tics.2003.10.004.
15. Hatfield E, Rapson RL. *Love, Sex and Intimacy. Their Psychology, Biology and History.* New York: Harper & Collins; 1993.
16. Hatfield E, Rapson RL. *Love and Sex. Cross-Cultural Perspectives.* Boston: Allyn and Bacon; 1996.
17. Berscheid E, Walster E. *Interpersonal Attraction*, 2nd ed. Reading, MA: Addison-Wesley; 1978.
18. Hatfield E, Sprecher S. Measuring passionate love in intimate relations. *J Adolesc.* 1986;9:383–410.
19. Clark M, Grote N. Why aren't indices of relationship costs always negatively related to indices of relationship quality? *Personal Soc Psychol.* 1998;2(1):2–17. http://journals.sagepub.com/doi/abs/10.1207/s15327957pspr0201_1. Accessed September 10, 2017.
20. Hatfield E, Rapson RL, Aumer-Ryan K. Social justice in love relationships: Recent developments. *Soc Justice Res.* 2008;21(4):413–31. doi:10.1007/s11211-008-0080-1.
21. James W. Theories about the hypnotic state. *The Principles of Psychology.* New York: Henry Holt & Co.; 1890. http://coursecontent.learn21.org/ENG3x-HS-U10/a/unit04/resources/docs/E3HS_4.C_OrgTable_Psychology.pdf. Accessed September 10, 2017.
22. Tooby J, Cosmides L. The psychological foundations of culture. *Psychol Gener Cult.* 1992:19–136. https://books.google.com/books?hl=en&lr=&id=dIGKDgAAQBAJ&oi=fnd&pg=PA19&dq=cosmides+tooby&ots=9Xm497SA9x&sig=YfwQV6_lmDiZIINh8qZen0hXtSU. Accessed September 10, 2017.
23. Hatfield E, Rapson RL. The neuropsychology of passionate love. In: Cuyler E, Ackhart M, eds. *Psychology of Relationships.* New York: Nova Science Publishers; 2009.
24. Berscheid E, Hatfield E. *Interpersonal Attraction.* New York: Addison-Wesley; 1969.
25. Hendrick SS, Hendrick C. Love and sexual attitudes, self-disclosure and sensation seeking. *J Soc Pers Relat.* 1987;4:281–97. http://journals.sagepub.com/doi/pdf/10.1177/0265407587004003003. Accessed September 13, 2017.

26. Ortigue S, Bianchi-Demicheli F, Patel N, Frum C, Lewis JW. Neuroimaging of love: fMRI meta-analysis evidence toward new perspectives in sexual medicine. *J Sex Med.* 2010;7(11):3541–52. doi:10.1111/j.1743-6109.2010.01999.x.
27. Cacioppo S, Bianchi-Demicheli F, Frum C, Pfaus JG, Lewis JW. The common neural bases between sexual desire and love: A multilevel kernel density fMRI analysis. *J Sex Med.* 2012;9(4):1048–54. http://www.ncbi.nlm.nih.gov/pubmed/22353205. Accessed October 28, 2012.
28. Cacioppo S. Neuroimaging of female sexual desire and hypoactive sexual desire disorder. *Sex Med Rev.* August 2017. doi:10.1016/j.sxmr.2017.07.006.
29. Cacioppo S, Bianchi-Demicheli F, Hatfield E, Rapson RL. Social neuroscience of love. *Clin neuropsychiatry.* 2012;9(1):3–13.
30. Insel TR, Young LJ. The neurobiology of attachment. *Nat Rev Neurosci.* 2001;2(2):129–36. doi:10.1038/35053579.
31. Bartels A, Zeki S. The neural basis of romantic love. *Neuroreport.* 2000;11(17):3829–34. http://www.ncbi.nlm.nih.gov/pubmed/11117499. Accessed September 10, 2017.
32. Ortigue S, Bianchi-Demicheli F, Hamilton AF, Grafton ST. The neural basis of love as a subliminal prime: An event-related functional magnetic resonance imaging study. *J Cogn Neurosci.* 2007;19(7):1218–30. doi:10.1162/jocn.2007.19.7.1218.
33. Bianchi-Demicheli F, Grafton ST, Ortigue S. The power of love on the human brain. *Soc Neurosci.* 2006;1(2):90–103. doi:10.1080/17470910600976547.
34. Ortigue S, Bianchi-Demicheli F, Patel N, Frum C, Lewis J. Neuroimaging of love: fMRI meta-analysis evidence towards new perspectives in sexual medicine. *J Sex Med.* 2010;7(11):3541–52.
35. Seghier ML. The angular gyrus: Multiple functions and multiple subdivisions. *Neuroscientist.* 2013;19(1):43–61. doi:10.1177/1073858412440596.
36. Aron A, Paris M, Aron EN. Falling in love: Prospective studies of self-concept change. *J Pers Soc Psychol.* 1995;69(6):1102–12.
37. Aron A, Aron EN, Tudor M, Nelson G. Close relationships as including other in the self. *J Pers Soc Psychol.* 1991;60:241–53.
38. Fisher H, Aron A, Brown LL. Romantic love: An fMRI study of a neural mechanism for mate choice. *J Comp Neurol.* 2005;493(1):58–62. doi:10.1002/cne.20772.
39. Beauregard M, Courtemanche J, Paquette V, St-Pierre EL. The neural basis of unconditional love. *Psychiatry Res.* 2009;172(2):93–98. doi:10.1016/j.pscychresns.2008.11.003.
40. Acevedo BP, Aron A, Fisher HE, Brown LL. Neural correlates of long-term intense romantic love. *Soc Cogn Affect Neurosci.* 2012;7(2):145–59. doi:10.1093/scan/nsq092.
41. Aron A, Fisher H, Mashek DJ, Strong G, Li H, Brown LL. Reward, motivation, and emotion systems associated with early-stage intense romantic love. *J Neurophysiol.* 2005;94(1):327–37. doi:10.1152/jn.00838.2004.
42. Song H, Zou Z, Kou J, et al. Love-related changes in the brain: A resting-state functional magnetic resonance imaging study. *Front Hum Neurosci.* 2015;9. doi:10.3389/fnhum.2015.00071.
43. De Felice M, Ossipov MH. Cortical and subcortical modulation of pain. *Pain Manag.* 2016;6(2):111–20. doi:10.2217/pmt.15.63.
44. Kanwisher N, Stanley D, Harris A. The fusiform face area is selective for faces not animals. *Neuroreport.* 1999;10(1):183–87.
45. McKone E, Kanwisher N, Duchaine BC. Can generic expertise explain special processing for faces? *Trends Cogn Sci.* 2007;11(1):8–15. doi:10.1016/j.tics.2006.11.002.
46. Kanwisher N, Yovel G. The fusiform face area: A cortical region specialized for the perception of faces. *Philos Trans R Soc Lond B Biol Sci.* 2006;361(1476):2109–28. doi:10.1098/rstb.2006.1934.
47. Gobbini MI, Haxby JV. Neural response to the visual familiarity of faces. *Brain Res Bull.* 2006;71:76–82. doi:10.1016/j.brainresbull.2006.08.003.

48. Arsalidou M, Barbeau EJ, Bayless SJ, Taylor MJ. Brain responses differ to faces of mothers and fathers. *Brain Cogn*. 2010;74(1):47–51. doi:10.1016/j.bandc.2010.06.003.
49. Dai J, Zhai H, Wu H, et al. Maternal face processing in Mosuo preschool children. *Biol Psychol*. 2014;99(1):69–76. doi:10.1016/j.biopsycho.2014.03.001.
50. Cacioppo SC, Couto B, Bolmont M, et al. Selective decision-making deficit in love following damage to the anterior insula. *Curr Trends Neurol*. 2013;7:15–19.
51. Cacioppo S, Weiss RM, Runesha HB, Cacioppo JT. Dynamic spatiotemporal brain analyses using high performance electrical neuroimaging: Theoretical framework and validation. *J Neurosci Methods*. 2014;238:11–34. doi:10.1016/j.jneumeth.2014.09.009.
52. Ortigue S, Patel N, Bianchi-Demicheli F. New electroencephalogram (EEG) neuroimaging methods of analyzing brain activity applicable to the study of human sexual response. *J Sex Med*. 2009;6(7):1830–45. doi:10.1111/j.1743-6109.2009.01271.x.
53. Cacioppo S, Grafton ST, Bianchi-Demicheli F. The speed of passionate love, as a subliminal prime: A high-density electrical neuroimaging study. *Neuroquantology*. 2012;10(4):715–24. http://www.neuroquantology.com/index.php/journal/article/view/509. Accessed September 26, 2013.
54. Ortigue S, Bianchi-Demicheli F. The chronoarchitecture of human sexual desire: A high-density electrical mapping study. *Neuroimage*. 2008;43(2):337–45.doi:10.1016/j.neuroimage.2008.07.059.
55. Birbaumer N, Lutzenberger W, Elbert T. Imagery and brain processes. In: Birbaumer N, Ohman A, eds. *The Structure of Emotion*. Toronto: Hogrefe and Huber; 1993:122–38.
56. Davidson RJ, Cacioppo JT. New developments in the scientific study of emotion: An introduction to the special section. *Psychol Sci*. 1992;3(1):21–22. doi:10.1111/j.1467-9280.1992.tb00250.x.
57. Cacioppo S, Cacioppo JT. Dynamic spatiotemporal brain analyses using high-performance electrical neuroimaging, Part II: A step-by-step tutorial. *J Neurosci Methods*. 2015;256:184–97. doi:10.1016/j.jneumeth.2015.09.004.
58. Başar E, Schmiedt-Fehr C, Oniz A, Başar-Eroğlu C. Brain oscillations evoked by the face of a loved person. *Brain Res*. 2008;1214:105–15. doi:10.1016/j.brainres.2008.03.042.
59. Vico C, Guerra P, Robles H, Vila J, Anllo-Vento L. Affective processing of loved faces: Contributions from peripheral and central electrophysiology. *Neuropsychologia*. 2010;48(10):2894–902. doi:10.1016/j.neuropsychologia.2010.05.031.
60. Donchin E, Heffley EF. Multivariate analysis of event-related potential data: A tutorial review. In: Otto D, ed. *Multidisciplinary Perspectives in Event-Related Brain Potential Research*. Washington, DC: U.S. Government Printing Office; 1978:555–72. http://apsychoserver.psych.arizona.edu/JJBAReprints/PSYC501A/Readings/Donchin, Heffley_Multivariate analysis of ERP_Multidiciplinary perspectives_1978.pdf. Accessed September 10, 2017.
61. Cacioppo S, Cacioppo JT. Research in social neuroscience: How perceived social isolation, ostracism, and romantic rejection affect our brain. In: Riva P, Eck J, eds. *The Many Faces of Social Exclusion*. Cham, Switzerland: Springer International Publishing; 2016:ix–xv. doi:10.1007/978-3-319-33033-4.
62. Cacioppo S. What happens in your brain during mental dissociation? A quest towards neural markers of a unified sense of self. *Curr Behav Neurosci Reports*. 2016;3(1):1–9. doi:10.1007/s40473-016-0063-8.
63. Fisher HE, Brown LL, Aron A, Strong G, Mashek D. Reward, addiction, and emotion regulation systems associated with rejection in love. *J Neurophysiol*. 2010;104(1):51–60. doi:10.1152/jn.00784.2009.
64. Mearns J. Coping with a breakup: Negative mood regulation expectancies and depression following the end of a romantic relationship. *J Pers Soc Psychol*. 1991;60(2):327–34. http://www.ncbi.nlm.nih.gov/pubmed/2016673. Accessed September 21, 2017.

65. Cacioppo S, Frum C, Asp E, Weiss RM, Lewis JW, Cacioppo JT. A quantitative meta-analysis of functional imaging studies of social rejection. *Sci Rep.* 2013;3(1):2027. doi:10.1038/srep02027.
66. Williams KD. *Ostracism: The Power of Silence.* New York: Guilford Press; 2001.
67. Williams KD. Ostracism: The kiss of social death. *Soc Personal Psychol Compass.* 2007;1(1):236–47. doi:10.1111/j.1751-9004.2007.00004.x.
68. Burklund LJ, Eisenberger NI, Lieberman MD. The face of rejection: Rejection sensitivity moderates dorsal anterior cingulate activity to disapproving facial expressions. *Soc Neurosci.* 2007;2(3–4):238–53. doi:10.1080/17470910701391711.
69. Eisenberger NI, Lieberman MD, Williams KD. Does rejection hurt? An fMRI study of social exclusion. *Science.* 2003;302(5643):290–92. doi:10.1126/science.1089134.
70. Eisenberger NI. The neural bases of social pain: Evidence for shared representations with physical pain. *Psychosom Med.* 2012;74(2):126–35. doi:10.1097/PSY.0b013e3182464dd1.
71. Somerville LH, Heatherton TF, Kelley WM. Anterior cingulate cortex responds differentially to expectancy violation and social rejection. *Nat Neurosci.* 2006;9(8):1007–8. doi:10.1038/nn1728.
72. Woo C-W, Koban L, Kross E, et al. Separate neural representations for physical pain and social rejection. *Nat Commun.* 2014;5:5380. doi:10.1038/ncomms6380.
73. Cacioppo S, Juan E, Monteleone G. Predicting intentions of a familiar significant other beyond the mirror neuron system. *Front Behav Neurosci.* 2017;11(August):1–12. doi:10.3389/fnbeh.2017.00155.
74. Ortigue S, Patel N, Bianchi-Demicheli F, Grafton ST. Implicit priming of embodied cognition on human motor intention understanding in dyads in love. *J Soc Pers Relat.* September 2010:1–15. doi:10.1177/0265407510378861.
75. Ortigue S, Bianchi-Demicheli F. Why is your spouse so predictable? Connecting mirror neuron system and self-expansion model of love. *Med Hypotheses.* 2008;71(6):941–94. http://www.ncbi.nlm.nih.gov/pubmed/18722062.
76. Frith U, Frith C. The biological basis of social interaction. *Curr Dir Psychol Sci.* 2001;10(5):151–55. doi:10.1111/1467-8721.00137.
77. Wlodarski R, Dunbar RIM. The effects of romantic love on mentalizing abilities. *Rev Gen Psychol.* 2014;18(4):313–21. doi:10.1037/gpr0000020.
78. Ruscher JB, Santuzzi AM, Hammer EY. Shared impression formation in the cognitively interdependent dyad. *Br J Soc Psychol.* 2003;42(Pt 3):411–25. doi:10.1348/014466603322438233.
79. Cutting JE, Kozlowski LT. Recognizing friends by their walk: Gait perception without familiarity cues. *Bull Psychon Soc.* 1977;9(5):353–56. http://www.haskins.yale.edu/Reprints/HL0213.pdf. Accessed September 21, 2017.
80. Cook R, Johnston A, Heyes C. Self-recognition of avatar motion: How do I know it's me? *Proceedings Biol Sci.* 2012;279(1729):669–74. doi:10.1098/rspb.2011.1264.
81. Jokisch D, Daum I, Troje NF. Self recognition versus recognition of others by biological motion: Viewpoint-dependent effects. *Perception.* 2006;35(7):911–20. doi:10.1068/p5540.
82. Baron-Cohen S, Wheelwright S, Hill J, Raste Y, Plumb I. The "Reading the Mind in the Eyes" test revised version: A study with normal adults, and adults with Asperger syndrome or high-functioning autism. *J Child Psychol Psychiatry.* 2001;42(2):241–51. doi:10.1111/1469-7610.00715.
83. Fazio RH, Sanbonmatsu DM, Powell MC, Kardes FR. On the automatic activation of attitudes. *J Pers Soc Psychol.* 1986;50(2):229–38. doi:10.1037//0022-3514.50.2.229.
84. Ortigue S, Bianchi-Demicheli F, Hamilton AF, Grafton ST. The neural basis of love as a subliminal prime: An event-related functional magnetic resonance imaging study. *J Cogn Neurosci.* 2007;19(7):1218–30. doi:10.1162/jocn.2007.19.7.1218.

85. Dehaene S, Naccache L, Le Clec'H G, Koechlin E, Mueller M, et al. Imaging unconscious semantic priming. *Nature*. 1998;395(6702):597–600.
86. Bargh JA, Chen M, Burrows L. Automaticity of social behavior: Direct effects of trait construct and stereotype-activation on action. *J Pers Soc Psychol*. 1996;71(2):230–44. http://www.ncbi.nlm.nih.gov/pubmed/8765481.
87. Fisher HE, Xu X, Aron A, Brown LL. Intense, passionate, romantic love: A natural addiction? How the fields that investigate romance and substance abuse can inform each other. *Front Psychol*. 2016;7:687. doi:10.3389/fpsyg.2016.00687.
88. Shea JA, Adams GR. Correlates of romantic attachment: A path analysis study. *J Youth Adolesc*. 1984;13(1):27–44. doi:10.1007/BF02088651.
89. Feygin DL, Swain JE, Leckman JF. The normalcy of neurosis: Evolutionary origins of obsessive–compulsive disorder and related behaviors. *Prog Neuro-Psychopharmacology Biol Psychiatry*. 2006;30(5):854–64. doi:10.1016/j.pnpbp.2006.01.009.
90. Tesser A, Paulhus DL. Toward a causal model of love. *J Pers Soc Psychol*. 1976;34(6): 1095–105. doi:10.1037/0022-3514.34.6.1095.
91. Isen AM. An influence of positive affect on decision making in complex situations: Theoretical issues with practical implications. *J Consum Psychol*. 2001;11(2):75–85. http://www.psychologie.uni-heidelberg.de/ae/allg/mitarb/ms/Isen_2001.pdf. Accessed September 12, 2017.
92. Isen AM, Means B. The influence of positive affect on decision-making strategy. *Soc Cogn*. 1983;2(1):18–31. doi:10.1521/soco.1983.2.1.18.
93. Cacioppo S, Hatfield E. Passionate love and sexual desire. In: Whelehan P and Bolin A, eds. *The International Encyclopedia of Human Sexuality*. Hoboken, NJ: Wiley-Blackwell; 2015:1–2. doi:10.4324/9780203888124.
94. Frascella J, Potenza MN, Brown LL, Childress AR. Shared brain vulnerabilities open the way for nonsubstance addictions: Carving addiction at a new joint? *Ann NY Acad Sci*. 2010;1187:294–315. doi:10.1111/j.1749-6632.2009.05420.x.
95. Olsen CM. Natural rewards, neuroplasticity, and non-drug addictions. *Neuropharmacology*. 2011;61(7):1109–22. doi:10.1016/j.neuropharm.2011.03.010.
96. van Steenbergen H, Langeslag SJE, Band GPH, Hommel B. Reduced cognitive control in passionate lovers. *Motiv Emot*. 2013;38(3):444–50. doi:10.1007/s11031-013-9380-3.
97. Song S, Zou Z, Song H, et al. Romantic love is associated with enhanced inhibitory control in an emotional stop-signal task. *Front Psychol*. 2016;7:1574. doi:10.3389/fpsyg.2016.01574.
98. Volkow ND, Baler RD, Goldstein RZ. Addiction: Pulling at the neural threads of social behaviors. *Neuron*. 2011;69(4):599–602. doi:10.1016/j.neuron.2011.01.027.
99. Parvaz MA, Alia-Klein N, Woicik PA, Volkow ND, Goldstein RZ. Neuroimaging for drug addiction and related behaviors. *Rev Neurosci*. 2011;22(6):609–24. doi:10.1515/RNS.2011.055.
100. Volkow ND, Fowler JS, Wang G-J. The addicted human brain: Insights from imaging studies. *J Clin Invest*. 2003;111(10):1444–51. doi:10.1172/JCI200318533.
101. Volkow ND, Koob GF, McLellan AT. Neurobiologic advances from the brain disease model of addiction. *New Engl J Med*. 2016;374(4):363–71. doi:10.1056/nejmra1511480.
102. Koob GF, Volkow ND. Neurocircuitry of addiction. *Neuropsychopharmacology*. 2010;35(1):217–38. doi:10.1038/npp.2009.110.
103. Liu Y, Wang ZX. Nucleus accumbens oxytocin and dopamine interact to regulate pair bond formation in female prairie voles. *Neuroscience*. 2003;121(3):537–44. http://www.ncbi.nlm.nih.gov/pubmed/14568015. Accessed September 15, 2017.
104. Gardner EL. Addiction and brain reward and antireward pathways. *Adv Psychosom Med*. 2011;30:22–60. doi:10.1159/000324065.

105. Hong SB, Zalesky A, Cocchi L, et al. Decreased functional brain connectivity in adolescents with internet addiction.. *PLoS One*. 2013;8(2):e57831. doi:10.1371/journal.pone.0057831.
106. Cacioppo JT, Cacioppo S, Capitanio JP, Cole SW. The neuroendocrinology of social isolation. *Annu Rev Psychol*. 2015;66(1):733–67. doi:10.1146/annurev-psych-010814-015240.
107. Cacioppo S, Capitanio JP, Cacioppo JT. Toward a neurology of loneliness. *Psychol Bull*. 2014;140(6):1464–504. doi:10.1037/a0037618.
108. Holt-Lunstad J, Smith TB, Baker M, Harris T, Stephenson D. Loneliness and social isolation as risk factors for mortality: A meta-analytic review. *Perspect Psychol Sci*. 2015;10(2):227–37. doi:10.1177/1745691614568352.
109. Cacioppo JT, Hawkley LC. Perceived social isolation and cognition. *Trends Cogn Sci*. 2009;13(10):447–54. doi:10.1016/j.tics.2009.06.005.
110. Cacioppo S, Cacioppo JT. Do you feel lonely? You are not alone: Lessons from social neuroscience. *Front Young Minds*. 2013;1(November):1–6. doi:10.3389/frym.2013.00009.
111. Cole SW, Hawkley LC, Arevalo JM, Sung CY, Rose RM, Cacioppo JT. Social regulation of gene expression in human leukocytes. *Genome Biol*. 2007;8(9):R189. doi:10.1186/gb-2007-8-9-r189.
112. Cacioppo JT, Berntson GG, Malarkey WB, et al. Autonomic, neuroendocrine, and immune responses to psychological stress: The reactivity hypothesis. *Ann NY Acad Sci*. 1998;840:664–73.
113. Cacioppo JT, Cacioppo S, Cole SW, Capitanio JP, Goossens L, Boomsma DI. Loneliness across phylogeny and a call for comparative studies and animal models. *Perspect Psychol Sci*. 2015;10(2):202–12. doi:10.1177/1745691614564876.
114. Capitanio JP, Hawkley LC, Cole SW, Cacioppo JT. A behavioral taxonomy of loneliness in humans and rhesus monkeys (*Macaca mulatta*). Bard KA, ed. *PLoS One*. 2014;9(10):e110307. doi:10.1371/journal.pone.0110307.
115. Norman G, DeVries A, Hawkley L, Cacioppo JT, Berntson GG. Social influences on cardiovascular processes: A focus on health. In: Wright R, Gendolla G, eds. *How Motivation Affects Cardiovascular Response: Mechanisms and Applications*. Washington, DC: American Psychological Association; 2012:307–25. doi:10.1037/13090-015.
116. Matthews T, Danese A, Gregory AM, Caspi A, Moffitt TE, Arseneault L. Sleeping with one eye open: Loneliness and sleep quality in young adults. *Psychol Med*. 2017;47(12)2177–86. doi:10.1017/S0033291717000629.
117. Caspi A, Harrington H, Moffitt TE, Milne BJ, Poulton R. Socially isolated children 20 years later: Risk of cardiovascular disease. *Arch Pediatr Adolesc Med*. 2006;160(8):805–11. doi:10.1001/archpedi.160.8.805.
118. Cacioppo JT, Patrick W. *Loneliness: Human Nature and the Need for Social Connection*. New York: W. W. Norton & Company; 2008.
119. Cacioppo JT, Chen HY, Cacioppo S. Reciprocal influences between loneliness and self-centeredness: A cross-lagged panel analysis in a population-based sample of African American, Hispanic, and Caucasian adults. *Personal Soc Psychol Bull*. 2017;43(8):1125–35. doi:10.1177/0146167217705120.
120. Cacioppo S, Balogh S, Cacioppo JT. Implicit attention to negative social, in contrast to nonsocial, words in the Stroop task differs between individuals high and low in loneliness: Evidence from event-related brain microstates. *Cortex*. 2015;70:213–33. doi:10.1016/j.cortex.2015.05.032.
121. Cacioppo S, Bangee M, Balogh S, Cardenas-Iniguez C, Qualter P, Cacioppo JT. Loneliness and implicit attention to social threat: A high-performance electrical neuroimaging study. *Cogn Neurosci*. 2016;7(1–4):138–59. doi:10.1080/17588928.2015.1070136.

122. Wilson RS, Krueger KR, Arnold SE, et al. Loneliness and risk of Alzheimer disease. *Arch Gen Psychiatry*. 2007;64(2):234–40. doi:10.1001/archpsyc.64.2.234.
123. Karelina K, Norman GJ, Zhang N, Morris JS, Peng H, DeVries AC. Social isolation alters neuroinflammatory response to stroke. *Proc Natl Acad Sci*. 2009;106(14): 5895–900. doi:10.1073/pnas.0810737106.
124. Karelina K, Norman GJ, Zhang N, DeVries AC. Social contact influences histological and behavioral outcomes following cerebral ischemia. *Exp Neurol*. 2009;220(2):276–82. doi:10.1016/j.expneurol.2009.08.022.
125. Venna VR, Weston G, Benashski SE, et al. NF-κB contributes to the detrimental effects of social isolation after experimental stroke. *Acta Neuropathol*. 2012;124(3):425–38. doi:10.1007/s00401-012-0990-8.
126. Cole SW, Hawkley LC, Arevalo JMG, Cacioppo JT. Transcript origin analysis identifies antigen-presenting cells as primary targets of socially regulated gene expression in leukocytes. *Proc Natl Acad Sci*. 2011;108(7);3080–5. doi:10.1073/pnas.1014218108.
127. Jaremka LM, Fagundes CP, Peng J, et al. Loneliness promotes inflammation during acute stress. *Psychol Sci*. 2013;24(7):1089–97. doi:10.1177/0956797612464059.
128. Steptoe A, Owen N, Kunz-Ebrecht SR, Brydon L. Loneliness and neuroendocrine, cardiovascular, and inflammatory stress responses in middle-aged men and women. *Psychoneuroendocrinology*. 2004;29(5):593–611. doi:10.1016/S0306-4530(03) 00086-6.
129. Hackett RA, Hamer M, Endrighi R, Brydon L, Steptoe A. Loneliness and stress-related inflammatory and neuroendocrine responses in older men and women. *Psychoneuroendocrinology*. 2012;37(11):1801–9. doi:10.1016/j.psyneuen.2012.03.016.
130. Berns GS, Brooks AM, Spivak M. Scent of the familiar: An fMRI study of canine brain responses to familiar and unfamiliar human and dog odors. *Behav Processes*. 2015;110:37–46. doi:10.1016/j.beproc.2014.02.011.
131. Cook PF, Prichard A, Spivak M, Berns GS. Awake canine fMRI predicts dogs' preference for praise vs food. *Soc Cogn Affect Neurosci*. 2016;11(12):1853–62. doi:10.1093/scan/nsw102.
132. Blanke O, Ortigue S, Landis T, Seeck M. Stimulating illusory own-body perceptions. *Nature*. 2002;419(6904):269–70. doi:10.1038/419269a.
133. Arzy S, Seeck M, Ortigue S, Spinelli L, Blanke O. Induction of an illusory shadow person. *Nature*. 2006;443(7109):287. doi:10.1038/443287a.
134. Bünning S, Blanke O. The out-of body experience: Precipitating factors and neural correlates. *Progress in Brain Research*. 2005;150:331–50,605–6. doi:10.1016/S0079-6123(05)50024-4.
135. Blanke O, Arzy S. The out-of-body experience: Disturbed self-processing at the temporo-parietal junction. *Neurosci*. 2005;11(1):16–24. doi:10.1177/1073858404270885.
136. Brugger P. Reflective mirrors: Perspective-taking in autoscopic phenomena. *Cogn Neuropsychiatry*. 2002;7(3):179–94. doi:10.1080/13546800244000076.
137. Ionta S, Martuzzi R, Salomon R, Blanke O. The brain network reflecting bodily self-consciousness: A functional connectivity study. *Soc Cogn Affect Neurosci*. 2014;9(12):1904–13. doi:10.1093/scan/nst185.
138. Rizzolatti G, Fogassi L. The mirror mechanism: Recent findings and perspectives. *Philos Trans R Soc Lond B Biol Sci*. 2014;369(1644):20130420. doi:10.1098/rstb.2013.0420.
139. Blanke O, Mohr C. Out-of-body experience, heautoscopy, and autoscopic hallucination of neurological origin: Implications for neurocognitive mechanisms of corporeal awareness and self-consciousness. *Brain Res Rev*. 2005;50(1):184–99. doi:10.1016/j.brainresrev.2005.05.008.
140. Heydrich L, Blanke O. Distinct illusory own-body perceptions caused by damage to posterior insula and extrastriate cortex. *Brain*. 2013;136(Pt 3):790–803. doi:10.1093/brain/aws364.

141. Blanke O, Landis T, Spinelli L, Seeck M. Out-of-body experience and autoscopy of neurological origin. *Brain*. 2004;127(Pt 2):243–58. doi:10.1093/brain/awh040.
142. Ainsworth MD. Infant-mother attachment. *Am Psychol*. 1979;34(10):932–7. http://www.ncbi.nlm.nih.gov/pubmed/7318528.
143. Ainsworth MD., Blehar MC, Waters E, Wall S. *Patterns of Attachment: A Psychological Study of the Strange Situation*. Hillsdale, NJ: Erlbaum; 1978.
144. Bowlby J. *Attachment and Loss. Vol I. Attachment*. New York: Basic Books; 1969. https://www.abebe.org.br/files/John-Bowlby-Attachment-Second-Edition-Attachment-and-Loss-Series-Vol-1-1983.pdf. Accessed September 23, 2017.
145. Bowlby J. The nature of the child's tie to his mother. *Int J Psychoanal*. 1958;39:350–73. http://www.psychology.sunysb.edu/attachment/online/nature%20of%20the%20childs%20tie%20bowlby.pdf. Accessed September 23, 2017.
146. Bowlby J. Attachment and loss: Retrospect and prospect. *Am J Orthopsychiatry*. 1982;52(4):664–78. doi:10.1111/j.1939-0025.1982.tb01456.x.
147. Bowlby J, Ainsworth M. Attachment theory. https://blogs.nursing.osu.edu/draudt_12/files/2011/12/Attachment-Theory.pdf. Accessed September 23, 2017.
148. Vrtička P, Vuilleumier P. Neuroscience of human social interactions and adult attachment style. *Front Hum Neurosci*. 2012;6:212. doi:10.3389/fnhum.2012.00212.
149. Fonagy P, Luyten P. A developmental, mentalization-based approach to the understanding and treatment of borderline personality disorder. *Dev Psychopathol*. 2009;21(4):1355–81. doi:10.1017/S0954579409990198.

References for Appendixes

Aron A, Fisher H, Mashek DJ, Strong G, Li H, Brown LL. Reward, motivation, and emotion systems associated with early-stage intense romantic love. *Journal of Neurophysiology*. 2005;94:327–37.

Aron A, Henkemeyer, L. Marital satisfaction and passionate love. *Journal of Social and Personal Relationships*. 1995;12:139–46.

Aron A, Norman CC, Aron EN, McKenna C, Heyman R. Couples' shared participation in novel and arousing activities and experienced relationship quality. *Journal of Personality and Social Psychology*. 2000;78:273–84.

Bartels A, Zeki S. The neural correlates of maternal and romantic love. *Neuroimage*. 2004;21:1155–66.

Beck JG, Bozman AW, Qualtrough T. The experience of sexual desire: Psychological correlates in a college sample. *Journal of Sex Research*. 1991;28:443–56.

Crowne DP, Marlowe D. *The Approval Motive: Studies in Evaluative Dependence*. New York: Wiley; 1964.

Fehr B. Prototype-based assessment of laypeople's views of love. *Personal Relationships*. 2005;1:309–31.

Gardner A, West SA. Spite. *Curr Biol*. 2006;16(17):662–64.

Graham DL, Rawlings EI, Ihms K, Latimer D, Foliano J, et al. A scale for identifying "Stockholm syndrome" reactions in young dating women: Factor structure, reliability, and validity. *Violence and Victims*. 1995;10:3–22.

Grote NK, Frieze IH. The measurement of friendship-based love in intimate relationships. *Personal Relationships*. 1994;1:275–300.

Hatfield E, Rapson RL. *Love and Sex: Cross-Cultural Perspectives*. Lanham, MD: University Press of America; 2005.

Hatfield E, Rapson RL, Martel, LD. Passionate love and sexual desire. In: Kitayama S, Cohen D, eds. *Handbook of Cultural Psychology*. New York: Guilford Press; 2007:760–79.

Hatfield E, Sprecher S. Measuring passionate love in intimate relations. *Journal of Adolescence.* 1986;9:383–410.

Hatfield E, Sprecher S. The passionate love scale. In: Fisher TD, Davis CM, Yaber WL, Davis SL, eds. *Handbook of Sexuality-Related Measures: A Compendium,* 3rd ed. Thousand Oaks, CA: Taylor & Francis; 2011:466–68.

Hatfield E, Sprecher S. In: Kluger J. Why we love. *Time Magazine.* 2004;(Jan 19):60. http://content.time.com/time/health/article/0,8599,1704355,00.html. Accessed June 15, 2019.

Hatfield E, Walster GW. *A New Look at Love.* Lanham, MD: University Press of America; 1978.

Hendrick C, Hendrick SS. Research on love: Does it measure up? *Journal of Personality and Social Psychology.* 1989;56:784–94.

James P. Effects of a communication training component added to an emotionally focused couples therapy. *Journal of Marital and Family Therapy.* 2007;17:263–75.

Landis D, O'Shea WA. Cross-cultural aspects of passionate love: An individual difference analysis. *Journal of Cross-Cultural Psychology.* 2000;31(6):754–79.

Sprecher S, Regan PC. Passionate and companionate love in courting and young married couples. *Sociological Inquiry.* 1998;68:163–85.

Tucker P, Aron A. Passionate love and marital satisfaction at key transition points in the family life cycle. *Journal of Social and Clinical Psychology.* 1993;12:135–47.

INDEX

Italic page references refer to figures or tables

activation likelihood estimate (ALE), 61–62
Adolphs, Ralph, 10b, 57, 93, 99, 128–29, 194
adrenocorticotropic hormone (ACTH), *49*
adrenomedullary system (SAM), 36, *49*
affective perspective taking, *82*, 87–95, 109–10
aggression, 25, 28, 60, 120, 169, 178
aging, 40, 121, 144, 216
Agricultural Revolution, 67
Ainsworth, Mary, 201b
altruism: Cacioppo Evolutionary Theory of Loneliness (ETL) and, 213–15; connecting forces and, 110; cooperation and, 166–69; group processes and, 152–57, 166–69; punitive, 29, 166–69; social brain and, 73, 81, 83; social connections and, 29–30, 39, 211; spite and, 1–2, 29–30, 39, 73, 170, 213, 215
Alzheimer's Disease, 44–45, 51
Amodio, David, 160, 164
amygdala: connecting forces and, 87, 91–93, *94*; eye perception and, 139–40, *142*, 145, 148–50; face perception and, 120, 123, 126, 128–29, 131–32; group processes and, 162–65; social behavior and, 10; social brain and, *57*, 58–59, 69–70; social connections and, *49*, 50–51, 200, 207; social influence and, 179
A/New Caledonia vaccination, 43
anger, 48, 120, 126–27, 131, 134
angular gyrus, 59, 186, 192–99, 206, 210
anterior cingulate cortex (ACC): connecting forces and, 91, *92*, 98b, 109; eye perception and, 150; face perception and, 121, 132; group processes and, 163–65; social connections and, 192, 194, 199, 206–7; social influence and, 187
anterior insula: connecting forces and, 87, 91, *92*; group processes and, 163, 168; social brain and, 80; social connections and, 192–94, 199, 201, 203–4, 206
antibodies, 43–44
anxiety: connecting forces and, 98; depression and, 25, 41, 48; loneliness and, 41, 48, 80; Passionate Love Scale (PLS) and, 217; social brain and, 80; social connections and, 25, 41, 45, 48, 201b
Ardipithecus ramidus, 54, 153
Aristotle, 182
Arsalidou, Marie, 200
artificial signals, 123–24
Asch, Solomon, 119, 174
Astolfi, Laura, 167b
astrocytes, 9
attachment styles, 200, 201b
attention: Cacioppo Evolutionary Theory of Loneliness (ETL) and, 214; connecting forces and, 87, 96, 109–10; eye perception and, 138, 140, 143b, 144, 145b, 147–48; face perception and, 118, 133; group processes and, 160; hyperattention, 6; joint, 138, 140, 144, 145b; social behavior and, 6–7, 11; social brain and, 59, 67, 81; social connections and, 28, 193, *196*, 198b, 207, 209–10; social influence and, 176–78; visual, 11, 133, 143b, 144, 145b, 191, 198b
auditory areas, 5b, 11–12, 47, *196*
automatic preparatory adjustments postulate, 215
aversive signals, 33, 214–15
axons, 67, 79–80, 125, 131

Babiloni, Fabio, 167b
baboons, *11*, 23, 37–38, *66*, 131
Baron-Cohen, Simon, 75, 138–39
Bartels, Andreas, 192
Başar, E., 205
bed nucleus of the stria terminalis (BNST), 50
Bennett, David, 51
Berns, Gregory, 195b
Bernston, G. G., 146–47
Berra, Yogi, 155–56
bi-parental caregiving, 8, 208, 212

Birbaumer, N., 205
birds, 54, 73, 79, 109, 172
Blakemore, Sarah J., 70b
blood pressure, 23, *32*, 36–38
bodily self, 59, 191, 198b
Boehm, Christopher, 179
bonobos, *66*, 70b, 111, 178
Borge, Victor, 70b
Bowlby, John, 201b
Bowles, Samuel, 169
brain-body interactions, 6
brain-derived neurotropic factor (BDNF), 51
brain size, 4–6, 9, 55b, 65, 68, 71, 73, 78–80, 85
brain stem, *12*, *94*, *140*; deception and, 146, 149; face perception and, 123–24; group processes and, 162; social brain and, *66*, *70*; social connections and, 50, *202*
Brewer, Marilynn, 158
Broca's area, 11
Brodmann area, 39, 195

Cacioppo, John T., 21, 64
Cacioppo, Stephanie, 91, 102, 192–95, 207, 209
Cacioppo Evolutionary Theory of Loneliness (ETL): altruism and, 213–15; attention and, 214; competition and, 214; cost-benefit analysis and, 213; depression and, 47–48; food and, 214; genetics and, 41–42; hormones and, 215; inflammatory substrate and, 44; mortality and, 22, 28; mutual benefit and, 213–14; pain and, 214; propositions of, 30–31, 213–16; salutary relationship and, 213–14; selfishness and, 213–15; short-term preservation and, 28; social behavior and, 213–15; social connections and, 22, 28–31, 33, 36, 41, 44, 47, 213–16; spite and, 213, 215; survival and, 213–15; sympathetic tonus and, *29*, *32*, 33, 36–38; transcriptome dynamics and, *29*, *32*, 33, 38–45
Calit2, 126
cardiovascular system, 23, *32*, 36–40, 44–45
caregiving, 8, 22, 72–73, 85–86, 212
Caruso, Eugene, 106
central nervous system (CNS), 41, 45, 146–49, 185
cerebellum, 10, *12*, *66*, *70*, *196*
cerebral cortex, *70*; deception and, 146; eye perception and, 146; social brain and, 65, *66*, *70*; structure of, 9–12
cerebral hemispheres, 12–13, *82*
chemokine monocyte chemotactic protein (MCP-1), 45
Chicago Health, Aging, and Social Relations Study, 40
chimpanzees: axonal projections and, 67, *68*; eye perception and, 141; face perception and, 114, 129, 131; group processes and, 153; guilt and, 178; laughter and, 70b; social behavior and, 5, 19; social brain and, *66–68*, 70b; social influence and, 178; social learning and, 109; TOM mentalizing and, 111; war and, 153
Chua, Hanna Faye, 185–86
Cicero, Marcus, 134–35, 182
Claidmère, Nicolas, 172
Cocaine and Amphetamine Regulated Transcript Protein (CART), 42
Coelho, A. M., 37–38
cognition: connecting forces and, *88*, 89–101, *108*, 110; deception and, 144–50; face perception and, 114, 118, 129; group processes and, 159–60, 164–65; love and, 25, 191–201, 205–12, 217; social brain and, 3, 5–8, 12–13, 16, 18–20, 54, 57, 59–62, 66–68, 71–80, 85; social connections and, 23, 28, 51–52, 191–201, 205–12; social influence and, 176, 179–80, 183–88
Cole, S. W., 42
colliculi, 10
communication: connecting forces and, 91, 191; deception and, 135, 146, 150–51; Elaboration Likelihood Model (ELM) and, 183–85; eye perception and, 134–35, 146, 150–51; face perception and, 112, 125–26; persuasion and, 179–87; social brain and, 7–8, 14, 67, 72, 75–76, 79–80; social connections and, 22; social influence and, 179–87
competition: Cacioppo Evolutionary Theory of Loneliness (ETL) and, 214; for food, 22, 72, 100, 178, 214; group processes and, 152–54, 158–59, 170; interspecific, 153–54; intraspecific, 153–54; social brain and, 8, 72, 76; social connections and, 22, 178
computers, 7–8, 86, 119, 126, 131, 206, 209
conceptual self, 189, *191*
conditioning, 108–9, 120, 150, 163
conformity: connecting forces and, 96; social brain and, 8, 75; social influence and, 172–77
connecting forces: affective perspective taking and, 87–95, 109–10; altruism and, 110; amygdala and, 87, 91–93, *94*; anterior insula and, 87, 91, *92*; anxiety and, 98; attention and, 87, 96, 109–10; cognition and,

88, 89–101, *108*, 110; communication and, 91, 191; conformity and, 96; contagion and, 86–95, 109; cooperation and, 87; cultural issues and, 96; egocentrism perspective and, 97–99; emotion and, 86–97, 105–6, 109–10; empathy and, 86–95, 109–10; evolution and, 100; fear and, 109–10; functional magnetic resonance imaging (fMRI) and, 91, 103, 106; gyrus and, 96, 100, 103, *104*; hormones and, *89*; hypothalamus and, *94*; identification and, 86–87, 93, 95–97, 103, 110; imitation and, 86–87, *88*, 95–97, *100*, 109–10; impressions and, 21, 99, 106; infants and, 86–87; intention and, 86–87, *88*, 91–95, 99–107, 110–11; macaques and, 98b, 111; memory and, 103, 105; mentalizing and, *92*, 93, *94*, 97–109, 111; multiple pathways and, 107–8; neuroimaging and, 90b, 91, 93, 98b, 99, 103; neurons and, 87, 98b, 99–104, 111; norms and, 96; orbital frontal cortex (OFC) and, 91, *92*, *94*, 98b, *162*; pain and, 93–94, 109–10; prefrontal cortex (PFC) and, 90b, 91–93, 98b, 109; simulation perspective and, 99–104; social learning and, 96, 108–10; social perception and, 90, 100; society and, 98; somatosensorimotor resonance and, 87, *88*; status and, 90, 107; superior temporal sulcus (STS) and, 91–92, *93–94*, 96, 103, 107; survival and, 91, 100, 103, 109; theory of mind (TOM) and, *92*, 97, 98b, 105–7, 109, 111
conservation postulate, 214
Conserved Transcriptional Response to Adversity (CTRA), 39–42
conspecifics, 2–3, 21–22, 26, 35, 39, 51, 72, 75, 97, 171, 214
contagion: affective perspective taking and, 87–95, 109–10; connecting forces and, 86–95, 109; emotional, 86–95, 109; social brain and, 8; social connections and, 21–22; somatosensorimotor resonance and, 87, *88*
cooperation: connecting forces and, 87; deception and, 135; group processes and, 153–56, 159, 166–70; punitive altruism and, 166–69; social brain and, 8, 14, 72, 76, 81; social connections and, 22, 208
coronary heart disease (CHD), 37
corticotropin releasing hormone (CRH), 10b, *49*
cost-benefit analysis: Cacioppo Evolutionary Theory of Loneliness (ETL) and, 213; extended immaturity and, 73, 85; eye perception and, 150; group processes and, 72, 110, 150, 153, 156–57, 168–69, 171, 187–88; love and, 207–11; social behavior and, 1–2, 213; social connections and, 29–30; social influence and, 171, 187–88; social learning and, 73
courtship, 143, 144b
Crowne, D. P., 219
cultural issues: connecting forces and, 96; deception and, 145, 149; group processes and, 152, 154, 157–58, 160, 165–66, 169; Passionate Love Scale (PLS) and, 219; persuasion and, 179–82; social brain and, 8–9, 14, 56, 67–68, 72–73, 85; social connections and, 22, *29*; social influence and, 171, 174, 179, 181–82, 187; social media and, 7, 181, *182*
cultural learning, 2, 73, 85, 96, 145, 158
cytokines, 15, 41, 45–46

Darwin, Charles, 8–9, 122
deception: central nervous system (CNS) organization and, 146–47; cerebral cortex and, 146; cognition and, 144–50; communication and, 135, 146, 150–51; cooperation and, 135; cultural issues and, 145, 149; detection of, 145–49; dorsolateral prefrontal cortex and, 148; emotion and, 137–38, 144, 146, 148; evolution and, 134–37, 145–51; expressions and, 135–38; eye perception and, 135–38, 145–49; fake smiles and, 135–36; fear and, 134; food and, 140; happiness and, 135–36, *142*; impressions and, 135, 148–50; intention and, 136, 138, 145b, 148; love and, 142, *143*; mutual benefit and, 149; neurons and, 137, 145–46, 148; production of, 145–49; selfishness and, 136; social behavior and, 149; status and, 142–43; survival and, 135, 139, 145; theory of mind (TOM) and, 148–49
Decety, Jean, 91
deleterious effects, *29*, *77*, 170, 216
dendrites, 80, 204
deoxyribonucleic acid (DNA), 15–16, 24, 38–39, 40b, 63
depression: anxiety and, 25, 41, 48; face perception and, 123, 133; loneliness and, 25, *32*, 33, 35, 41, 47–48, 51; Passionate Love Scale (PLS) and, 221; social connections and, 21, 25, *29*, *31–32*, 33, 35, 41, 47–48, 51, 206
desert locust (*Schistocerca gregaria*), 2–5, 78
DeVries, A. C., 46
diabetes, 44

Digital Revolution, 85
discrimination: group processes and, 22, 60, 75–77, 90b, 129, 133, 154, 160, 162–64, 170, 195b, 214; outgroups and, 157–60, 163–64, 166, 170; prejudice and, 158, 160, 162–66, 170; stereotyping and, 159–65, 170
divorce, 143, 189, *190*, 203
doctrine of multilevel analysis, 14–18
dogs, 15b, 36, 78, 194
dopamine, 10b, 81; face perception and, 120–21; love and, 192–93, 195, 199–200, 211–12; social connections and, 35, 42, 50–51
dorsal striatum, 163, 168, 200
dorsolateral prefrontal cortex: connecting forces and, 90b, 109; deception and, 148; eye perception and, 148; face perception and, 132; group processes and, 165, 168; social brain and, 77
dorsomedial prefrontal cortex (DMPFC): connecting forces and, 78, 90b, 91, 97–99, 106, 109; group processes and, 160, 162, 164; social connections and, 199
down-regulation, *32*, 39–42, 64, 176
Drosophila melanogaster, 5b
drugs, 81, 211
Dunbar, Robin, 73

egocentric mentalizing, 10, 97–99, 105, 149, 198b
Eichmann, Adolf, 176–77, 179
Einstein, Albert, 59
Einstein (robot), 126
Ekman, Paul, 125–26, 205
Elaboration Likelihood Model (ELM), 183–85
electroencephalograms (EEGs), *18*, 34, 56, 63, 91, 167b, 204
embarrassment, 131–32
emotion: affective perspective taking and, 87–95, 109–10; anger, 48, 120, 126–27, 131, 134; anxiety, 25, 41, 45, 48, 98, 201b; characteristics of core processes of, *88*; connecting forces and, 86–97, 105–6, 109–10; contagion and, 86–95, 109; deceptive, 135–38, 145–49; depression, 21, 25, *29*, *31–32*, 33, 35, 41, 47–48, 51, 123, 133, 206, 221; embarrassment, 131–32; empathy, 88b, 89–91, *92*; expressions of, 125–26, *127*; face perception and, 118–33; fear, 65 (*see also* fear); gaze and, *88–89*, 142–45; grief, 21, 25, 122; group processes and, *162*, 163, 168; guilt, 86, 105, 131–32, 178; laughter, 70b, 133; loneliness, 22 (*see also* loneliness); love, 58–59 (*see also* love); negative moral, 131–32; omega sign and, 122; Passionate Love Scale (PLS) and, 217–19, *220*, *222*; sadness, 28, 47, 123, 126, *127*, 131, 137; shame, 131–32; social brain and, 6, 19, 57–58, 68–70, 77; social connections and, 189, 192–95, 199–212; social influence and, 181; somatosensorimotor resonance and, 87, *88*; surprise, 126, *127*, 131, 144
empathy, 8; affective perspective taking and, 87–95, 109–10; cognitive, *88*, 89, *92*; connecting forces and, 86–95, 109–10; development of, *89*, 90–91; emotional, 88b, 89–91, *92*; functional magnetic resonance imaging (fMRI) and, 57, 91; gaze and, *89*; group processes and, 159, 180; network of, 57; pain and, 57, 91–95; social behavior and, 91, 95; social brain and, 57, 72, 80–81, 85; social connections and, 22, 195; social influence and, 179; somatosensorimotor resonance and, 87, *88*
encephalization, 180, 185
endorphins, 10b
endotoxin, 45
epigenetic processes, 6, 63–65
Epstein-Barr Virus (EBV), 43
event-related potentials (ERPs), 56, 204–5
evolution: *Ardipithecus ramidus* and, 54, 153; Cacioppo Evolutionary Theory of Loneliness (ETL) and, 22, 28–31, 33, 36, 41, 44, 47, 213–16; connecting forces and, 100; Darwin and, 8–9, 122; deception and, 134–37, 145–51; face perception and, 125; group processes and, 152–54, 158; Hamilton's rule and, 2, 30; *Homo sapiens* and, 54, 65–71, 84, *141*, 185; human ascendancy and, 54, 55b; loneliness and, 28–31; natural selection and, 1, 3, 30, 38–39, 68, 85, 108, 134, 139, 145, 169, 171; persuasion and, 179–85; social behavior and, 1–6, 9–10; social brain and, 54, 55b, 65–74, 81; social connections and, 22, 28–31, 33, 39, 43–44, 47, 85, 191, 197, 201b; social influence and, 172, *173*, 178–85
executive functioning, 7, 11, *21*, 77, 90, 105, 109, 148, 150
Expressions of Emotions in Man and Animals, The (Darwin), 121–22
extended self, 189, *191*
extrastriate body area (EBA), 117
extroversion, 120
eye perception: amygdala and, 139–40, *142*, 145, 148–50; attention and, 138, 140, 144b,

144, 145b, 147–48; cerebral cortex and, 146; chimpanzees and, 141; communication and, 134–35, 146, 150–51; cost-benefit analysis and, 150; deception and, 125–28, 145–49; fear and, 138–40, *142*, 144–45; fusiform gyrus and, 144; gaze and, 138–45; happiness and, 23, 135–36, *142*; human brain and, 137, 145; identity and, 144; impressions and, 44, 135, 144, 148–50; infants and, 141; intention and, 134, 136, 138–39, 142, 143, 145b, 148; isolation and, 146–47, 151; macaques and, 144; memory and, 151; mentalizing and, 138, 144, 148–49; mindreading and, 138–39; neocortex and, 146; neuroimaging and, 140, 144–45, 148; neurons and, 137, 142, 144–46, 148; physical influences and, 149–50; prefrontal cortex (PFC) and, 144, 148–50; reading eyes and, 138–39; social influence and, 149–50; superior temporal sulcus (STS) and, 144, 148; theory of mind (TOM) and, 138, 144, 148

face perception: amygdala and, 120, 123, 126, 128–29, 131–32; animal research in, 128–32; artificial signals and, 123–24; attention and, 118, 133; axons and, 125, 131; chimpanzees and, 114, 129, 131; cognition and, 114, 118, 129; communication and, 112, 125–26; depression and, 123, 133; distinction of face and, 112–18; Ekman-Friesen system and, 205; emotion and, 118–33; evolution and, 125; eye perception and, 134–51; fear and, 120, 126–31; functional magnetic resonance imaging (fMRI) and, 114–15, 120, 123, 126; fusiform gyrus and, 58, 114–15, *116–17*, 126, 144, 148, 160, 163, 200; gaze and, 117; happiness and, 120–21, 126, 131; hippocampus and, 117, 126; holistic, 114; human brain and, *113*, *116–17*, 126; hypothalamus and, 123; identification and, 117–18; identity and, 118b, 119; imitation and, 123; impressions and, 112, 119–22; infants and, 120; intention and, 119, 123, 125, 131, 133–34; International Affective Picture System and, 205; love and, 199–200; macaques and, 118b, 131, 144; memory and, 114; mentalizing and, 112, 114, *116*, 132; neuroimaging and, 114, 118, 120, 132; neurons and, *116*, 124–25, 131, 133, 137; omega sign and, 122; orbital frontal cortex (OFC) and, 121; patient studies in, 128–32; power/dominance and, 119; prefrontal cortex (PFC) and, 121, 126, 132; rapid signals and, 124–32; slow signals and, 121–23; static signals and, 118–21; superior temporal sulcus (STS) and, 116–17, 126, *130*, 131; theory of mind (TOM) and, 114, *116*, 132; valence/trustworthiness and, 119

falsehood, 13, 59, 106, 136, 148, 181

Farah, Martha, 114

fascia, 113, 124

fear: connecting forces and, 109–10; deception and, 134; eye perception and, 138–40, *142*, 144–45; face perception and, 120, 126–31; Passionate Love Scale (PLS) and, 221; social brain and, 65; social connections and, 45

fish, 52–54, 79, 109, 171–72

fitness, 1, 28–31, 47, *69*, 152, 169, 187, 213–15

food: Cacioppo Evolutionary Theory of Loneliness (ETL) and, 214; competition for, 22, 72, 100, 178, 214; desert locust and, 3; eye perception and, 140; group processes and, 156; limited resources of, 22; organisms and, 14; sharing of, 65, 84; social brain and, 54, 55b, 65, *69*, 72, 84; social connections and, 100, 195, 211; social determinants and, 15; social influence and, 172, 178–79; variety of, 54, 55b

forebrain, 10, *94*

Freiwald, Winrich, 98b

Friesen, W. V., 205

frontal lobe, 11, *12*, *66*, 71, 85

functional magnetic resonance imaging (fMRI): activation likelihood estimate (ALE) and, 61–62; BOLD, 58; brain matrix from, 57–58; causal inferences and, 59–60; connecting forces and, 91, 103, 106; cumulative science approach and, 61–62; empathy and, 57, 91; face perception and, 114–15, 120, 123, 126; intention and, 103–4; loneliness and, 80–83; love and, 192, 200, 204, 209; mentalizing network and, 57; moderator variables and, 60–61; multilevel kernel density analysis (MKDA) and, 61–62; Passionate Love Scale (PLS) and, 192; persuasion and, 185; simplicity and, 58–59; social behavior and, *18*; social brain and, 56–63, 70b, 80–83; social connections and, 192, 200, 204, 209; social influence and, 185; social perception network and, 58; time course and, 62–63; uses of, 56–63

fusiform face area (FFA), 114–15, *116–17*, 126, 148, 200

fusiform gyrus: eye perception and, 144; face perception and, 58, 114–15, *116–17*, 126, 144, 148, 160, 163, 200; gaze and, 58, 144; group processes and, 160, 163; social brain and, 58–59; social connections and, *194*, 199

Gamma-aminobutyric acid [GABA], 10b
Gardner, A., 213
gaze: automatic mimicry and, *88*; communication and, 142; courtship and, 143, 144b; direction of, 138–45; emotion and, *88–89*; eye perception and, 138–45; face perception and, 117; fusiform gyrus and, 58, 144; imitation and, *88*; infants and, 141; love and, 142–44; mentalizing and, 144; prefrontal cortex (PFC) and, 76; superior temporal sulcus (STS) and, 80; survival and, 139; theory of mind (TOM) and, 144
Gazzaniga, Michael, 9
gender, 6, 21, 26, 52, 76, 83, 118, 161, 164
genetics: Cacioppo Evolutionary Theory of Loneliness (ETL) and, 41–42; CTRA and, 39–42; DNA and, 15–16, 24, 38–39, 40b, 63; epigenetic processes and, 6, 63–65; genotype and, 38–39, 134; Hamilton's rule and, 2, 30; HPA axis and, 40b; inclusive fitness and, 152, 169; maternal behavior and, 15–16; phenotypes and, 38, 41, 47–48, 134; relatedness and, 30; RNA and, 35, 38, 42, 63; social influence and, 171; transcriptome dynamics and, *29*, *32*, 33, 38–45; up/down regulation and, *32*, 39–42, 64, 176
genotypes, 38–39, 134
Gintis, Herbert, 169, 179
glial cells, 9, 45–46, 124
glucocorticoids (GCs), 34, 36, 39–42, *49*, 64
Gobbini, M. I., 118b
golden triangle, 14–18
gorillas, *66*, 70b, *141*
gray matter, 76b, 79–80, 93
grief, 21, 25, 122
Grossmann, Tobias, 75
group processes: altruism and, 152–57, 166–69; amygdala and, 162–65; anterior insula and, 163, 168; attention and, 160; bias mitigation and, 164–66; chimpanzees and, 153; cognition and, 159–60, 164–65; competition and, 152–54, 158–59, 170; cooperation and, 153–56, 159, 166–70; cost-benefit analysis and, 72, 110, 150, 153, 156–57, 168–69, 171, 187–88; cultural issues and, 152, 154, 157–58, 160, 165–66, 169; discrimination and, 22, 60, 75–77, 90b, 129, 133, 154, 160, 162–64, 170, 195b, 214; emotion and, *162*, 163, 168; empathy and, 159, 180; evolution and, 152–54, 158; fitness and, 152, 169; food and, 156; gyrus and, 160, 162–63, 165; hormones and, 162; human brain and, 159; hyperscanning and, 167b; hypothalamus and, 162; identification and, 157–59, 163; impressions and, 160–62, 164; ingroups and, 96, 157–60, 163–64, 166, 170; kinship and, 2, 67, 152, 157; macaques and, 153; memory and, 160–61; mentalizing and, *162*; neuroimaging and, 160, 164, 166, 168; norms and, 157, 164–65, 169; orbital frontal cortex (OFC) and, 162–63; outgroups and, 157–60, 163–64, 166, 170; pain and, 163; pair-bonds and, 153; persuasion and, 179–88; prefrontal cortex (PFC) and, 160–64, *165*, 168; prejudice and, 158, 160, 162–66, 170; reciprocity and, 154–57, 159, 168–70, *210*, 217; selfishness and, 166–68; social behavior and, 153, 156, 168; social learning and, 169; social perception and, 170; society and, 153, 158, 160; spite and, 170; stereotyping and, 159–62, 164–65, 170; survival and, 154, 157, 159, 170; teams and, 2, 90, 163; ventral striatum and, 162, 168
guilt, 86, 105, 131–32, 178
Gürek, Özgür, 166
gyrus, 76b; angular, 59, 186, 192–99, 206, 210; connecting forces and, 96, 100, 103, *104*; eye perception and, 144; face perception and, 115, *116–17*, 126; fusiform, 58–59, 115, *116–17*, 126, 144, 160, 163, *194*, 199; group processes and, 160, 162–63, 165; gyrification and, *11*; love and, 195, 197; social brain and, 58–59, *68*, 80; social connections and, *31*, 192–99, 206–7, 210; social influence and, 186–87; supramarginal, *31*, 207

Hamilton's rule, 2, 30
Hanson Robotics, 126
happiness: deception and, 135–36, *142*; eye perception and, 135–36, *142*; face perception and, 120–21, 126, 131; Passionate Love Scale (PLS) and, 220, 222; social brain and, 76; social connections and, 23, 52, 189
Harrari, Yuval, 65
Hatfield, Elaine, 217–19
Haxby, J. V., 118b
Health and Retirement Study, 33
Heatherton, Todd, 83
Hillis, Argye, 91
hindbrain, 9–10, *66*

hippocampus: face perception and, 117, 126; social brain and, 68, 79; social connections and, *49*, 192, *193*, *196*, 209, 211
Holocaust, 176–77, 179
Homo sapiens, 54, 65–71, 84, *141*, 185
hormones: Cacioppo Evolutionary Theory of Loneliness (ETL) and, 215; connecting forces and, *89*; group processes and, 162; social behavior and, 2, 8, 16–19, 23; social brain and, 54, 56, 78b; social connections and, 25, 28, 33–34, 36, 38, 40b, 47, *49*
human brain: animal research and, 10b; brain-body interactions and, 6; capabilities of, 2; cognitive functions of, 6–8 (*see also* cognition); cranial nerves of, *113*; current model of, 6; design features of, 2; emotion and, 91 (*see also* emotion); eye perception and, 137, 145; face perception and, *113*, *116–17*, 126; functional connectivity of, 31; gray matter and, 76b, 79–80, 93; group processes and, 159; human identity and, 9–14; imaging techniques for, 56; impressions and, 7; language and, 8–9, 11, 67, *68*, 109; love and, 204–6; methods for studying, 55–65; number of neurons in, 6; perceived social isolation and, 48–52; plasticity of, 6–7, 25, 51, 79; size of, 4–6, 9, 55b, 65, 68, 71, 73, 78–80, 85; social connections and, 53–54, 197, 204–6; structure of, 6 (*see also* specific structure); subtractive method and, 23–24; TOM mentalizing and, 107, *116*; uniqueness of, 8–14. *See also* social brain
human immunodeficiency virus (HIV), 44
hyperscanning, 167b
hypothalamic-pituitary-adrenocortical (HPA) axis, *32*, 33–36, 38, 40b, *49*, 51
hypothalamus: connecting forces and, *94*; face perception and, 123; group processes and, 162; social behavior and, 10; social brain and, 70; social connections and, 35, 40b, *49*, 200

identification: connecting forces and, 86–87, 93, 95–97, 103, 110; face perception and, 117–18; group processes and, 157–59, 163; imitation and, 95–97; ingroups and, 96, 157–60, 163–64, 166, 170; self-attribution and, 90b; social connections and, 198b, 209; visual word form area (VWFA) and, 117
identity: bodily self and, 59, 191, 198b; constructs of self and, 189, 191; eye perception and, 144; face perception and, 118b, 119; group processes and, 157, 159; social connections and, 189, 191, 209; social influence and, 181
imitation: connecting forces and, 86–87, *88*, 95–97, *100*, 109–10; face perception and, 123; gaze and, *88*; human/animal, *100*; identification and, 95–97; mimicry and, 87, *88–89*, 95, 99, 112–13, 124–25, 137–38; social brain and, 198b
immunity: antibodies and, 43–44; HIV and, 44; SIV and, 43–44; social behavior and, 15; social connections and, *29*, *32*, 33–34, 39, 43–44, 46; vaccination and, 43
immunoglobulin, 43–44
implicit vigilance, 28, 30, 47, 53, 81, 215
impressions: connecting forces and, 21, 99, 106; deception and, 44, 135, 148–50, 160–64; eye perception and, 135, 144, 148–50; face perception and, 112, 119–22; false, 106, 136; formation of, 7; group processes and, 160–62, 164; human brain and, 7; social brain and, 7; social connections and, 21, 198b
Industrial Revolution, 67, 84
infanticide risk hypothesis, 74
infants: connecting forces and, 86–87; eye perception and, 141; face perception and, 120; gaze and, 141; HPA activation and, 35; language and, 74–76; mother attachments and, 2, 8, 14; neural commitment and, 75; pair-bonds and, 2, 8, 14, 35, 74, 78b; prefrontal cortex (PFC) and, 75–76; rodent DNA and, 15–16; social brain and, 74–77; social connections and, 189; social influence and, 172–73, 178, 185
inflammatory substrate, 15, *32*, 33, 39–46, 52, 64
Information Revolution, 67
ingroups, 157–59
intention: connecting forces and, 86–87, *88*, 91–107, 110–11; deception and, 136, 138, 145b, 148; eye perception and, 134, 136, 138–39, 141, 143, 145b, 148; face perception and, 119, 123, 125, 131, 133–34; functional magnetic resonance imaging (fMRI) and, 103–4; social brain and, 58, 62, 66, 76–77, 85; social connections and, 3, 47, *191*, 208–9; social influence and, 171, 179, 182, 187; sports and, 101–3
interleukin-1 receptor antagonist (IL-1Ra), 45
interleukin-6 (IL-6), 45
International Affective Picture System, 205
interpersonal self, 191
introversion, 21, 25, 120

Irlenbusch, Bernd, 166
isolation: consequences of, 31–48; eye perception and, 146–47, 151; measuring, 24–28; objective, 24–28, *31*, 33; outgroups and, 157–60, 163–64, 166, 170; pathways of, 31–48; perceived, 24–28, 33, 38, 47–53, 83; social behavior and, 3, 5b, 10b, 15, 17; social brain and, 48–52, 64–65, *66*, *70*, 77–83; social connections and, 24–53, 206; social control and, 24–25
Italian Renaissance, 182

Kanai, Ryota, 79–80
Kanwisher, Nancy, 114
Karelina, K., 46
Keltner, Dacher, 131–32
Kennedy, D. P., 57
Kiecolt-Glaser, J. K., 43
kinship, 2, 67, 152, 157
Klucharev, Vasily, 175
knockout mice, 24
Korsakoff's syndrome, 185
Kuhl, Patricia, 75

language: axonal projections and, 67; body, 76, *89*; Broca's area and, 11; connecting forces and, 109; CTRA and, 41; gene analysis and, 41; genetics and, 171; human brain and, 8–9, 11, 67, *68*, 109; infants and, 75; mentalizing and, 76; social behavior and, 8; social brain and, 67, *68*, 72–76; social connections and, 210; social influence and, 171, *182*
laughter, 70b, 133
Legare, Christine, 96
leukocytes, *29*, *32*, 33, 38–46
Levine, S., 42
lifespan, 6, 19, 22, 28, 33, 36–37, 79, 90, 108, 118, 189, 216
loneliness: anxiety and, 41, 48, 80; Cacioppo Evolutionary Theory of Loneliness (ETL) and, 22, 28–31, 33, 36, 41, 44, 47, 213–16; cardiovascular system and, 23, *32*, 36–40, 44; CART and, 42; chronic, 6, 42, 64; contagion and, 21–22; cost-benefit analysis and, 213; dejection and, 47; depression and, 25, *32*, 33, 35, 41, 47–48, 51; evolutionary theory of, 28–31; functional magnetic resonance imaging (fMRI) and, 80–83; hypothalamic-pituitary-adrenocortical (HPA) axis and, *32*, 33–36, 38, 40b, *49*, 51; immunity and, *29*, *32*, 33–34, 39, 43–44, 46; inflammatory substrate and, 15, *32*, 33, 39–42, 44–46, 52, 64; leukocytes and, *29*, *32*, 33, 38–46; macaques and, 41, 43–44; marriage and, 22–23, 25, 189; mortality and, 22–25, 28, *29*, 32–48, 189, 215; objective, 24–28; pain and, 214; Passionate Love Scale (PLS) and, 222; perceived, 24–28; prepotent responding and, *29*, *32*, 33, 46–47, 81; repair/replacement postulate and, 215; scale for, 27b; short-term preservation and, 28; sleep quality and, *29*, 32–34, 52; social behavior and, 6; social brain and, 64, 77, 79–84; social connections and, 21–22, 26–33, 36–52, 206; social control and, 24–25; sympathetic tonus and, *29*, *32*, 33, 36–38; transcriptome dynamics and, *29*, *32*, 33, 38–45
love: affect system and, 194; angular gyrus and, 195, 197; attachment styles and, 200, 201b; biological drive and, 199–204; brain network of, *194*; Brodmann area 39 and, 195; cognition and, 25, 191–201, 205–12; constructs of self and, 191; deception and, 142, *143*; desire and, 201–2; dopamine and, 192–93, 195, 199–200, 211–12; Ekman-Friesen system and, 205; face perception and, 199–200; familial, 199–200; functional magnetic resonance imaging (fMRI) and, 192, 200, 204, 209; gaze and, 142–44; human brain and, 204–6; International Affective Picture System and, 205; maternal, 199; other types of, 199–204; passionate, 81, 191–212; Passionate Love Scale (PLS) and, 192, 197, 199, 217–23; quest for union and, 191–92; rejection and, 206–7; reward/motivation system and, 193–94; romantic, 81, 142–43, 191–99, 204–12, 220, 222; self-expansion model and, 192–93, 197, 198b, 209; social brain and, 58–59, 81; social connections and, 191–212; soul mates and, 191; speed of in human brain, 204–6; unrequited, 25, 217; ventral tegmental area (VTA) and, 200
lower motoneurons (LMNs), 124–25, 137
Lynch, J., 37

macaques: axonal projects and, *67*, *68*; connecting forces and, 98b, 111; CTRA and, 41; dedicated brain networks and, 98b; eye perception and, 144; face perception and, 118b, 131, 144; group processes and, 153; loneliness and, 41, 43–44; social brain and, *66*, 67, *68*; TOM mentalizing and, 111
magneto-encephalography (MEG), *18*, 56

Marlowe, D., 219
marriage, 22–23, 25, 143, 189, *190*, 203
Matthews, Gillian, 50–51
McEnroe, John, 101–2
Mearns, Jack, 206
medial prefrontal cortex, 126, *194*; connecting forces and, 91, *92*, 98b; deception and, 144, 148–49; group processes and, 164–66; social brain and, 77–78; social connections and, *49*, 59; social influence and, 186, 188; ventromedial prefrontal cortex (VMPFC), 52, 77, 78, 90b, 91–92, *93*, 96–97, 99, 107, 121, 132, 148, 162–64, 207
medulla, *4*, 10, 36, *49*
memory: connecting forces and, 103, 105; eye perception and, 151; face perception and, 114; group processes and, 160–61; losing, 5b; social behavior and, 5b, 7, 12; social brain and, 59, 79–80; social connections and, 32, 191, 193, *196*; social influence and, *184*, 185
mentalizing: body language and, 76; cerebral hemispheres and, 11–12; connecting forces and, *92*, 93, *94*, 97–109, 111; egocentric, 97–99, 105, 149, 198b; eye perception and, 138, 144, 148–49; face perception and, 112, 114, *116*, 132; functional magnetic resonance imaging (fMRI) and, 57; gaze and, 144; group processes and, *162*; internal states and, 57; mindreading and, 138–39; multiple pathways and, 107–8; network of, 57, 93, *94*, 98b, 99, 105–7, 111, *116*, 132; simulation perspective and, 99–104; social behavior and, 11–12; social brain and, 57, 76; social connections and, 195, 207–8; social influence and, 171; theory of mind (TOM), *92*, 97, 98b, 105–7, 114, *116*, 132, 138
mice, 5b, 15, 23–24, 34, 44–45, 50–52, 79
midbrain, 5b, 10, 78
Milgram, Stanley, 177–78
mimicry, 87, *88–89*, 95, 99, 112–13, 124–25, 137–38
mindreading, 138–39
mirror neurons, 87, 98b, 99–103, *104*, 111, *116*, 195, 198b, 209
Moll, Jorge, 93
Montague, P. R., 62–63
moral emotion, 131–32
mortality: immunity and, *29*, *32*, 33–34, 39, 43–44, 46; inflammatory substrate and, 15, *32*, 33, 39–42, 44–46, 52, 64; leukocytes and, *29*, *32*, 33, 38–46; loneliness and, 22–25, 28, *29*, 32–48, 189, 215; prepotent responding and, *29*, *32*, 33, 46–47, 81; sleep quality and, *29*, 32–34, 52; social connections and, 22–25, 28, *29*, 32–48, 189, 215; sympathetic tonus and, *29*, *32*, 33, 36–38; transcriptome dynamics and, *29*, *32*, 33, 38–45
Mosso, Angelo, 56
mother-infant attachments, 2, 8, 14
motor neurons, 124–25, 133, 137, 145–46, 148
multilevel kernel density analysis (MKDA), 61–62
Mundy, P., 145b
mutual benefit: Cacioppo Evolutionary Theory of Loneliness (ETL) and, 213–14; deception and, 149; social behavior and, 1–2; social brain and, 73, 83; social connections and, 29–30, 39, 197, 208
myelin, 51, 79–80

natural selection, 1, 3, 30, 38–39, 68, 85, 108, 134, 139, 145, 169, 171
Neisser, U., 189
neocortex, 5, *69*, *71*, 73, 145
neural commitment, 75
neuroimaging: connecting forces and, 90b, 91, 93, 98b, 99, 103; cortical structures and, 10b; eye perception and, 140, 144, 148; face perception and, 114, 118, 120, 132; group processes and, 160, 164, 166, 168; hyperscanning and, 167b; smells and, 195b; social behavior and, 10b, 16–19; social brain and, 55b, 56, 59–62, 77, *82*, 83; social connections and, 52, 192, 195b, 197–200, 201b, 205, 207, 211–12; social influence and, 174, 185–86. *See also* specific technique
neurons: connecting forces and, 87, 98b, 99–104, 111; deception and, 137, 145–46, 148; eye perception and, 137, 142, 144–46, 148; face perception and, *116*, 124–25, 131, 137; mirror, 87, 98b, 99–103, *104*, 111, *116*, 195, 198b, 209; motor, 124–25, 133, 137, 145–46, 148; number of in human brain, 6; social behavior and, 6, 9, 10b, *18*; social brain and, 58, 67, 70, 79–81; social connections and, 35, 46, 50–51, 195, 198b, 204, 209; synapses and, 6, 9, 50, 146, 204
Newell, L., 145b
Nielsen, Mark, 96
Nisbett, Richard, 13
norepinephrine, 10b
norms: connecting forces and, 96; group processes and, 157, 164–65, 169; social brain and, 8, 14, 72; social influence and, 172–74, 187

obedience to authority, 176–79
obesity, 32, 44
omega sign, 122
Opie, Christopher, 74
orangutans, *66*, 70b, 111
orbital frontal cortex (OFC): connecting forces and, 91, *92*, *94*, 98b, *162*; face perception and, 121; group processes and, 162–63; social connections and, *49*, 199, 206, 211
outgroups, 157–60, 163–64, 166, 170
out-of-body experiences (OBE), 198b
oxytocin, 10b, 35, 77, 78b

pain: Cacioppo Evolutionary Theory of Loneliness (ETL) and, 214; connecting forces and, 93–94, 109–10; empathy and, 57, 91–95; group processes and, 163; social behavior and, 134; social brain and, 57–58; social connections and, 30, 33, 199, 207
pair-bonds: group processes and, 153; infants and, 2, 8, 14, 35, 74; marriage and, 22–23, 25, 189; social brain and, 2, 8, 10b, 14, 18, 72–74, 77–78; social connections and, 22, 35, 50, 209, 211
Parfitt, D. N., 37
parietal lobe, 11, *12*, *66*, 67
Parr, Lisa, 129
Passionate Love Scale (PLS): anxiety and, 217; Bartels and, 192; behavioral components of, 218; cognitive components of, 217–18; cultural issues and, 219; depression and, 221; emotion and, 217–19, *220*, *222*; fear and, 221; functional magnetic resonance imaging (fMRI) and, 192; happiness and, 220, 222; Hatfield and, 217–19; loneliness and, 222; scoring of, 218; Social Desirability Scale and, 219; Song and, 197, 199; Sprecher and, 217–19; version A, 220–21; version B, 222–23; Zeki and, 192
peahens, 143, 144b
Peromyscus leucopus, 44
perspective taking, *82*, 87–95, 109–10
persuasion: communication and, 179–87; cultural issues and, 179–82; Elaboration Likelihood Model (ELM) and, 183–85; evolution and, 179–85; functional magnetic resonance imaging (fMRI) and, 185; malevolent uses of, 181–82; motivation and, 186; social influence and, 179–88
phenotypes, 38, 41, 47–48, 134
plasticity, 6–7, 25, 51, 79
Plato, 182
pons, 10, 125
positron emission tomography (PET), 56, 114
posterior cingulate cortex, 77, 132, 186, 197
prairie voles, 10b, 35, 37, 52, 78b, 79
predation, 5b, 39, 52–53, 65, *69*, 78, 84, 100, 109, 139, 171
prefrontal cortex (PFC): connecting forces and, 90b, 91–93, 98b, 109; eye perception and, 144, 148–50; face perception and, 121, 126, 132; gaze and, 76; group processes and, 160–65, 168; medial, *49*, 59, 77–78, 91, *92*, 98b, 126, 144, 148–49, 164–66, 186, 188, *194*; social brain and, 55b, 59, 67, *68*, 75–80, 84; social connections and, *49*, 50–52, *194*, 199, 207; social influence and, 179, 186, 188; ventromedial prefrontal cortex (VMPFC), 52, 77–78, 90b, 91–92, *93*, 96–97, 99, 107, 121, 132, 148, 162–64, 207
prejudice, 158, 160, 162–66, 170
prepotent responding, *29*, *32*, 33, 46–47, 81
principle of multiple determinism, 14–15
principle of nonadditive determinism, 15
principle of reciprocal determinism, 15–16
Prisoner's Dilemma game, 167b
private self, 189, *191*
proinflammatory gene expression, 40b, 52
psychopathy, 20, 93

rapid-eye movement, 34
rapid signals, 124–32
Reading the Mind in the eyes (RME) test, 138
reciprocity, 154–59, 168–70, *210*, 217
Regan, P. C., 218–19
rejection, 206–7
repair/replacement postulate, 215
ribonucleic acid (RNA), 35, 38, 42, 63
Rizzolatti, Giacomo, 99
robots, 126
Rockenbach, Bettina, 166
romance, 81, 142–43, 191–99, 204–12, 220, 222
Ross, Marina, 70b
Rueggeberg, Rebecca, 34

sadness, 28, 47, 123, 126, *127*, 131, 137
salubrity, 3, *32*, 33–34
salutary relationship postulate, 213–14
Scientific Revolution, 67
self-attribution, 90b
self-centeredness, 21, 28, 30–31, 81, 97, 192, 215
self expansion, 192–93, 197, 198b, 209
selfishness: Cacioppo Evolutionary Theory of Loneliness (ETL) and, 213–15; deception

and, 136; group processes and, 166–68; social behavior and, 1–2; social brain and, 72–73; social connections and, 29–30, 39; social influence and, 187; spite and, 1–2, 29–30, 39, 73, 170, 213, 215
self-preservation, 28, 34, 44, 47, 53, 77, 84, 214–16
self-representation, 11, 59, 107, 193, 201
Semendeferi, Katerina, 68–71
serotonin, 10b
sexual behavior, 15, 142, *143*, 153, 164, 200–4, 211, 217–19
shame, 131–32
Shankar, A., 36
siblings, 24, 35, 77, 205, 214
Silk, Joan, 23
Silwa, Julia, 98b
simian immunodeficiency virus (SIV), 43–44
simulation perspective, 99–104
sleep, *29*, 32–34, 52
slow signals, 121–23
smells, 195b
Smith, L., 74
social behavior: altruism and, 1–2, 8; amygdala and, 10; attention and, 6–7, 11; Cacioppo Evolutionary Theory of Loneliness (ETL) and, 213–15; cerebral cortex and, 9–12; chimpanzees and, 5, 19; comprehensive accounts of, 16; cost-benefit analysis and, 1–2, 213; deception and, 149; dessert locust and, 2–5, 78; empathy and, 91, 95; evolution of, 1–2; extroversion and, 120; functional magnetic resonance imaging (fMRI) and, *18*; group processes and, 153, 156, 168; Hamilton's rule and, 2, 30; hormones and, 2, 8, 16–19, 23; hypothalamus and, 10; immunity and, 15; inappropriate, 132–33; introversion and, 21, 25, 120; kinship and, 2, 67, 152, 157; language and, 8; loneliness and, 6; memory and, 5b, 7, 12; mentalizing and, 11–12; mutual benefit and, 1–2; neocortex and, 5; neuroimaging and, 10b, 16–19; older subcortical structures and, 10; pain and, 134; persuasion and, 179–88; principle of multiple determinism and, 14–15; selfishness and, 1–2; social brain and, 56–60, 67, 73, 85; social connections and, 23, 28–30, 39, 51–52; spite and, 1–2; thalamus and, 10
social brain: altruism and, 73, 81, 83; amygdala and, *57*, 58–59, 69–70; anterior insula and, 80; anxiety and, 80; attention and, 59, 67, 81; axons and, 67, 79–80; cerebral cortex and, 65, *66*, *70*; chimpanzees and, *66*, 67, *68*, 70b; cognition and, 3, 5–8, 12–13, 16, 18–20, 54, 57, 59–62, 66–68, 71–80, 85; communication and, 7–8, 14, 67, 72, 75–76, 79–80; competition and, 8, 72, 76; conformity and, 8, 75; contagion and, 8; cooperation and, 8, 14, 72, 76, 81; cultural issues and, 8–9, 14, 56, 67–68, 72–73, 85; development of from infancy, 74–77; emotion and, 6, 19, 57–58, 68–70, 77; empathy and, 57, 72, 80–81, 85; energy expense of, 5b; epigenetic processes and, 6, 63–65; evolution and, 1–6, 9–10, 54, 55b, 65–74, 81; fear and, 65; food and, 54, 55b, 65, *69*, 72, 84; frontal cortex and, 69–70; functional magnetic resonance imaging (fMRI) and, 56–58, 61–62, 70b, 80–83; gyrus and, 58–59, 68, 80; happiness and, 76; hippocampus and, 68, *79*; hormones and, 54, 56, 78b; hypothalamus and, 70; hypothesis of, 71–74; imitation and, 198b; impressions and, 7; infants and, 74–77; intention and, 58, 62, 66, 76–77, 85; isolation and, 3, 5b, 10b, 15, 17, 48–52, 64, 65, *66*, *70*, 77–83; language and, 67, *68*, 72–76; loneliness and, 64, 77, 79–84; love and, 58–59, 81; macaques and, *66*, 67, *68*; memory and, 59, 79–80; mentalizing and, 57, 76; methods for study of, 55–65; mother-infant attachments and, 2, 8, 14; multimodal integration centers and, 5b; mutual benefit and, 73, 83; neocortex and, *69*, *71*, 73; neuroimaging and, 55b, 56, 59–62, 77, *82*, 83; neurons and, 6, 9, 10b, *18*, 58, 67, 70, 79–81; norms and, 8, 14, 72; pain and, 57–58; pair-bonds and, 2, 8, 10b, 14, 18, 72–74, 77–78; prefrontal cortex (PFC) and, 55b, 59, 67, *68*, 75–80, 84; salutary relationships and, 77–84; selfishness and, 72–73; social behavior and, 56–60, 67, 73, 85; social learning and, 8, 14, 66–67, *69*, 73, 85; social perception and, 7, 12, 54, 58, 76, 79–80; spite and, 73; status and, 8, 72; superior temporal sulcus (STS) and, 58, 78–80; survival and, 72, 74, 84–85; theory of mind (TOM) and, 81; ventral striatum and, 76–77, 81–83
social connections: altruism and, 29–30, 39, 211; amygdala and, *49*, 50–51, 200, 207; anterior insula and, 192–94, 199, 201, 203–4, 206; anxiety and, 25, 41, 45, 48, 201b; attention and, 28, 193, *196*, 198b, 207, 209–10; Cacioppo Evolutionary Theory of

social connections (cont.)
Loneliness (ETL) and, 22, 28–31, 33, 36, 41, 44, 47, 213–16; cardiovascular system and, 23, *32*, 36–40, 44; cognition and, 23, 28, 51–52, 191–201, 205–12; communication and, 22; competition and, 22, 178; constructs of self and, 189, 191; contagion and, 21–22; cooperation and, 22, 208; cost-benefit analysis and, 29–30; cultural issues and, 22, *29*; depression and, 21, 25, *29*, *31–32*, 33, 35, 41, 47–48, 51, 206; emotion and, 189, 192–95, 199–212; empathy and, 22, 195; evolution and, 22, 28–31, 33, 39, 43–44, 47, 85, 191, 197, 201b; fear and, 45; food and, 100, 195, 211; functional magnetic resonance imaging (fMRI) and, 192, 200, 204, 209; gyrus and, *31*, 192–99, 206–7, 210; happiness and, 23, 52, 189; hippocampus and, *49*, 192, *193*, *196*, 209, 211; hormones and, 25, 28, 33–34, 36, 38, 40b, 47, *49*; hypothalamic-pituitary-adrenocortical (HPA) axis and, *32*, 33–36, 38, 40b, *49*, 51; hypothalamus and, 35, 40b, *49*, 200; identification and, 198b, 209; identity and, 189, 191, 209; immunity and, *29*, *32*, 33–34, 39, 43–44, 46; impressions and, 21, 198b; infants and, 189; inflammatory substrate and, 15, *32*, 33, 39–42, 44–46, 52, 64; intention and, 3, 47, *191*, 208–9; isolation and, 24–53, 206; language and, 210; leukocytes and, *29*, *32*, 33, 38–46; loneliness and, 21–22, 26–33, 36–52, 206; love and, 25, 191–212; macaques and, 41, 43–44; marriage and, 22–23, 25, 143, 189, *190*, 203; memory and, 32, 191, 193, *196*; mentalizing and, 195, 207–8; mortality and, 22–25, 28, *29*, 32–48, 189, 215; mutual benefit and, 29–30, 39, 197, 208; neuroimaging and, 52, 192, 195b, 197–200, 201b, 205, 207, 211–12; neurons and, 35, 46, 50–51, 195, 198b, 204, 209; orbital frontal cortex (OFC) and, *49*, 199, 206, 211; pain and, 30, 33, 199, 207; pair-bonds and, 22, 35, 50, 209, 211; prefrontal cortex (PFC) and, *49*, 50–52, *194*, 199, 207; prepotent responding and, *29*, *32*, 33, 46–47, 81; quest for union and, 191–92; rejection and, 206–7; salubrious, 3; salutary, 8, 22–24, 28, 33, 52, 77–85, 87, 103, 189–215; selfishness and, 29–30, 39; sleep quality and, *29*, 32–34, 52; social behavior and, 23, 28–30, 39, 51–52; social learning and, 22; society and, 200; spite and, 29–30, 39; status and, 23, 25, 53; survival and, 22–23, 28, *29*, 33–34, 39, 48, 52–53; sympathetic tonus and, *29*, *32*, 33, 36–38; transcriptome dynamics and, *29*, *32*, 33, 38–45; ventral striatum and, 200
social control, 24–25, 108, 168, 179, 182
Social Desirability Scale, 219
social influence: advertising and, 181; amygdala and, 179; attention and, 176–78; cognition and, 176, 179–80, 183–88; communication and, 179–81, 183–87; conformity and, 172–77; cost-benefit analysis and, 171, 187–88; cultural issues and, 171, 174, 179, 181–82, 187; emotion and, 181; empathy and, 179; evolution and, 172, *173*, 178–85; food and, 172, 178–79; functional magnetic resonance imaging (fMRI) and, 185; genetics and, 171; gyrus and, 186–87; Holocaust and, 176–77, 179; identity and, 181; infants and, 172–73, 178, 185; intention and, 171, 179, 182, 187; language and, 171, *182*; memory and, *184*, 185; mentalizing and, 171; Milgram experiment and, 177–78; neuroimaging and, 174, 185–86; norms and, 172–74, 187; obedience to authority and, 176–79; persuasion and, 179–88; prefrontal cortex (PFC) and, 179, 186, 188; selfishness and, 187; social learning and, 171, 174, 179, 187; society and, 180; status and, 179; ventral striatum and, 174–75
social learning: chimpanzees and, 109; connecting forces and, 96, 108–10; cost-benefit analysis and, 73; group processes and, 169; social brain and, 8, 14, 66–67, *69*, 73, 85; social connections and, 22; social influence and, 171, 174, 179, 187; types of, *108*
social media, 7, 181, *182*
social networks, 8, 19, 21–22, 80
social perception: connecting forces and, 90, 100; deception and, 134–51; face perception and, 112–33; functional magnetic resonance imaging (fMRI) and, 58; group processes and, 170; social brain and, 7, 12, 54, 58, 76, 79–80
somatosensory states, 11, 87, *88*, 90b, 163, 200
Song, Hongwen, 197, 199
spider monkeys, 71
spinal cord, 124, 146, 149, 151
spite: altruism and, 1–2, 29–30, 39, 73, 170, 213, 215; Cacioppo Evolutionary Theory of Loneliness (ETL) and, 213, 215; group processes and, 170; social behavior and, 1–2; social brain and, 73; social connections and, 29–30, 39

Sprecher, Susan, 217–19
squirrel monkeys, 35
Stanley, Damian, 10b, 93, 99, 194
static signals, 118–21
status: connecting forces and, 90, 107; deception and, 142–43; social brain and, 8, 72; social connections and, 23, 25, 53; social influence and, 179
Steptoe, A., 37n, 45
stereotyping, 159–62, 164–65, 170
steroids, 51
stroke, 37, 44–45, 52, 203
substantia nigra, 10
subtractive method, 23–24
suicide, 206
sulci, 11, 76b
superior temporal sulcus (STS): connecting forces and, 91–92, *93–94*, 96, 103, 107; empathy and, 92; eye perception and, 144, 148; face perception and, 116–17, 126, *130*, 131; gaze and, 80; social brain and, 58, 78–80
surprise, 126, *127*, 131, 144
survival: Cacioppo Evolutionary Theory of Loneliness (ETL) and, 213–15; connecting forces and, 91, 100, 103, 109; deception and, 135, 139, 145; eye perception and, 139; group processes and, 154, 157, 159, 170; social brain and, 72, 74, 84–85; social connections and, 22–23, 28, *29*, 33–34, 39, 48, 52–53
Suzuki, Shinsuke, 107
sympathetic adrenomedullary system (SAM), 36, *49*
sympathetic nervous system (SNS), *49*
sympathetic tonus, *29*, *32*, 33, 36–38
synapses, 6, 9, 50, 146, 204

teams, 2, 90, 163
Technau, G. M., 5b
tectum, 10
tegmentum, 10, *94*
temporal cortex, *68*, 76–77, *92*, 93, 144, 150, *193*
temporal lobe, 11, *12*, *66*, *68*, 93, *94*, 118, 150, 160–62, 185
temporoparietal junction (TPJ), 78, 90b, 106–7, 109, *116*, *194*, 198b, 199
thalamus, 10, 121, *140*, 192–93, *194*, 207
theory of mind (TOM): Caruso on, 106; connecting forces and, *92*, 97, 98b, 105–7, 109, 111; deception and, 148–49; egocentric perspective and, 97–99; eye perception and, 138, 144, 148–49; face perception and, 114, *116*, 132; gaze and, 144; mentalizing and, *92*, 97, 98b, 105–7, 114, *116*, 132, 138; mindreading and, 138–39; multiple pathways and, 107–8; social brain and, 81
titi monkeys, 35
Todorov, Alex, 119
Tomasello, Michael, 145b
tonic preparatory response, 32, 36, 50
transcranial magnetic stimulation (TMS), 19, 56, 60, 161, 175–76, 198b
transcriptome dynamics, *29*, *32*, 33, 38–45
tumor necrosis factor, 45

upper motoneurons (UMNs), 124–25, 133, 137, 145–46, 148
up-regulation, *32*, 39–42, 64

vaccination, 43
Valtorta, N. K., 37
van Schaik, Carel, 179
vasopressin, 10b, 199
ventral striatum: group processes and, 162, 168; social brain and, 76–77, 81–83; social connections and, 200; social influence and, 174–75
ventral tegmental area (VTA), 50, 192–93, *194*, 200, 209, 211
ventromedial prefrontal cortex (VMPFC), 52, 207; connecting forces and, 90b, 91–92, *93*, 96–97, 99, 107; deception and, 148; face perception and, 121, 132; group processes and, 162–64; social brain and, 77–78
Vico, C., 205
visuospatial attention, 11, 191, 198b
Vrticka, Pascal, 201b

Walster, E., 217
waste, 14
West, S. A., 213
When You Come to a Fork in the Road, Take It! (Berra), 155–56
Whitehall cohort, 45
Whiten, Andrew, 172
Wilson, Rob, 51
Wilson, Timothy, 13
Woo, Choong-Wan, 207

Yorzinski, J. L., 143b

Zebrowitz, Leslie, 120
Zeki, Semir, 192